Advanced Oxidation Processes for Water and Wastewater Treatment

Advanced Oxidation Processes for Water and Wastewater Treatment

Edited by Simon Parsons

Publishing

Published by IWA Publishing, Alliance House, 12 Caxton Street, London SW1H 0QS, UK

Telephone: +44 (0) 20 7654 5500; Fax: +44 (0) 20 7654 5555; E-mail: publications@iwap.co.uk
Web: www.iwapublishing.com

First published 2004; reprinted 2005

Typeset by Gray Publishing, Tunbridge Wells, UK

British Library Cataloguing in Publication Data
A CIP catalogue record for this book is available from the British Library

Library of Congress Cataloging-in-Publication Data
A catalog record for this book is available from the Library of Congress

ISBN: 1 84339 017 5

Contents

Preface

Advances in chemical water and wastewater treatment have lead to a range of processes termed advanced oxidation processes being developed. The suitability of advanced oxidation processes (AOPs) for aqueous pollutant degradation was recognised in the early 1970s and much research and development work has been undertaken to commercialise some of these processes. AOPs have shown great potential in treating pollutants at low and high concentrations and have found applications as diverse as groundwater treatment, municipal wastewater sludge destruction and VOCs control.

This book is an overview of those advanced oxidation processes currently used or proposed for the remediation of water, wastewater, odours and sludges. It is divided into two introductory chapters covering an introduction to advanced oxidation processes and a background to UV photolysis, seven chapters focus on individual AOPs and three on specific applications. Given that the subject is very extensive, the amount of detail is largely determined by the target readership which in this case is hoped that those readers interested in water and wastewater treatment processes, including problem holders and equipment suppliers as well as students and academics studying in this area, will find this book useful.

Simon Parsons

List of contributors

THE EDITOR

Simon Parsons

Dr Parsons is a Senior Lecturer in The School of Water Sciences at Cranfield University with 10 years' experience of industrial and academic R&D. In his current post he has conducted research into a wide range of physical and, advanced oxidation processes as applied to water and wastewater treatment. Topics have including photocatalysis, Fenton's for disinfection by-product precursor removal, electrochemical oxidation of textile dye waste and ammonia liquors and using chemical membrane reactors for water recovery. He has contributed to a number of books on both water and wastewater treatment and has published extensively in peer-reviewed journals. Dr Parsons manages research activities pertaining to potable water treatment, advanced oxidation processes and scaling and corrosion and is the organiser of the UK Network on Potable Water Treatment and Supply, a research council sponsored initiative to encourage dissemination of information between the industrial and academic communities in the UK.

Contact: s.a.parsons @cranfield.ac.uk

CONTRIBUTORS

Idil Arslan-Alaton

Istanbul Technical University, Faculty of Civil Engineering, Environmental Engineering Department, 34469 Maslak – Istanbul, Turkey

http://www.ins.itu.edu.tr/cevre/personel/alaton/

Tel: + 90 212 286 65 76; Fax: + 90 212 285 65 87; Email: arslanid@itu.edu.tr

Merzak Belkhodja
Veolia Water Systems, L'Aquarène – 1 pl. Montgolfier, F-94417 Saint Maurice cedex, France
Tel: +33 1 45 11 55 55; Fax: +33 1 45 11 58 10;
Email: Merzak.belkhodja@veoliawater.com

Tony Byrne
Research Associate, Nanotechnology Research Institute, 25B10, University of Ulster at Jordanstown, Northern Ireland, BT37 0QB
http://www.engj.ulst.ac.uk/nibec/photocatalysis/photocatalysis.htm
Tel: +44 28 90368941; Email: j.byrne@ulster.ac.uk

William M. Copa
USF – Zimpro, 301 West Military Road, Rothschild, WI, 54474, US
Tel: +1 715 359 7211; Fax: +1 715 355 3221; Email: Copab@usfilter.com

Paul Christensen
School of Chemical Engineering & Advanced Materials, University of Newcastle on Tyne, NE1 7RU, UK

Didier Cretenot
Veolia Water Systems, L'Aquarène – 1 pl. Montgolfier, F-94417 Saint Maurice cedex, France
Tel: +33 1 45 11 55 55; Fax: +33 1 45 11 58 10;
Email: Didier.cretenot@veoliawater.com

Terry Egerton
School of Chemical Engineering & Advanced Materials, University of Newcastle on Tyne, NE1 7RU, UK
Email: T.A.Egerton@ncl.ac.uk

Francis Luck
Kompetenzzentrum Wasser Berlin gGmbH, Cicerostrasse 24, D-10709 Berlin, Germany
www.kompetenz-wasser.de
Tel: + 49 30 536 53 802; Fax: + 49 30 536 53 888;
Email: francis.luck@kompetenz-wasser.de

Timothy J Mason
Professor of Chemistry, School of Science and the Environment, Coventry University, Coventry CV1 5FB, UK
Email: t.mason@coventry.ac.uk

Clayton B. Maugans
USF – Zimpro, 301 West Military Road, Rothschild, WI, 54474, US
Tel: +1 715 359 7211; Fax: +1 715 355 3221; Email: Maugansc@usfilter.com

Lucie Patria
Veolia Water/Anjou Recherche, Chemin de la Digue, BP 76, F-78603 Maisons-Laffitte cedex, France
Tel: +33 1 34 93 31 31; Fax: +33 1 34 93 31 10;
Email: Lucie.patria@generale-des-eaux.net

Christian Pétrier
Laboratoire de Chimie Moléculaire et environnement, ESIGEC-Université de Savoie, 73 376 Le Bourget du Lac, France
Tel: +33 4 79 75 88 04; Fax: +33 4 79 75 87 72;
Email: Christian.petrier@univ-savoie.fr

Mihela I Stefan
Trojan Technologies Inc., 3020 Gore Road, London, Ontario, N5V 4T7, Canada
Tel: (519) 457 3400; Fax: (519) 457 3030; Email: mstefan@trojanuv.com

Richard Stuetz
Centre for Water & Waste Technology, School of Civil & Environmental Engineering, The University of New South Wales, Sydney, NSW 2052, Australia
www.cwwt.unsw.edu.au
Email: r.stuetz@unsw.edu.au

Tuula Tuhkanen
Tampere Univerity of Technology
tuula.tuhkanen@tut.fi

Kris Van Craeynest
EnVOC Research Group, Organic Chemistry Department, Faculty of Agricultural and Applied Biological Sciences, Ghent University, Coupure Links 653, 9000 Ghent, Belgium
Tel: +32 9 264 5998; Fax: +32 9 264 6243; Email:
Kris.VanCraeynest@UGent.be

Herman Van Langenhove
EnVOC Research Group, Organic Chemistry Department, Faculty of Agricultural and Applied Biological Sciences, Ghent University, Coupure Links 653, 9000 Ghent, Belgium
Tel: +32 9 264 5950; Fax: +32 9 264 6243;
Email: Herman.VanLangenhove@Ugent.be

Susan Wadley
School of Civil & Environmental Engineering, The University of New South Wales, Sydney, NSW 2052, Australia
Tel: 61 2 9385 5793; Fax: 61 2 9385 6139; Email: wadley@civeng.unsw.edu.au

T David Waite
Centre for Water & Waste Technology, School of Civil & Environmental Engineering, The University of New South Wales, Sydney, NSW 2052, Australia
Tel: 61 2 9385 5060; Fax: 61 2 9385 6139; Email: d.waite@unsw.edu.au

Mike Williams
RMCS, Cranfield University

Christian Williamson
Trojan Technologies Inc., 510 E. Newport Dr., Tucson, AZ 85704, USA
Tel: (520) 544 5121; Fax: (520) 544 5184; Email: cwilliamson@trojanuv.com

1

Introduction

Simon A. Parsons and Mike Williams

1.1 INTRODUCTION

Advances in chemical water and wastewater treatment have led to a range of processes termed advanced oxidation processes (AOPs) being developed. The processes have shown great potential in treating pollutants of low or high concentrations and have found applications as diverse as groundwater treatment, municipal wastewater sludge destruction and volatile organic compounds (VOCs) treatment. AOPs have been developed and studied over the past 30 years and the scientific literature surrounding their development and application is extensive (Figures 1.1 and 1.2). Suty *et al.* (2003) reviewed the literature from 1975 until 2000 and showed the diversity of AOPs reported in the literature and how the number of papers had significantly increased decade after decade. This trend was seen for all AOPs and those papers referring directly to water treatment.

This book aims to give the reader an overview of the range of AOPs available and cover both the underpinning science behind the processes and where its current or future applications may lie. In this chapter we will look at a background to oxidation and where opportunities for AOPs exist.

 ISBN: 1 84339 017 5

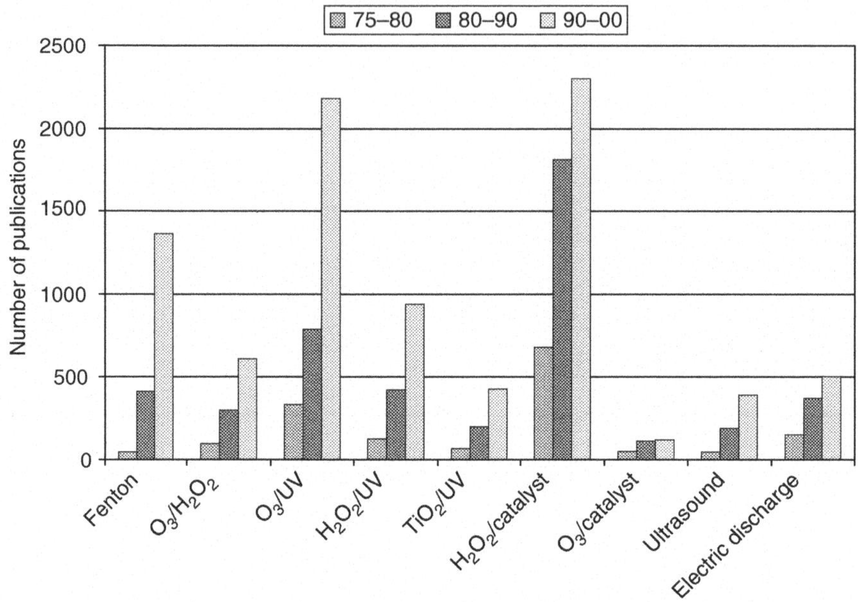

Figure 1.1 Number of AOP publications over the 1975–2000 period (adapted from Suty *et al.*, 2003).

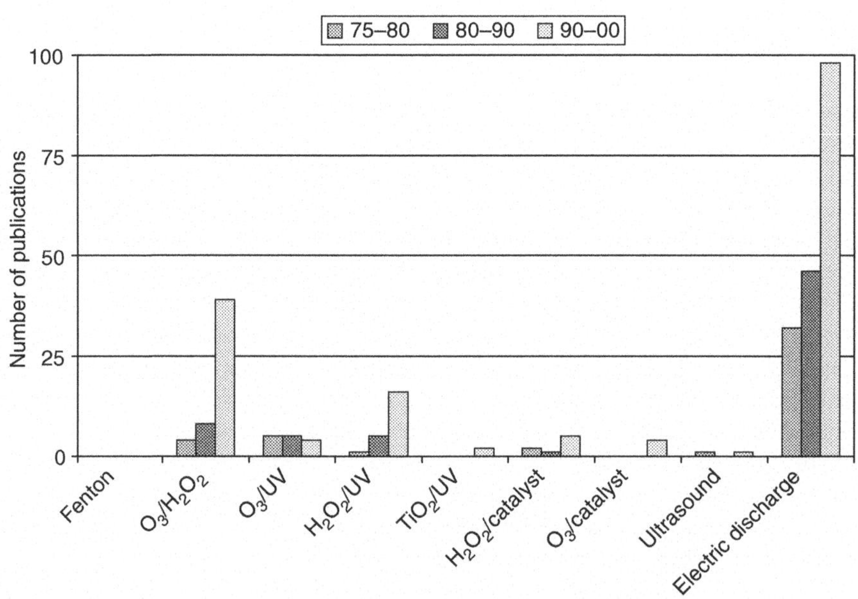

Figure 1.2 Number of water treatment AOP publications over the 1975–2000 period (adapted from Suty *et al.*, 2003).

1.1.1 Oxidation

The ultimate aim of the oxidation of pollutants in water is to 'mineralise', that is to convert the constituents of an organic pollutant into simple, relatively harmless and inorganic molecules:

- carbon to carbon dioxide,
- hydrogen to water,
- phosphorous to phosphates or phosphoric acids,
- sulphur to sulphates,
- nitrogen to nitrates,
- halogens to halogen acids.

Oxidation can be generalised as follows:

- *Inorganic compounds*. The removal of electrons to produce a higher oxidation state Fe^{2+} going to Fe^{3+}.
- *Organic compounds*. The combination of carbonaceous material with oxygen to produce a more heavily oxygenated compound; for instance, in wine, ethanol to acetic acid. 'Forced' oxidation of hydrocarbons for water purification purposes should, ultimately, form carbon dioxide and water.

Each oxidation is accompanied by a corresponding reduction so that the charge balance is preserved. The driving force for oxidation is the stability of the ultimate products. Heats of formation (in $kJ\,mol^{-1}$) are:

Carbon dioxide (g)	−393
Water (l)	−285
Phosphate (aq)	−1276
Sulphate (aq)	−908
Nitrate (aq)	−207
Chloride (aq)	−167

In contrast, the heats of formation (in $kJ\,mol^{-1}$) of organic molecules are typically:

Ethylamine (C_2H_7N)	−74
Benzene (C_6H_6)	+49

So conversion of ethylamine to carbon dioxide (2 per molecule), water (3.5 per molecule) and nitrate (1 per molecule) is thermodynamically favourable, because nearly 2000 $kJ\,mol^{-1}$ of energy is released.

Strong oxidising agents are formed from the electronegative elements of the top right-hand corner of the periodic table (F, Cl, Br, O, S). The relative strength of oxidants is shown in Table 1.1. Here conventional oxidants such as chlorine, permanganate and ozone are compared to the hydroxyl radical. It is this radical that gives many of the AOPs covered in this book their power.

Table 1.1 Oxidation potential of common species.

Species	Oxidation potential (V)
Fluorine	3.03
Hydroxyl radical	*2.80*
Atomic oxygen	2.42
Ozone	2.07
Hydrogen peroxide	1.78
Perhydroxyl radical	1.70
Permanganate	1.68
Hypobromous acid	1.59
Chlorine dioxide	1.57
Hypochlorous acid	1.49
Chlorine	1.36

Table 1.2 Comparison of rate constants for ozone and hydroxyl radicals for a range of compounds.

Organic compound	Rate constant ($M^{-1}s^{-1}$)	
	O_3	$HO^\bullet$
Benzene	2	7.8×10^9
Toluene	14	7.8×10^9
Chlorobenzene	0.75	4×10^9
Trichloroethylene	17	4×10^9
Tetrachloroethylene	<0.1	1.7×10^9
n-Butanol	0.6	4.6×10^9
t-Butanol	0.03	0.4×10^9

1.1.2 Free radicals

Many strong oxidants are 'free radicals', of which the hydroxyl radical $HO^\bullet$ is the most powerful oxidising species after fluorine. The aim of most AOPs is to produce the hydroxyl radical in water. For example, the hydroxyl radical is able to oxidise a wide range of organic compounds significantly (10^9) faster than ozone (Table 1.2). A free radical is not an ionic species but is formed from an equal cleavage of a two-electron bond.

$$HO{:}OH \rightarrow HO^\bullet + {}^\bullet OH \quad \text{photolysis}$$

Each $HO^\bullet$ is uncharged and two $HO^\bullet$ will recombine to form HOOH (also uncharged). The symbol $^\bullet$ indicates the radical centre and represents a single unpaired electron.

Once a free radical reaction has been initiated by photolysis, ozone, hydrogen peroxide, heat, etc. then a series of simple reactions will ensue. The complexity of the chemistry of such systems is due to the very large number of reactions that are possible. With such complicated mechanisms it is very difficult to predict all of the products of an oxidation.

The rate of oxidation will depend on radical concentration, oxygen concentration and pollutant concentration. Many factors can affect the radical concentration, such as pH, temperature, the presence of ions, the type of pollutant, as well as the presence of scavengers such as the bicarbonate ion, which is an effective radical trap.

1.2 AOPs FOR WATER AND WASTEWATER TREATMENT

Developments in chemical treatment have made a range of AOPs suitable for water and wastewater applications. Table 1.3 lists those AOPs that have been developed and whilst the list is not comprehensive it does highlights the range and variety of processes developed or under development that could have applications in water and wastewater treatment. A number of these processes are commercially available and in some instances widely used. Ultraviolet (UV) for instance has over 3000 applications in Europe as a disinfection processes and is used in the US for treating groundwater pollutants such as methyl tertiary-butyl ether (MTBE) and N-nitrosodimethylamine (NDMA). Process such as the chemical combinations of H_2O_2, O_3 and UV, Fenton's reagent, wet air oxidation, supercritical water oxidation and electron beam have all been used at full scale. Other processes such as photocatalysis and ultrasound have been evaluated at pilot scale but many of the process are being developed at the laboratory bench. While the list includes individual processes much research has been undertaken using a combination of the processes, such as UV, ultrasound and ozone together, which offer significant kinetic and performance benefits over each of the processes alone.

The literature on applications is too varied to cover all topics so here those range of applications covered at the recent AOP3 conference have been reviewed (Vogelpohl, 2003). Firstly generic areas have been identified and it can been seen that AOPs have found applications in all areas of water and wastewater treatment (Table 1.4). AOPs are available as an alternative processes option for removing pollutants from liquid, solid and gaseous streams. Examples of specific contaminants and wastes treated by AOPs covered at the conference are shown in Table 1.5. While some of the applications are more viable than others the overall conclusion

Table 1.3 AOPs evaluated for water and wastewater treatment (italicised processes are covered in the following chapters).

Catalysis	*Pulsed plasma*
Electrochemical	Supercritical water oxidation
Fenton's reagent	*Ultrasound*
Ferrate	*UV*
Ionising radiation	*UV/H_2O_2*
Microwave	*$UV/H_2O_2/O_3$*
Photo Fenton's reagent	*Vacuum UV*
Photocatalysis	*Wet air oxidation*

Table 1.4 Areas of opportunities for AOPs.

Groundwater	Industrial wastewater
Odour and VOCs	Industrial sludge's
Surface water	Municipal wastewater
Swimming pools	Leachates
Water recycling	Municipal sludge's
Disinfection	Ultrapure

Table 1.5 Examples of contaminants and wastes treated with AOPs (Vogelpohl, 2003).

Amino acids	MTBE
Antibiotics	Tannery wastewater
Arsenic	Municipal sludge
Chromium	Pesticide wastewaters
Coliforms	VOCs
Disinfection by-products	Paper mill effluent
Distillery wastewater	Landfill leachate
Drug residues	*Cryptosporidium*
Glass fibre wastewater	Taste and odour causing compounds
Hospital wastewater	Grey water
Insecticide	Rubber process wastewater
Kraft bleaching wastewater	Chemical specialities wastewater
Natural organic matter	Humic materials
Nickel plating wastewater	Stillbeen derivatives
Oilfield wastewater	Cyanide
Olive mill wastewater	*Escherichia coli*
Parasites	Municipal wastewater treatment plants effluent
Phenolic wastewater	Urine
Printing wastewater	Pesticides
Seed corn wastes	Endocrine disruptors
Trinitrotoluene (TNT)	Phenolic resins
X-ray contrast media	Spent caustic

from the literature is that as processes develop and the need for greater, faster removal of pollutants is required then AOPs will find their place in water and wastewater treatment flowsheets.

REFERENCES

Suty, H., De Traversay, C. and Coste, M. (2003) Applications of advanced oxidation processes: present & future. In: *Proceedings of the 3rd Conference on Oxidation Technologies for Water and Wastewater Treatment*, Goslar, Germany, May 18–22, p. 8.

Vogelpohl (2003) *Proceedings of the 3rd Conference on Oxidation Technologies for Water and Wastewater Treatment*, Goslar, Germany, May 18–22.

2

UV photolysis: background

Mihaela I. Stefan

2.1 INTRODUCTION

Over the past two decades, photochemical technologies have become more attractive than the conventional treatment technologies, such as air stripping, adsorption on the activated carbon and biodegradation, for pollutant removal from the contaminated environment. Photochemical technologies are simple and clean, cost-effective in many applications, and, often, give the end-user the dual benefit of both environmental contaminant treatment and disinfection. Ultraviolet (UV)-driven advanced oxidation processes (AOPs) are primarily based on the generation of powerful oxidising species, such as the hydroxyl radical ($^{\bullet}OH$), through the direct photolysis of hydrogen peroxide (H_2O_2), or, through photo-induced processes as in the photo-Fenton type reactions or photocatalysis. In UV direct photolysis, the contaminant to be destroyed must absorb the incident radiation and undergo degradation starting from its excited state. The typically low concentrations of pollutants in the contaminated water and, in general, the low efficiency of photodissociation resulting from the light absorption event, limit the industrial applications of the UV photolysis process as compared to the hydroxyl radical-driven technologies, where the light absorption by the target pollutant is not absolutely required. However, there are many cases when the target pollutants are strong absorbers of the UV radiation, and therefore, their UV photolysis may become a significant component during the treatment by any UV-driven AOP. Sometimes, a contaminant can be treated by both direct photolysis and $^{\bullet}OH$ radical-induced processes. In such cases, the optimisation of the treatment process is driven by economic considerations.

There is growing interest in the application of UV light to environmental contaminant treatment, interest particularly driven by the overall public concern about the potential carcinogenic and toxic effects of contaminants in water, and by the progressively more restrictive requirements regarding the admissible levels of such contaminants in aquatic environment imposed by the regulatory agencies. As a result of this increased interest, numerous research studies have been undertaken on the UV light-induced degradation of various classes of organic and inorganic compounds of environmental concern, and a rich published literature has become available in this domain.

In this chapter, we shall briefly discuss the principles of UV direct photolysis, as well as the characteristics of the UV lamps used in UV light-based water remediation, and some process efficiency parameters. In the chapter, we have followed IUPAC recommendations, 1996 (http://www.unibas.ch/epa/) with regard to terms and definitions used in photochemistry.

2.2 FUNDAMENTALS OF UV PHOTOLYSIS

2.2.1 Electromagnetic radiation and light absorption by chemical species

In 1666, Isaac Newton was first to observe the diffraction of a white light beam when passing through a prism, opening the investigation of the properties of the light components. Almost 150 years later (1800–1801), Sir William Herschel and J.W. Ritter discovered that there was radiant energy beyond the two ends of the visible light, which were identified as infrared and UV regions, and in 1804, Thomas Young showed that the invisible chemically active radiations beyond the violet end of spectrum were subject to the laws of interference. Such an observation, along with further extended investigations, indicated that the visible, UV and infrared chemical effects were all properties of the same electromagnetic radiation, differing from each other only in frequency.

2.2.1.1 Electromagnetic radiation and bond dissociation energies

Light has both wave and particle properties (Calvert and Pitts, 1966; Parker, 1968). Light is a form of electromagnetic radiation, which explains its wave properties, such as reflection, refraction, diffraction, interference and polarisation. The relationship between the wavelength λ (m, usually given in nm in photochemical application) and the frequency ν (s^{-1}) of radiation is

$$\lambda = \frac{c}{\nu} \tag{2.1}$$

where $c = 3.0 \times 10^8 \, m\,s^{-1}$ is the velocity of light in a vacuum. The particle properties of light involve the absorption/emission of light, and can explain the photoelectric effect and photochemical reactions, based on Planck's quantum theory of radiation. Light is absorbed or emitted in discrete units of energy E, called *quanta* or *photons* ($h\nu$), which are related to the frequency of radiation through Equation (2.2):

$$E = h\nu = \frac{hc}{\lambda} = hc\overline{\nu} \tag{2.2}$$

where $h = 6.6256 \times 10^{-34}\,J\,s$ is a proportionality constant (Planck's constant), and $\overline{\nu}$ is wave number (m^{-1}, usually given in cm^{-1}). Equation (2.2) is known as *Planck's Law of Radiation* and expresses the wave/particle duality of light, that is, light comes in discrete packets of energy called photons or quanta, which have a frequency, and thus, wavelength, associated with them. One mole (6.02×10^{23}) of photons (quanta) is called an einstein, therefore, the energy of 1 einstein of wavelength λ (nm) is

$$E = 6.02 \times 10^{23} \frac{hc}{\lambda} = \frac{1.19626 \times 10^5}{\lambda} \text{kJ einstein}^{-1} \tag{2.3}$$

UV radiation is usually defined as the electromagnetic radiation of wavelength between 4 and 400 nm (Koller, 1965) [according to other sources, between 10 and 400 nm (CRC Handbook, 1991)], the spectral domain that covers the gap between the X-ray and visible regions. Figure 2.1 shows the electromagnetic spectrum from 100 to 1000 nm.

The UV spectral range of interest for the UV photolysis applications in water is the UVC [200–280 nm (Phillips, 1983)], where both the pollutants and the

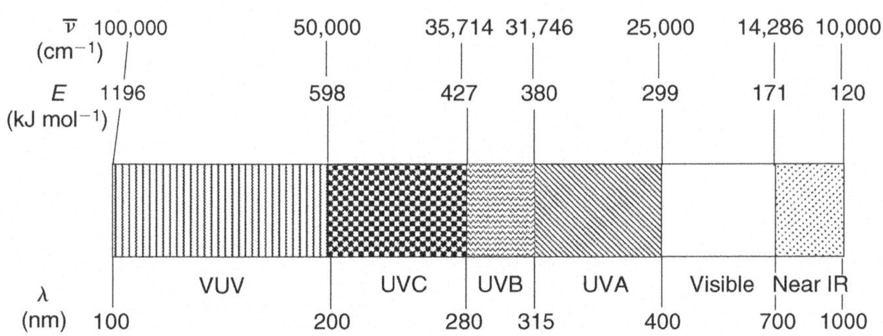

Figure 2.1 Spectrum of the electromagnetic radiation (Phillips, 1983).

water constituents (dissolved organic and inorganic compounds) absorb the radiation. Some authors consider the UVB within 280–320 nm range (Sonnemann, 1992), others define it from 290 to 320 nm (Moore, 1996). The vacuum UV (VUV) region within the 100–200 nm range is of a particular interest for AOPs, since these radiations are absorbed by water, according to reaction (2.4), thereby generating highly reactive species that further induce oxidative degradation of the dissolved organic pollutants (Adam and Oppenländer, 1986; Jakob *et al.*, 1993; Oppenländer, 2002; Oppenländer and Gliese, 2000).

$$H_2O + h\nu(\lambda < 190\,\text{nm}) \rightarrow H^\bullet + {}^\bullet OH \tag{2.4}$$

Organic compounds also absorb in the VUV spectral range, but when dissolved in aqueous solution, water is the main absorber, as it is present at a concentration of about 55.5 M, which is typically a million times or more the concentration of any contaminant present. VUV photolysis has commercial applications, particularly in the total organic carbon removal in the ultrapure water.

It is interesting to examine the energies carried by UV radiations, and to compare them with the energies required for chemical bond dissociation. Table 2.1 lists some examples of single bond dissociation energy ΔE° (kJ mol^{-1}) and the corresponding 'threshold' wavelength λ_D (nm), defined as the maximum wavelength for which the photon energy matches the bond energy required to result in a homolytic bond cleavage.

Theoretically, any radiation of $\lambda < \lambda_D$, therefore, carrying energies $E_\lambda > E_{\lambda_D}$ can break the corresponding chemical bond, as a result of light absorption. However, whether or not such a process occurs is determined by two factors: the probability of the light absorption event, which depends on the optical properties

Table 2.1 Bond dissociation energies and corresponding light 'threshold' wavelengths [(Calvert and Pitts, 1966), otherwise as indicated].

Bond	$\Delta E^\circ_{298\,K}$ (kJ mol^{-1})	λ_D (nm)
n-C_3H_7—H	407	294
C_6H_5—H	428	279
H_3C—CH_3	349	343
C_2H_5—OH	382	313
H_3C—C(O)CH_3	328	365
H_3C—Cl	340	352
C_2H_5—NH_2	324	370
C_2H_5—ONO	244	490
C_2H_5O—NO	156	765
$(CH_3)_2N$—NO	170 (Burgess and Lavanish, 1964)	703
H_2N—NH_2	236	507
OCl—O	265	451
HO—Cl	244	490
HO—OH	211	568

of the compound that are quantified in its absorption spectrum, and the probability that the excited state reached through the light absorption process proceeds to a chemical reaction, which is expressed as the quantum yield.

2.2.1.2 *Interaction of light with molecules in solution: absorption, electronic spectra and water background impact*

The *first law of photochemistry* (Grotthus–Draper law) states that only the light that is absorbed by a molecule can be effective in producing a photochemical change in that molecule (Calvert and Pitts, 1966). The quantitative aspects of light absorption in the solution by a particular constituent are expressed by the *Beer–Lambert law*, which states that the fraction of light absorbed by the system does not depend on the incident spectral radiant power P^o_λ (W m^{-1}, commonly used unit W nm^{-1}), and the amount of light absorbed is proportional to the number of the constituent molecules absorbing the radiation. The law expression is given in Equation (2.5):

$$P_\lambda = P^o_\lambda \times 10^{-(\alpha_\lambda + \varepsilon_\lambda C)l} \tag{2.5}$$

where P_λ and P^o_λ are the transmitted and the incident spectral radiant power, respectively, C (M) is the concentration of the constituent of interest in the irradiated solution, l (m, common unit cm) is the irradiation pathlength, and α_λ and ε_λ are the decadic attenuation coefficient of the medium (if water, commonly referred as the 'water background') (m^{-1}, commonly used unit cm^{-1}) and the decadic molar absorption coefficient (m^2 mol^{-1}, commonly used unit M^{-1} cm^{-1}) of the constituent in that medium, respectively. The sum $\alpha_\lambda + \varepsilon_\lambda C$ is called absorption coefficient a (m^{-1}, frequently cm^{-1}) of medium, accounting for both constituent and water background absorption per unit of irradiation pathlength.

When more than one absorbing species is present in the homogeneous solution, the exponential term in Equation (2.5) accounts for their contribution to the total absorption. Using the Beer–Lambert expression, two other common terms in photochemistry are introduced, such as *absorbance*, A, and *transmittance*, T, both dimensionless, which are interrelated in the following equations:

$$A_\lambda = \log\frac{P^o_\lambda}{P_\lambda} = (\alpha_\lambda + \varepsilon_\lambda C)l \quad \text{or} \quad A_\lambda = \sum_i \varepsilon_{\lambda,i} C_i l \tag{2.6}$$

$$T_\lambda = \frac{P_\lambda}{P^o_\lambda} = 10^{-A_\lambda} \quad \text{or} \quad A_\lambda = -\log T_\lambda \tag{2.7}$$

As shown in Equation (2.6), absorbance is an additive property. The *molar absorption coefficient* of a pure absorbing species is a constant at a given wavelength, and is a measure of the probability that the quantum–molecule interaction results in a light absorption event. The molar absorption coefficient is a fundamental physical property of a compound. It depends on the wavelength, solvent and pH, if the chromophore structure of the molecule is pH-dependent. Therefore, the absorption spectrum of the compound of interest will depend on the same factors.

The Beer–Lambert law assumes that the interactions between the solute molecules are negligible, which, in fact is the case in the AOP applications, where the concentrations of the organic pollutants are very low. In such cases, the dissolved pollutants absorb the radiation independently, and, therefore, their degradation processes can be examined independently as well. Deviations from this law are expected at high concentrations of solute, when the interactions between the molecules are important and dimers or other aggregates can form. In the latter case, the linear relationship between the absorbance and solute concentration [Equation (2.6)] does not hold.

The photochemical process induced by the radiation in a pollutant P depends on the fraction of light absorbed by that specific compound. That is only a part of the total fraction of light absorbed in the irradiated matrix. For a monochromatic radiation, the expression of fraction of light absorbed by P in the irradiated system is

$$\chi_{P,\lambda} = \frac{\varepsilon_{P,\lambda} C_P}{a_\lambda}(1 - 10^{-a_\lambda l}) \tag{2.8}$$

When a polychromatic light source is used, the fraction of light absorbed will be integrated over the emission spectrum of the light source:

$$\chi_P = \frac{\displaystyle\int_{\lambda_1}^{\lambda_2} N_\lambda (1 - 10^{-a_\lambda l}) \frac{\varepsilon_{P,\lambda} C_P}{a_\lambda} \, \mathrm{d}\lambda}{\displaystyle\int_{\lambda_1}^{\lambda_2} N_\lambda \, \mathrm{d}\lambda} \tag{2.9}$$

where N_λ represents the relative incident photon flux (relative spectral emission from the lamp), corrected for the quartz transmission.

The absorption spectrum characteristics depend on the molecular structure of the absorber, and on the interactions between that compound and the solvent. Most UV light absorbers contain double bonds or conjugated double bonds, involving carbon, nitrogen or oxygen atoms, and are characterised by delocalised π-electrons. Such systems are called *chromophores*. The environmental pollutants containing chromophoric structures include alkenes, aromatic and heterocyclic compounds, aldehydes, ketones, and carboxylic acids, textile dyes, nitro- and nitroso-derivatives, etc. The saturated compounds containing the same atoms do not absorb the light with $\lambda > 205$–$210\,\text{nm}$, and therefore, cannot be economically treated by a direct UV photolysis technology, but can often be treated by other AOPs. Table 2.2 gives a few examples of chromophores, whereas Figure 2.2 shows the electronic absorption spectra of some common substances, which are environmental pollutants (in figure, DOC = dissolved organic carbon; NOM = natural organic matter).

Even though some of the compounds illustrated in Figure 2.2 are strong absorbers of UV light [atrazine (ATZ), N-nitrosodimethylamine (NDMA), tetrachloroethylene (PCE)], since their concentrations in the aquatic environment is low (μM level), the fraction of light absorbed by those pollutants is very small as

Table 2.2 Chromophoric characteristics of typical organic functional groups (Larson and Weber, 1994, and references therein).

Chromophore	λ_{max} (nm)	ε_{max} (approx.) ($M^{-1} cm^{-1}$)	Transition
C—H or C—C	<180	1000	$\sigma \rightarrow \sigma^*$
C=C	180	10,000	$\pi \rightarrow \pi^*$
C=C—C=C	220	20,000	$\pi \rightarrow \pi^*$
Benzene	260	200	$\pi \rightarrow \pi^*$
Naphtalene	310	200	$\pi \rightarrow \pi^*$
Anthracene	350	10,000	$\pi \rightarrow \pi^*$
Phenol	275	1500	$\pi \rightarrow \pi^*$
Aniline	290	1500	$\pi \rightarrow \pi^*$
RS–SR	300	300	$n \rightarrow \pi^*$
C=O	280	20	$n \rightarrow \pi^*$
C=C—C=O	320	50	$n \rightarrow \pi^*$
C=N	<220	20	$n \rightarrow \pi^*$
N=N	350	50	$n \rightarrow \pi^*$
N=O	300	100	$n \rightarrow \pi^*$
Ar—NO_2	280	7000	$\pi \rightarrow \pi^*$
Indole	290	5000	$\pi \rightarrow \pi^*$
(*s*)-Triazine	230	40,000	$\pi \rightarrow \pi^*$

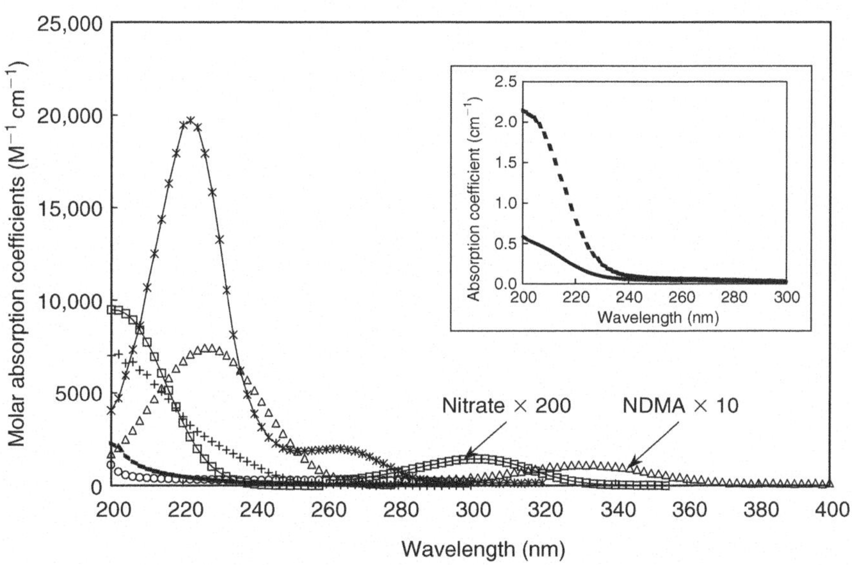

Figure 2.2 Examples of absorption spectra of environmental pollutants: PCE (+), ATZ (×0.5, −*−), NDMA (△), MTBE (×1000, ○), 1,4-dioxane (×100, ---), nitrate (□). Insert, two different surface water qualities: nitrate-rich water (– – –), water containing 3.5 ppm DOC as NOM and 3 ppm NO_3^- (—).

compared to that of the water background that is primarily due to the dissolved organic matter and inorganic constituents, such as nitrate ions. Therefore, very often, the water background constituents are strong competitors with the environmental pollutants for the light, and background absorbance should be carefully considered when evaluating the treatment efficiency of a UV technology, both with respect to the lamp type and oxidant alternatives. The inner filter effect of the dissolved natural organic matter or inorganic constituents (e.g. nitrate) narrows the range of the effective radiation in pollutant treatment with medium-pressure mercury (Hg) lamps to $\lambda \geqslant 235$ nm.

Organic pollutants, such as methyl *tert*-butyl ether (MTBE) and 1,4-dioxane, are very weak absorbers of the UV radiation, therefore, they cannot be treated by UV photolysis. However, hydroxyl radical-driven AOPs have proven very effective in removal of such pollutants from contaminated waters.

2.2.1.3 Excited states: generation, chemical and physical deactivation

A light absorption event is a one-quantum process per absorbing molecule [*second law of photochemistry*, Starck–Einstein law (Calvert and Pitts, 1966)], and results in the generation of an electronically excited state (singlet or triplet). High power light source techniques (flash photolysis and lasers) that can generate high photon flows may lead to bi- (multi-)photonic absorption events per absorbing molecule within the lifetime of the electronically excited state (μs–ns). Multiple photon processes require a large concentration of the excited states, and are very unlikely to occur at the low concentrations of pollutants that are generally observed in the aquatic environment. The electronic transitions in the unsaturated or heterocyclic compounds are of $\pi \rightarrow \pi^*$ and $n \rightarrow \pi^*$ types, where either a π or a n electron from bonding and non-bonding orbitals, respectively, are promoted to higher energy π antibonding orbitals. In general, the $\pi \rightarrow \pi^*$ electronic transition bands have larger molar absorption coefficients than the $n \rightarrow \pi^*$ transition bands, therefore, higher probability to occur (see also Table 2.2). The $n \rightarrow \pi^*$ transition is expected to occur in molecules containing atoms with non-bonding electron pairs, such as oxygen and nitrogen, and is responsible for the longer wavelength-absorption band of the pollutant absorption spectrum. For example, the absorption spectrum of NDMA (Figure 2.2) exhibits two absorption bands, a strong one at $\lambda_{max} = 228$ nm ($\varepsilon_{max} = 7378\,M^{-1}cm^{-1}$), and a weak one, of lower energy, at $\lambda_{max} = 352$ nm ($\varepsilon_{max} = 109\,M^{-1}cm^{-1}$), which are regarded as a $\pi \rightarrow \pi^*$ intramolecular charge transfer and as a $n \rightarrow \pi^*$ transition, respectively (Toussaint and Würkert, 1978). When irradiated in those absorption bands, the N—N bond in the NDMA molecule breaks, resulting in two radicals that undergo further reactions; the extent of such an event depends on the quantum yield.

It is important to mention that given the low concentrations and short lifetimes of the electronically excited states (ns–μs) reached through the light absorption process, the bimolecular chemical reactions proceeding from such species have a very low probability, unless they occur at very fast diffusion-controlled rates. Most of the energy of such excited states is dissipated as heat in the surrounding matrix, when the molecule in the excited state returns to its ground (stable) state through non-radiative processes, such as internal conversion or intersystem crossing. Fluorescence (from excited singlet state) and phosphorescence (from excited triplet state) may occur as radiative deactivation processes from the excited states. For more details on the photophysical processes of the excited states see *Jablonski diagram* (Parker, 1968). The radiative deactivation processes reduce the probability of chemical reactions from the excited states, and, implicitly, impact the applicability of the UV direct photolysis as a treatment option.

2.2.2 Quantum yield

2.2.2.1 *Definition and importance in the AOPs*

The *quantum yield* is one of the most important parameters in evaluating the efficiency of a photochemical process. There are several types of quantum yield terms used in photochemistry, such as primary quantum yields (φ), product quantum yields (Φ), quantum yields of fluorescence, decomposition, rearrangement, etc. (Calvert and Pitts, 1966). In general, for direct photochemical processes in the environmental applications, it is not necessary to determine all the individual quantum yields corresponding to various pathways that the excited state can follow, but only the *reaction quantum yield*, which encompasses all reactions in a given set of conditions that lead to the decay of the parent pollutant *P*, as a result of light absorbed by *P* in the system. The quantum yield for direct UV photolysis of *P* is defined as

$$\Phi_r(\lambda) = \frac{\text{total number of moles of } P \text{ transformed}}{\text{total number of moles of photons of } \lambda \text{ absorbed by } P \text{ in the system}} \tag{2.10}$$

The same quantum yield can be defined in kinetic terms, such as the rates of decay and of light absorption by *P* in the system:

$$\Phi_r(\lambda) = \frac{-\dfrac{d[P]}{dt}(\text{mol L}^{-1}\,\text{s}^{-1})}{I_a(\lambda)(\text{mol L}^{-1}\,\text{s}^{-1})} \tag{2.11}$$

where $I_a(\lambda)$ ($\text{mol L}^{-1}\text{s}^{-1}$) is the rate of light absorption of *P*. Quantum yield is a unitless parameter [Equation (2.11)]. Sometimes, it is expressed in units of mol einstein^{-1} as derived from Equation (2.10), but given the definition of einstein, Φ_r is in fact unitless.

According to the second law of photochemistry, the primary quantum yields cannot exceed unity. However, the reaction quantum yields may be greater than one, if dark (thermal) chain reactions involving radicals that were generated in the photochemical step occur with the ground state of the compound under consideration [e.g. the quantum yield of Fe(II) generation in the ferrioxalate system irradiated with $\lambda = 254$ nm, $\Phi = 1.25$ (Hatchard and Parker, 1956)]. Such chain reactions imply bimolecular reactions, whose probability is low in the aquatic environment where the concentrations of the organic pollutants are low. Some water constituents may inhibit the chain reactions. Therefore, most of the quantum yields reported in the literature for the UV photolysis of various compounds, are very small. Table 2.3 compiles a few literature data on the quantum yields of various chemicals of environmental interest.

The application of direct UV photolysis process is therefore most suitable to aquatic contaminants, such as NDMA, PCE, and some pesticides, that exhibit reasonably large quantum yields. Since the quantum yield of the highly reactive $^{\bullet}OH$ radical generation through the direct UV photolysis of H_2O_2 is near unity and wavelength independent, the UV/H_2O_2 process is regarded as an efficient AOP and is increasingly applied to water remediation, with both monochromatic and polychromatic light sources.

2.2.2.2 Factors affecting the quantum yield

The reaction quantum yield is defined for monochromatic radiation, and assumes monophotonic absorption. It may depend on wavelength, pH, pollutant concentration, temperature, solvent and concentration of dissolved oxygen. The quantum yields are determined at specific wavelengths and at very low concentrations under 'vanishing absorption' conditions (von Sonntag and Schuchmann, 1992; von Sonntag *et al.*, 1993) where the interactions between solute molecules are not important. In practice, the determination of the quantum yield should be made at very short irradiation times, for example, low conversion of pollutant concentration, such that the change in the fraction of light absorbed during the irradiation is negligible.

Over the wavelength range of a given absorption band (same electronic transition), the quantum yield is expected to be wavelength independent (Schwarzenbach *et al.*, 1993; Zepp, 1978, and references therein). When polychromatic light sources are used for pollutant treatment, and the target compound exhibits more than one

Table 2.3 Quantum yields of UV photolysis of selected compounds.

Organic compound	Φ_r	Experimental conditions
PCE	0.84	$\lambda = 253.7$ nm (Mertens and von Sonntag, 1995)
NDMA	0.300 ± 0.006	$\lambda = 253.7$ nm, pH 3–10 (Stefan *et al.*, 2002)
Simazine	0.083	$\lambda = 253.7$ nm (Nick and Schoeler, 1995)
ATZ	0.05	$\lambda = 253.7$ nm (Nick and Schoeler, 1995)
	0.035	$\lambda = 253.7$ nm (Bolton and Stefan, 2002)
4-Chlorophenol	0.24 ± 0.04	Xe lamp, pH 3–8 (Ho *et al.*, 1996)
	0.25	$\lambda = 296$ nm (Boule *et al.*, 1982)

absorption band, it is necessary to determine the quantum yield at various wavelengths. Such a curve representing the variation of quantum yield of a photochemical reaction, as a function of wavelength constitutes the *action spectrum* of that specific compound (Braun *et al.*, 1991).

For polychromatic light, Braun *et al.* (1991) define a reaction *quantum efficiency* as the ratio of the number of molecules of product formed to the number of photons absorbed (N_a), in the spectral region used, during the reaction period:

$$\eta = \frac{\int_{\lambda_1}^{\lambda_2} \Phi(\lambda) N_a(\lambda)\, d\lambda}{\int_{\lambda_1}^{\lambda_2} N_a(\lambda)\, d\lambda} \tag{2.12}$$

The quantum efficiency η equals the quantum yield Φ only if the quantum yield is wavelength independent.

Examples of wavelength-dependent quantum yields for the UV photolysis of some organic compounds are: chlorpyrifos: $\Phi_{254\,nm} = 0.016$; $\Phi_{313\,nm} = 0.052$ (Wan *et al.*, 1994), 2-nitrophenol: $\Phi_{254\,nm} = 1.7 \times 10^{-3}$; $\Phi_{365\,nm} = 2.4 \times 10^{-6}$ (Tixier *et al.*, 2000), diazinon: $\Phi_{254\,nm} = 0.082$; $\Phi_{313\,nm} = 0.012$ (Wan *et al.*, 1994). Similarly, the quantum yield for UV photolysis of nitrate (quantum yield of nitrite generation) is wavelength-dependent: pH $\sim$ 6: $\Phi_{254\,nm} \sim 0.09$, $\Phi_{310\,nm} \sim 0.065$, $\Phi_{313\,nm} \sim 0.063$; pH 11.5: $\Phi_{207\,nm} = 0.25$, $\Phi_{254\,nm} = 0.17$, $\Phi_{282\,nm} = 0.024$ (Mack and Bolton, 1999; Mark *et al.*, 1996).

The quantum yield of the NDMA UV photolysis is pH dependent: $\Phi \sim 0.30$ (pH 3–10); $\Phi = 0.24$ (pH 11.5); $\Phi = 0.16$ (pH 12) (Stefan *et al.*, 2002).

Table 2.4 gives two examples of reaction quantum yield dependence of the *pollutant concentration*.

Table 2.4 Quantum yields of the UV photolysis of ATZ and metazachlor ($\lambda = 254$ nm): concentration dependence. $I_a = 7.1\ (\pm 0.2) \times 10^{-7}$ mol L^{-1} s^{-1} (Hessler *et al.*, 1993a).

Atrazine			Metazachlor		
C_o (µmol L^{-1})	A_{254}	Φ_r	C_o (µmol L^{-1})	A_{254}	Φ_r
3	0.012	0.047 (±0.006)[2]	4	0.002	0.44 (±0.022)[3]
6	0.023	0.046 (±0.004)[2]	12	0.005	0.41 (±0.08)
17	0.061	0.042 (±0.004)	17	0.007	0.37 (±0.06)
23[1]	0.078	0.037 (±0.003)	37	0.016	0.34 (±0.03)
33[1]	0.116	0.038 (±0.003)	73	0.032	0.32 (±0.02)
41[1]	0.144	0.033 (±0.003)	250[1]	0.111	0.26 (±0.02)
66[1]	0.23	0.034 (±0.003)	360[1]	0.159	0.23 (±0.02)
78[1]	0.274	0.031 (±0.003)			
110[1]	0.388	0.029 (±0.003)			
160[1]	0.567	0.028 (±0.003)			

[1]Non-vanishing absorption; [2]$I_a = 6.4 \times 10^{-7}$ mol L^{-1}s^{-1}; [3]$I_a = 7.7 \times 10^{-7}$ mol L^{-1}s^{-1}.

Table 2.5 Quantum yield of monochloroacetic acid photohydrolysis.

Φ_r (Cl^-)	Conditions
1.1 ± 0.1	$\lambda_{excitation}$ = 184.9 nm; T = 303 K; water
0.5 ± 0.1	$\lambda_{excitation}$ = 213.9 nm; T = 303 K; water
0.6 ± 0.1	$\lambda_{excitation}$ = 228.8 nm; T = 303 K; water
0.35 ± 0.05	$\lambda_{excitation}$ = 253.7 nm; T = 303 K; water
0.67	$\lambda_{excitation}$ = 253.7 nm; T = 303 K; $X_{CH_3OH} = 0.55$[1]
0.31	$\lambda_{excitation}$ = 253.7 nm; T = 299 K; water
0.69	$\lambda_{excitation}$ = 253.7 nm; T = 342 K; water

[1]X is the mole fraction of methanol in the aqueous solution used in experiment.

As mentioned above, the reaction quantum yield may be sensitive to *temperature*, even though the photochemical reactions require very low activation energies. An example is the quantum yield of $^{\bullet}OH$ radical generation from nitrite ion irradiated in its second absorption band (λ = 351 nm) at pH 8 (Zellner *et al.*, 1990): $\Phi_{278\,K} \sim 0.027$, $\Phi_{288\,K} \sim 0.038$, $\Phi_{298\,K} \sim 0.046$, $\Phi_{308\,K} \sim 0.058$, $\Phi_{328\,K} \sim 0.091$.

The quantum yield of chloride ion released during the UV photolysis of monochloroacetic acid is wavelength- (Neumann-Spallart and Getoff, 1979), temperature- (Smith *et al.*, 1939) and solvent-dependent (Neumann-Spallart and Getoff, 1979), as summarised in Table 2.5. Monochloroacetic acid is recommended as an actinometer for the 254 nm radiation, since the photochemical reaction of chloride ion release and the corresponding quantum yield are well characterised (Kuhn *et al.*, 1989). The quantum yield dependence on the quantum energy is explained by the cage effects that governs the homolytic scission of C—Cl bond, whereas the *solvent* influences the polarity of the transition state during the hydrolysis of the excited state (Neumann-Spallart and Getoff, 1979).

UV photolysis of acetone in aqueous solution occurs with low, wavelength-independent quantum yields, but affected by the level of *dissolved oxygen* [$\Phi_{270\,nm} = 0.061$ (Anpo and Kubokawa, 1977)]; $\Phi_{200-300\,nm} = 0.061 \pm 0.002$ and 0.14 ± 0.01 in air-saturated and oxygen-free solutions, respectively (Stefan and Bolton, 1996). The latter effect is related to the quenching of triplet acetone $^3(n \rightarrow \pi^*)$ excited states by the oxygen molecules.

The *photochemistry of dissolved natural organic matter* may play a role in the pollutant treatment, enhancing or diminishing the photochemical yield of pollutant degradation. Some inorganic components of the natural waters, such as nitrate and nitrite ions, metal complexes, etc. may have similar effects. Indirect photo-induced processes are mainly due to the reactivity of the photochemically generated excited states of natural organic matter constituents (humic and fulvic acids, flavins, polyheterocyclic compounds, etc.), that can be involved in energy and/or electron transfer processes, either directly to the ground state of the organic pollutant, or to the dissolved oxygen, leading to highly reactive species, such as singlet oxygen $^1O_2$ and dioxygen radical anion $^{\bullet}O_2^-$. Another reactive species generated photochemically in natural systems, particularly in the nitrate- or iron-rich waters, is the hydroxyl radical •OH, which increases the reaction quantum yield of pollutants in environmental systems.

In general, such photosensitised processes are insignificant, and the quantum yields determined in pure water are nearly the same as those obtained in natural waters. This is explained by the fact that the concentrations of the natural water constituents that undergo such reactions is very low, and also, by the natural organic matter itself being a 'sink' for the generated reactive species, competing efficiently with the target pollutants (Fasnacht and Blough, 2002; Zepp and Baughman, 1978). The chemistry of photosensitised processes is very complex and has been extensively reported in the published literature (Cooper *et al.*, 1989; Hoigné *et al.*, 1989; Larson and Weber, 1994; Leifer, 1988; Mabury and Crosby, 1994; Tsao and Eto, 1994; Zepp, 1988).

2.2.2.3 Determination of quantum yields: chemical actinometry

Quantum yields are determined experimentally, either directly, based on Equation (2.11), or indirectly, relative to a known quantum yield of the UV photolysis of a 'reference' compound that is added to the irradiated solution. In the latter case, the photolysis rates of the two compounds are compared, and should be proportional to the corresponding quantum yields and molar absorption coefficients. When a polychromatic light source is used for an UV photolysis application, the quantum yield should be known at several wavelengths of the spectral emission of the lamp. Therefore, a weighted-average quantum yield should be used in determining the rate of decay of target pollutant, unless, sophisticated models that can incorporate the wavelength-dependent quantum yields are available. Several studies are available in the literature regarding the experimental methods for the quantum yields determination (Beltran *et al.*, 2000; Bolton and Stefan, 2002; Draper, 1987; Braun *et al.*, 1991; Johns, 1971; Leifer, 1988; Murov *et al.*, 1993; Schwarzenbach *et al.*, 1993; Zepp, 1978; von Sonntag and Schuchmann, 1992).

For direct quantum yield determinations, the light absorbed by the solute is calculated based on the incident photon irradiance (E_p, mol $cm^{-2} s^{-1}$) on the solution surface, and an account is taken for the attenuation of the light beam while it traverses the solution (Bolton and Linden, 2003; Bolton and Stefan, 2002; Braun *et al.*, 1991; Schwarzenbach *et al.*, 1993). The incident photon irradiance is measured by either physical (radiometry) or chemical (actinometry) methods. A chemical actinometer is a chemical system (or single compound), which undergoes a well-characterised photochemical reaction with a known quantum yield. Ideally, the quantum yield should be wavelength independent, insensitive to the actinometer concentration, to the radiation rate (no multiple photonic absorption), and to temperature. When a polychromatic light source is used, the action spectrum of the actinometer should be known. Among the chemical actinometers, the most frequently used ones are: potassium ferrioxalate, potassium persulphate, uranyl oxalate, iodide/iodate and uridine. Kuhn *et al.* (1989) and Taylor (1971) published detailed compendia of actinometers and their use, but recent studies on new actinometers are available in the literature as well (Mark *et al.*, 1990; Nicole *et al.*, 1990; Pfeifer *et al.*, 1998; Rahn, 1997; Schulz *et al.*, 2001; Stefan *et al.*, 2001a). Some chemical actinometers, such as the iodide/iodate and the uridine actinometers have been used for the determination of fluence rate distribution in laboratory-scale UV reactors with both monochromatic and polychromatic radiation sources (Schulz *et al.*, 2001; Stefan *et al.*, 2001a). For practical applications, UV sensors (UV sensitive photocells, most commonly based on silicon carbide) are used in full-scale installation UV reactors to determine the irradiance E_o ($W\,m^{-2}$, often $mW\,cm^{-2}$) at given location relative to the UV lamps. The sensor irradiance readings are related to the lamp performance and *aging*, sleeve fouling and water transmittance on a local basis in a given reactor.

2.2.3 Kinetics of photochemical reactions

Knowledge of the kinetic parameters of the photochemical reactions of organic pollutants is important in order to both select and optimise the appropriate UV treatment technology. The kinetic parameters are required to predict the pollutant decay in a specific reactor geometry and for a given light source, flow and water quality, such that the treatment achieves the desirable discharge levels of the target pollutant.

The theory of photochemical reaction kinetics of the environmental aquatic pollutants has been extensively discussed by Leifer (1988), Schwarzenbach *et al.* (1993), Zepp *et al.* (1981a,b, 1985). The general expression of the *rate* of direct photolysis of a pollutant P under monochromatic radiation λ, is given by Equation (2.13):

$$-\left(\frac{d[P]}{dt}\right)_\lambda = \Phi_r(\lambda) I_{a,P}(\lambda) \tag{2.13}$$

where $\Phi_r(\lambda)$ and $I_a(\lambda)$ ($mol\,L^{-1}\,s^{-1}$) are the reaction quantum yield and the rate of light absorption by P, respectively. $I_a(\lambda)$ is a function of the fraction of light absorbed by P [see Equation (2.8)], therefore, the above equation becomes:

$$-\left(\frac{d[P]}{dt}\right)_\lambda = \Phi_r \frac{N_o(\lambda)}{V} \frac{\varepsilon_P(\lambda)[P]}{a(\lambda) + \varepsilon_P(\lambda)[P]} (1 - 10^{-[a(\lambda)+\varepsilon_P(\lambda)[P]]l}) \tag{2.14}$$

In the above equation we use $N_o(\lambda)$ as a symbol for the incident photon flow ($mol\,s^{-1}$), instead of the IUPAC symbol Φ_p in order to avoid any confusion with the quantum yield. This equation is equivalent to the photon fluence rate-based rate expression of the UV photolysis of P:

$$-\left(\frac{d[P]}{dt}\right)_\lambda = \frac{\Phi_r(\lambda) E_p^o(\lambda) \varepsilon_P(\lambda)}{1000} \frac{1 - 10^{-a(\lambda)l}}{a(\lambda)l} [P] \tag{2.14a}$$

where E_p^o ($mol\,cm^{-2}\,s^{-1}$) is the photon fluence rate, 1000 is a conversion factor from L (dm^3) to cm^3, and the other parameters are as defined previously. Equation (2.14a) assumes a uniform radiation field across the solution surface.

From a general perspective, Equation (2.14) describes the interaction of the two 'reagents', namely, the chemical pollutant and the light. If the light absorption by the pollutant P is very strong, due to a large concentration of P and a large molar absorption coefficient at wavelength λ, such that $\varepsilon_P(\lambda)[P] \gg \alpha(\lambda)$, the limiting reagent is the light, and the degradation process follows a *zero-order kinetics* law [Equation (2.15)].

$$[P_o] - [P_t] = \Phi_r(\lambda)\frac{N_o(\lambda)}{V}t \tag{2.15}$$

Such cases are not encountered in natural water photochemistry, and are generally applicable to the actinometric systems.

As mentioned earlier, the concentrations of pollutants in the aquatic environment are generally low, and the main absorber of the UV radiation is the water background (humic substances, metal complexes, nitrate/nitrite ions, etc.). In such cases, the limiting reagent is the pollutant concentration. As $\alpha(\lambda) \gg \varepsilon_P(\lambda)[P]$, therefore, $(\alpha(\lambda) + \varepsilon_P(\lambda)[P]) \cong \alpha(\lambda)$, there are two extreme situations that can be discussed with regard to the kinetic Equation (2.14) and the fraction of light absorbed in the system. When the total absorbance is very low $[\alpha(\lambda) + \varepsilon_P(\lambda)[P]]$ $l \leqslant 0.02$, that is, 'vanishing' absorption condition] and most of the incident light is transmitted through the water, the term $10^{-(\alpha(\lambda)+\varepsilon_P(\lambda)[P])l}$ can be expanded in Taylor series and the simplified Equation (2.14) integrates to

$$\ln\frac{[P]_o}{[P]_t}\bigg|_\lambda = 2.303\Phi_r(\lambda)\frac{N_o(\lambda)}{V}\varepsilon_P(\lambda)lt \tag{2.16}$$

which is the expression of the *first-order kinetics* law, with the first-order kinetics rate constant $k_1^\lambda(\mathrm{s}^{-1})$ dependent on the photochemical characteristics of the pollutant and the incident photon flow:

$$k_1^\lambda = 2.303\Phi_r(\lambda)\frac{N_o(\lambda)}{V}\varepsilon_P(\lambda)l \tag{2.17}$$

When nearly all of the light is absorbed in the water $[(\alpha(\lambda) + \varepsilon_P(\lambda)[P])l \geqslant 2]$, the term $1 - 10^{-(\alpha(\lambda)+\varepsilon_P(\lambda)[P])l} \cong 1$, and the integrated form of Equation (2.14) is

$$\ln\frac{[P]_o}{[P]_t}\bigg|_\lambda = \Phi_r(\lambda)\frac{N_o(\lambda)}{V}\frac{\varepsilon_P(\lambda)}{\alpha(\lambda)}t \tag{2.18}$$

In this case, the reaction pathlength does not have any effect on the rate since all light is absorbed in the water. Thus, the optimisation of the reaction pathlength is an important parameter in reactor design.

Often, the light absorption lies between the two cases addressed above, and the first-order kinetics expression [Equation (2.19)] should account for the attenuation of photon flow passing through the water:

$$\ln\frac{[P]_o}{[P]_t}\bigg|_\lambda = \Phi_r(\lambda)\frac{N_o(\lambda)}{V}\frac{\varepsilon_P(\lambda)}{\alpha(\lambda)}[1 - 10^{-\alpha(\lambda)l}]t \tag{2.19}$$

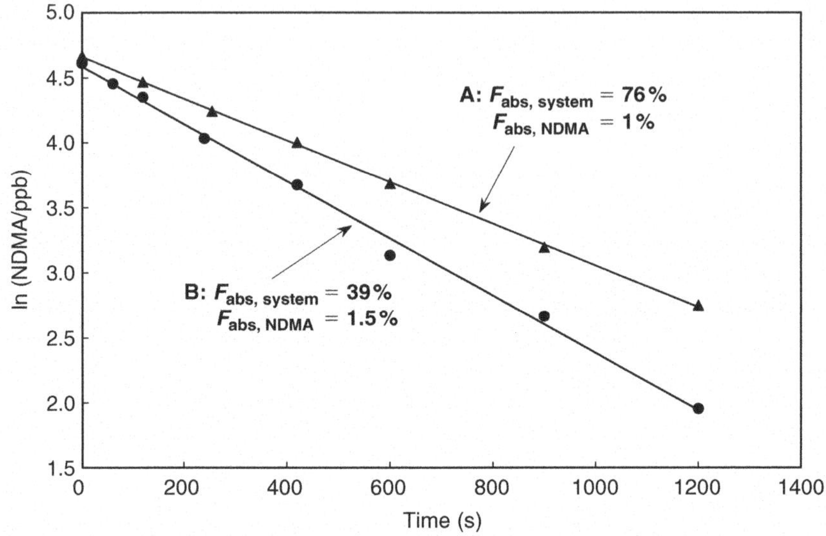

Figure 2.3 NDMA photolysis in a quasi-collimated beam.

When polychromatic light sources are used, the overall rate of degradation is integrated over the spectral distribution of the light emitted by the source within the wavelength range of interest (usually 200–300 nm), thereby accounting for the wavelength dependence of the quantum yield.

Figure 2.3 gives an example of the impact of water background on the pollutant photodegradation (Stefan *et al.*, 2002). In this specific example, the competition for light between NDMA and water background results in different fractions of light absorbed by NDMA, therefore, different rates of decay. The increase of amount of dissolved organic matter in the water from 2 mg L^{-1} (A) to 12 mg L^{-1} (B), translated to an increase of fraction of light absorbed by water background and a decrease of fraction of light absorbed by NDMA, from 39% to 76%, and from 1.5% to 1%, respectively.

One concludes that the photochemical reaction kinetics depends on the light absorption conditions of the system and the reaction quantum yield of the pollutant under consideration. Therefore, any water matrix characteristics that can affect the two reaction efficiency parameters (quantum yield and fraction of light absorbed), and implicitly the generation and evolution of the excited states, may affect the reaction kinetics and the reaction pathway. Such characteristics are water constituents (absorbers, photosensitisers and suspended solids), pH, ionic strength, concentration of dissolved oxygen and temperature (Mill and Mabey, 1985). Water temperature does not usually have a significant impact on photochemical reaction kinetics, as the activation energies for the photochemical processes are fairly low (10–20 kJ mol^{-1}), as encountered in diffusion-controlled reactions (Mill and Mabey, 1985). For example, Stefan *et al.* (2002) calculated an activation energy of

$10\,\text{kJ mol}^{-1}$ for the UV photodegradation of NDMA in simulated surface water. Along with the water characteristics, the spectral emission of the light source affects the reaction kinetics (Stefan *et al.*, 2001b).

Therefore, for AOP applications, it is important to determine the photo-induced reaction kinetic parameters in a representative water, using a light source with similar spectral characteristics as those chosen for that particular full-scale application.

In general, the UV photolysis processes are of low efficiency when compared to the $^{\bullet}$OH radical-induced AOPs. However, there are pollutants, such as NDMA, for which UV photolysis treatment is more efficient than the $^{\bullet}$OH radical-driven AOPs.

2.2.4 Figures-of-merit for the technical application of AOPs

Bolton *et al.* (2001) introduced two figures-of-merit in order to assess the technology efficiency with respect to the electrical energy consumption associated with the contaminant removal through a UV-driven AOP. The two figures-of-merit are the electrical energy per order of magnitude of removal (E_{EO}) and the electrical energy per mass of removal (E_{EM}), which characterise two different kinetic order regimes, namely, first-order (low contaminant concentrations) and zero-order (high contaminant concentrations) kinetics, respectively. These figures-of-merit can be used to compare various AOPs with respect to the electrical energy efficiency, but also in sizing the UV equipment to generate the required electrical energy dose for specific treatment goals. These parameters vary greatly between different types of UV reactors, and, therefore, will be reactor specific.

2.2.4.1 Electrical energy per order (E_{EO})

Electrical energy per order (E_{EO}) is defined as the electrical energy in kilowatt hours (kWh) required to degrade a contaminant C by one order of magnitude in a unit volume [e.g. m^3 (1000 L) or, frequently in North America, 1000 US gallons (1 US gal = 3.785 L)]. The E_{EO} values (kWh m^{-3} per order) are calculated with the following formulas (Bolton *et al.*, 2001):

$$E_{\text{EO}} = \frac{Pt1000}{V\lg(c_i/c_f)} \quad \text{batch operation} \tag{2.20}$$

$$E_{\text{EO}} = \frac{P}{F\lg(c_i/c_f)} \quad \text{flow-through operation} \tag{2.21}$$

where P, V, t, c_i and c_f, are the rated power (kW) of the UV system, the volume (L) of water treated in the time t (h), the initial and final concentrations (mol L^{-1}) of the contaminant, respectively, and F is the water flow rate ($\text{m}^3\,\text{h}^{-1}$). Since the flow rate is inversely proportional to the rate of pollutant decay through the same factor, the residence time, E_{EO} should not depend on F; however, in practical applications it has been observed that E_{EO} decreases when F increases, which can be explained by hydraulic effects leading to a better mixing of the water volume at higher flow rates. Equation (2.20) can be expressed in terms of the first-order kinetics rate constant k_1 (min^{-1}):

$$E_{EO} = \frac{38.4P}{Vk_1} \tag{2.22}$$

Thus, E_{EO} is both contaminant- and reactor-specific. E_{EO} is inversely proportional to the reaction quantum yield and the molar absorption coefficients of the target pollutant over the emission spectrum of the lamp, therefore, it depends on the amount of light absorbed by the pollutant. The absorption coefficient of the water background has a negative impact on E_{EO}. E_{EO} is also a function of the spectral radiant power over a specific wavelength range (usually 200–300 nm for polychromatic-, or 253.7 nm for monochromatic-light sources), therefore it is lamp spectrum dependent. E_{EO} decreases as the reaction pathlength increases; therefore, optimisation of UV reactor performance with regard to the synergetic effects of pollutant photochemistry, hydraulics, and reactor optics is critical.

The relationship between fluence (UV dose) and E_{EO} has been theoretically established (Bolton, 2002), and can be used for laboratory-scale UV reactors, where, the investigated systems can be regarded as 'ideal systems', consistent with the assumptions made in the kinetic and fluence rate distribution models. However, large-scale systems present significant complexities to modelling. In order to predict the reactor performance for contaminant destruction, one must have thorough knowledge of many empirical conditions, such as the detailed nature of the fluid flow and mixing of the fluid throughout the reactor, the spatial distribution of light within the reactor (this will be affected by the lamp types, power level and age of the lamps, transmission of the quartz sleeves, fouling of the sleeves, the water quality in terms of light absorption and light scattering properties), and, in the UV/oxidant processes, the oxidant concentrations as a function of location within the reactor. Almost all of these variables are reactor specific. In evaluating the effectiveness of the technology, the scientific questions, such as theoretical calculations of required fluence for a given application, are soon overtaken by the difficulty of engineering a practical solution and ensuring, through measurement, that the equipment performs properly. Hence, an engineering parameter such as E_{EO}, is a very practical parameter which can be easily measured on equipment of all sizes (including laboratory-scale equipment which, if designed well, should approach theoretical maximum efficiency and subsequently the lowest possible E_{EO}) and can be used to easily compare the effectiveness of systems regardless of their configurations. It is a very easy, efficient, and ultimately robust parameter to normalise comparisons between the sizing and performance of full-scale installations independent of their configuration, and it does not rely on any mathematical or modelling assumptions that can so easily lead to poor conclusions. Hence, significant effort is taken to minimise the E_{EO} for full-scale technologies through attacking all of the variables listed above in the equipment design process.

2.2.4.2 *Electrical energy per mass (E_{EM})*

Electrical energy per mass E_{EM} (kWh kg^{-1}) is the electrical energy in kWh required to bring about the degradation of a unit mass (e.g. 1 kg) of a contaminant C in polluted water. Used for high concentrations of pollutant, E_{EM} depends on the concentration, as the degradation rate law does (zero-order kinetics). Equations (2.23)–(2.24) give the expressions of E_{EM} (Bolton *et al.*, 2001):

$$E_{\mathrm{EO}} = \frac{Pt1000}{VM(c_i - c_f)} \quad \text{batch operation} \tag{2.23}$$

$$E_{\mathrm{EO}} = \frac{P}{FM(c_i - c_f)} \quad \text{flow-through operation} \tag{2.24}$$

where all symbols are as above, and M is the molar mass of C (g mol^{-1}).

This figure-of-merit is mostly used as an efficiency indicator in the laboratory-scale studies, particularly for the total organic carbon removal.

2.2.5 Reaction mechanisms in the UV photolysis

2.2.5.1 *Major reaction pathways in the UV photolysis*

The electronically excited state RX* generated through the light absorption process [first step of the reaction mechanism, Equation (2.25)] is highly energetic and can either deactivate to the ground state of the molecule through physical processes (such as fluorescence, phosphorescence, or non-radiative deactivation) or undergo 'dark' (thermal) chemical reactions. The most common chemical reaction pathways followed by the excited state are summarised in the following set of equations:

$$\mathrm{RX} + h\nu \rightarrow \mathrm{RX}^* \tag{2.25}$$

$$\mathrm{RX}^* \rightarrow (\mathrm{R}^\bullet \cdots {}^\bullet\mathrm{X})_{\mathrm{cage}} \rightarrow \mathrm{R}^\bullet + {}^\bullet\mathrm{X} \tag{2.26}$$

$$(\mathrm{R}^\bullet \cdots {}^\bullet\mathrm{X})_{\mathrm{cage}} \rightarrow \mathrm{RX} \tag{2.27}$$

$$\mathrm{RX}^* \rightarrow (\mathrm{R}^+ \cdots \mathrm{X}^-)_{\mathrm{cage}} \rightarrow \mathrm{R}^+ + \mathrm{X}^- \tag{2.28}$$

$$\mathrm{RX}^* + \mathrm{O}_2 \rightarrow \mathrm{RX}^{+\bullet} + \mathrm{O}_2^{-\bullet} \tag{2.29}$$

$$\mathrm{RX}^* + {}^3\mathrm{O}_2 \rightarrow \mathrm{RX} + {}^1\mathrm{O}_2 \tag{2.30}$$

The homolytic bond scission [Equation (2.26)] is the predominant chemical pathway, and occurs in the solvent cage. Escaped from the solvent cage, the radicals undergo further oxidation/reduction reactions, depending on their structure; for example, the carbon-centred radicals ($R^{\bullet}$) react with dissolved oxygen leading to peroxyl ($R{-}O_2^{\bullet}$) and oxyl radicals ($R{-}O^{\bullet}$), whose chemistry has been extensively investigated (Alfassi *et al.*, 1997; von Sonntag and Schuchmann, 1997). Organic pollutants may undergo different reaction mechanisms in the presence and the absence of dissolved oxygen, resulting in different end products. Recombination of the primary radicals in the solvent cage [Equation (2.27)] leading to the parent molecule occurs with high probability, which explains the low quantum yields frequently observed with organic compounds in condensed media as compared to those in the gas phase.

In polar solvents, such as water, heterolytic bond scissions [intramolecular electron transfer, Equation (2.28)] have been observed, particularly in aromatic halogenides (Choudhry *et al.*, 1986), atrazine (Nick *et al.*, 1992) and chlorinated dioxins (Nohr *et al.*, 1994).

Energy and electron-transfer processes [Equations (2.29) and (2.30)] to oxygen are also possible, but require a relatively long-lived excited state, such as the triplet state. Dissolved oxygen is the most likely reaction partner of the excited state of an organic pollutant, particularly if the concentrations of the other water constituents are low. Very reactive species, such as dioxygen radical anion [commonly known as superoxide radical, Equation (2.29)] and singlet oxygen [Equation (2.30)] are generated, along with the pollutant radical cation and the parent molecule in the ground state.

Proton and hydrogen atom transfers from the excited state to surrounding molecules are also possible (Larson and Weber, 1994), but with a much lower probability than the reactions (2.26)–(2.30).

In any AOP, dissolved oxygen plays an important role in the pollutant oxidative degradation, affecting the degradation reaction rates, the type and distribution of the reaction intermediates, and the level of mineralisation of the organic carbon. In UV photolysis processes an insignificant degree of mineralisation usually is observed.

2.2.5.2 *Examples of reaction mechanisms of UV photolysis of organic pollutants*

The next few pages depict the reaction mechanisms for the UV photolysis of NDMA, PCE and ATZ. These are organic pollutants of high environmental concern, and can be treated by UV direct photolysis, with either monochromatic or polychromatic light sources. However, as mentioned before, the water quality characteristics and the action spectrum of each of these organic compounds are the decision-making factors with regard to the lamp technology. Also, it is important to mention that both PCE and ATZ degradation rates are larger in the UV/H_2O_2 process than in the UV direct photolysis (Hessler *et al.*, 1993b; Mertens and von Sonntag, 1994, 1995; Stefan *et al.*, 2001b), whereas no significant contribution of the $^{\bullet}OH$ radical-induced processes was observed in the case of NDMA (Stefan *et al.*, 1999). Thus, in UV/oxidant applications, PCE and ATZ will degrade by a combination of direct photolysis and $^{\bullet}OH$ radical-induced oxidation.

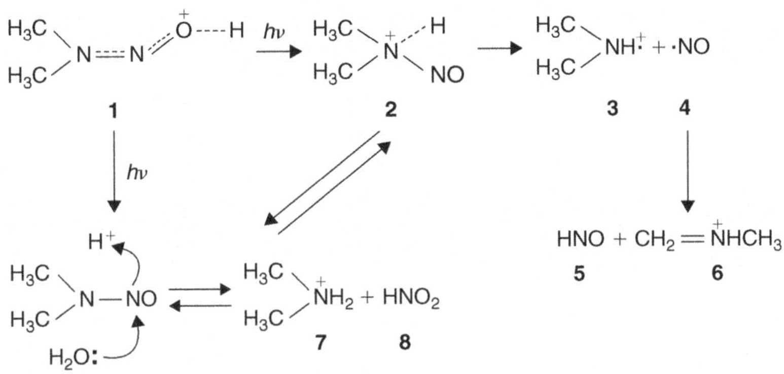

Figure 2.4 Proposed reaction mechanism for the UV photolysis of NDMA (Polo and Chow, 1976b).

NDMA photodegradation NDMA is released in the human environment from several industries, and has been found in a number of groundwater aquifers and some treated surface waters as well. NDMA is also a disinfection by-product (Mitch and Sedlak, 2002), and UV photolysis is considered to be the only economical treatment technology. NDMA is restricted in environmental remediation, such as drinking water and wastewater type applications, with effluent target treatment goals on the order of 2–10 ng L^{-1}. Figure 2.4 summarises the primary reactions in the photolysis of NDMA in aqueous solution (Polo and Chow, 1976b).

Two distinct pathways are proposed for the decay of the excited state of NDMA (**1**). One implies the hydrolysis of the excited state, leading to the protonated form of dimethylamine (DMA) **7** ($pK_a = 10.73$) and nitrous acid **8**. The second assumes the rearrangement of the excited state of NDMA to the N-nitrosodimethylammonium ion **2** followed by the N—N bond homolytic scission to the dimethylaminium radical **3** and nitric oxide **4**. Hyponitrous acid **5** and N-methylidenmethylamine **6** are generated as a result of the reaction of radicals **3** and **4** in the solvent cage (Burgess and Lavanish, 1964). The degradation products of NDMA photolysis are: di- and monomethylamine, nitrite and nitrate ions, formaldehyde and formic acid (Stefan and Bolton, 2002). Provided the low concentration levels of NDMA in the aquatic environment, none of these degradation products represents a by-product issue, therefore, are always below the regulatory action levels.

DMA is not a photosensitive compound and is not considered a dangerous substance, but recent studies indicated that it could interfere with the chloramination process in drinking water forming NDMA (Mitch and Sedlak, 2002). However, the occurrence of such a process depends on several factors, it is not always an issue, and must be evaluated from case to case.

PCE photodegradation Tetrachloroethylene (perchloroethylene, PCE) is a nonflammable solvent that is chemically inert and largely used industrially as a degreaser and a dry-cleaning solvent. It is usually found in groundwater at low levels (μg L^{-1}). The PCE limit in drinking water is set at 10 μg L^{-1} in Germany, and 5 μg L^{-1} in North America. Mertens and von Sonntag (1995) have investigated the photolysis ($\lambda = 253.7\,\text{nm}$) of PCE in aqueous solution. The major primary (photochemical) processes include C—Cl bond homolytic scission (i), solvent addition (ii), and electron transfer (heterolytic C—Cl cleavage) (iii):

$$CCl_2{=}CCl_2 + h\nu \rightarrow CCl_2{=}CCl^{\bullet} + Cl^{\bullet} \tag{i}$$

$$CCl_2{=}CCl_2 + h\nu + H_2O \rightarrow CCl_2(OH)CCl_2H \tag{ii}$$

$$CCl_2{=}CCl_2 + h\nu \rightarrow CCl_2{=}CCl^{+} + Cl^{-} \tag{iii}$$

The chlorine atom is one of the representative reactive species in water treatment. Being a strong electrofile, the chlorine atom reacts by addition to the unsaturated bonds, by H-atom abstraction, and by participating in electron-transfer processes with water constituents, such as carbonate–bicarbonate ions. One of the chlorine atom reactions in the PCE treatment is the addition to the double bond of the parent pollutant, leading to an intermediate radical responsible for the generation of trichloroacetic acid (TCA) as one of the degradation products (iv):

$$CCl_2{=}CCl_2 + Cl^{\bullet} \rightarrow CCl_3{-}CCl_2^{\bullet} \xrightarrow{O_2} CCl_3{-}CCl_2O_2^{\bullet} \rightarrow\rightarrow \mathbf{CCl_3{-}COOH} + HCl \tag{iv}$$

The chlorinated ethylic alcohol produced in reaction (ii) and the carbocation in reaction (iii) degrade rapidly to dichloroacetic acid (DCA), but with a lower reaction quantum yield than TCA:

$$CCl_2(OH){-}CCl_2H \xrightarrow{-HCl} C(O)Cl{-}CHCl_2 \xrightarrow{H_2O} \mathbf{Cl_2HC{-}COOH} \tag{v}$$

$$CCl_2{=}CCl^{+} + H_2O \rightarrow CCl_2{=}C(OH)Cl \rightarrow \mathbf{Cl_2HC{=}COOH} \tag{vi}$$

The peroxyl radical (trichlorovinylperoxyl radical) resulting from the carbon-centred radical generated in the photochemical reaction (i) is an intermediate species in reactions leading eventually to final products such as chloride ions, CO_2, CO and oxalic acid.

The fate of this radical follows a second-order reaction kinetics by self- or cross-termination reactions with peroxyl radicals generated in the system, through a tetraoxide structure. The reactions below exemplify only the self-termination route and some of the final products:

$$2CCl_2{=}CClO_2^{\bullet} \rightarrow CCl_2{=}CCl{-}O_4{-}CCl{=}CCl_2 \rightarrow 2CCl_2{=}CClO^{\bullet} + O_2 \tag{vii}$$

$$CCl_2{=}CClO^{\bullet} \Leftrightarrow {}^{\bullet}CCl_2{-}C(O)Cl \xrightarrow{O_2}\rightarrow {}^{\bullet}OCCl_2{-}C(O)Cl \xrightarrow{-Cl^{\bullet}} C(O)Cl{-}C(O)Cl \tag{viii}$$

$$C(O)Cl{-}C(O)Cl + 2H_2O \rightarrow \mathbf{HOOC{-}COOH} + \mathbf{2HCl} \quad \text{(ix)}$$

$$^{\bullet}OCCl_2{-}C(O)Cl \rightarrow O{=}CCCl_2 + {}^{\bullet}C(O)Cl \quad \text{(x)}$$

$$Cl_2C{=}C{=}O + H_2O \rightarrow \mathbf{CO_2} + 2HCl \quad \text{(xi)}$$

$$^{\bullet}C(O)Cl \rightarrow \mathbf{CO} + Cl^{\bullet} \quad \text{(xii)}$$

$$^{\bullet}C(O)Cl + O_2 \rightarrow\rightarrow {}^{\bullet}OC\,(O)\,Cl \rightarrow CO_2 + Cl^{\bullet} \quad \text{(xiii)}$$

The laboratory studies indicated that PCE treatment by UV photolysis in aerated aqueous solutions results in the high yield of mineralisation (chloride ions and CO_2 are the major final products), and low levels of haloacetic acids (TCA and DCA). In natural waters, which contain dissolved natural organic matter, radicals such as chlorine are scavenged by NOM, and also by carbonate and bicarbonate ions with large rate constants [($k(Cl^{\bullet} + HCO_3^-) = 2.2 \times 10^8 M^{-1}s^{-1}$, $k(Cl^{\bullet} + CO_3^{2-}) = 5.0 \times 10^8 M^{-1}s^{-1}$ (Mertens and von Sonntag, 1995)], preventing the formation of TCA from potentially present PCE in water.

ATZ photodegradation Atrazine is a widely used herbicide that became an environmental concern. The admissible levels of atrazine in drinking water are set at $0.1\,\mu g\,L^{-1}$ in the EC countries, and at $3\,\mu g\,L^{-1}$ by the EPA in the USA.

ATZ exhibits strong absorption in the UVC range and photolyzes under both monochromatic and polychromatic radiations. Figure 2.5 presents the major reaction products released during the ATZ photolysis (Torrents *et al.*, 1997).

Both ATZ and its degradation products undergo UV photolysis, with low quantum yields (Table 2.3). Hydroxyatrazine is the major primary degradation product of ATZ photolysis at $\lambda = 253.7$ nm, accounting for 90% of the ATZ degraded (Nick *et al.*, 1992). Nick *et al.* (1995) investigated the primary processes involving and following the photochemical step (generally accepted for the photolysis of chloroaromatic compounds), in the particular case of ATZ. These reactions are as follows:

$$ACl + h\nu \rightarrow ACl^* \quad \text{(a)}$$

$$ACl^* \rightarrow (A^{\bullet} \cdots {}^{\bullet}Cl)_{cage} \rightarrow A^{\bullet} + {}^{\bullet}Cl \quad \text{(b)}$$

$$(A^{\bullet} + {}^{\bullet}Cl)_{cage} \rightarrow (A^+ \cdots Cl^-)_{cage} \rightarrow A^+ + \mathbf{Cl^-} \quad \text{(c)}$$

$$A^+ + H_2O \rightarrow \mathbf{AOH} + H^+ \quad \text{(d)}$$

$$(A^+ + Cl^+)_{cage} \rightarrow ACl \quad \text{(e)}$$

$$ACl^* \rightarrow A^+ + Cl^- \quad \text{(f)}$$

The heterolytic scission of C—Cl bond (f) is favoured over the homolytic cleavage (b), in polar solvents, especially in water. As both mechanisms lead to the same product, hydroxy derivative **AOH**, the authors used methanol as a radical scavenger, to assess the contribution of each pathway to the product formation. In

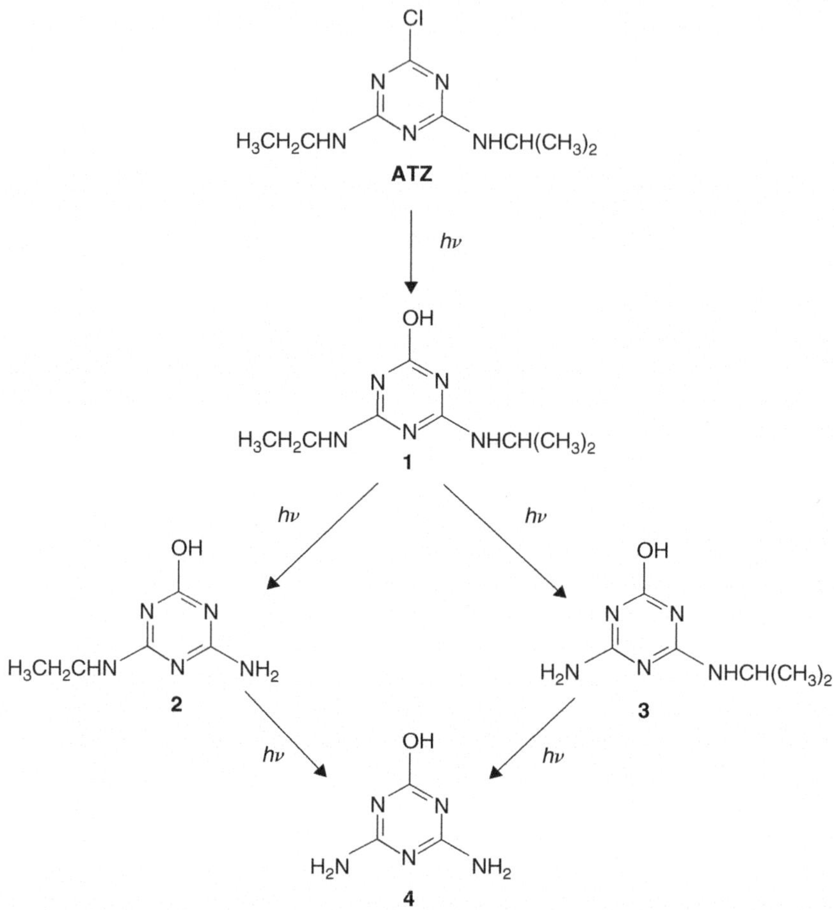

Figure 2.5 Products of the UV photolysis of ATZ: **1** hydroxy ATZ; **2** desisopropylhydroxy ATZ; **3** desethylhydroxyatrazine; **4** desethyldesisopropylatrazine.

the presence of methanol, the hydrogenated ATZ AH should be formed, as a result of hydrogen-abstraction reaction (g):

$$A^{\bullet} + CH_3OH \rightarrow AH + {}^{\bullet}CH_2OH \tag{g}$$

The analysis revealed only 0.5% of dehalogenated ATZ as being formed at a degradation level of ATZ of 90%, indicating that the major chemically deactivation of the excited state is the heterolytic bond scission [either of reactions (c) and (f)]. The same study showed that the saturation of ATZ solutions with oxygen reduced the quantum yield of ATZ destruction by 30% as compared to oxygen-free photolysis. This indicated that O_2 is a quencher of the excited state that can be a short-lived triplet state.

ATZ is a good example of chemical compound that can be treated by both UV light-induced processes, both in the absence (UV photolysis) and in the presence of oxidants, such as H_2O_2 and O_3 ($^{\bullet}OH$ radical-induced oxidation). Of the two indicated oxidants, H_2O_2 would be the option, provided the formation of bromate, a tightly restricted by-product, in bromide-containing waters. The extent of UV photolysis component in the UV/oxidant treatment of ATZ depends on several factors, such as water quality, ATZ and H_2O_2 concentrations, UV light source, such that the optimum treatment strategy lies in the consideration of all these factors.

One can conclude that the mechanistic studies on the UV light-induced degradation of chemical pollutants in the aquatic environment provide relevant information to be considered in the full-scale application of UV technologies, particularly with regard to the reaction intermediates and their probable levels in the treated water, ensuring that those do not exceed the regulatory standards. As most product studies reported in the literature have been conducted in either deionised or 'simulated' natural water, the relative concentration ratios of the parent pollutant to its degradation intermediates, as well as between various intermediates, do not necessarily reflect the degradation product yields for a given contaminant treated by a UV technology in real waters, such as surface, ground and drinking water. Therefore, due to the complexity of chemical reactions in the aquatic environment, the by-product formation from a given contaminant may or may not be an issue, and the evaluation should be performed on case-to-case basis.

2.2.5.3 Indirect (sensitised) photolysis of organic pollutants

Additional degradation pathways of environmental pollutants in natural waters are triggered by the photosensitisation processes from natural organic matter, or from photolysis of water constituents. Such processes might be of a particular interest in the polychromatic light applications, considering that most of the studies indicated photosensitisation reactions occurring as a result of simulated sunlight irradiation of NOM ($\lambda > 300\,\text{nm}$). The photochemistry of natural organic matter present in the surface waters is extensively addressed in the published literature (Canonica and Freiburghaus, 2001; Cooper *et al.*, 1989; Hoigné *et al.*, 1989; Mabury and Crosby, 1994, and references therein; Thomas-Smith and Blough, 2001). The long-lived (triplet) excited states of water organic constituents can induce chemical transformation of the organic pollutants either through direct energy transfer or by generating reactive species that further may react with pollutant. Zepp *et al.* (1985) assessed the steady-state concentration of humus-derived triplets in the 10^{-15}–$10^{-13}\,\text{mol L}^{-1}$ range in the upper levels of sunlight irradiated natural waters, being proportional to the UV absorption coefficients of waters. The authors determined the rate constant of the energy transfer reactions in water from the triplet state of humic substances $k_{et} = 3 \times 10^9\,\text{L mol}^{-1}\,\text{s}^{-1}$, with a probability of unity, and identified the classes of compounds that can be photosensitised by energy transfer, such as: polycyclic aromatic hydrocarbons, nitroaromatics, conjugated dienes. The humic substances triplet states have energies of minimum $250\,\text{kJ mol}^{-1}$.

The reactive species generated through the light-induced process in natural organic matter are: molecular singlet oxygen ($^1O_2$), peroxyl radicals ($RO_2^{\bullet}$), hydroperoxyl and superoxide anion radicals $HO_2^{\bullet}/O_2^{-\bullet}$, hydroxyl radicals ($OH^{\bullet}$), hydrated

Table 2.6 Concentration of reactive species in waters (Mill, 1990).

Species	Concentration (mol L^{-1})
Singlet molecular oxygen	10^{-14}–10^{-13}
Peroxyl radicals	10^{-11}–10^{-10}
Hydroxyl radical	10^{-18}–10^{-19}
Hydrated electron	10^{-17}–10^{-15}
Hydroperoxyl/superoxide	10^{-9}–10^{-8}

electrons (e^-_{aq}), carbonate radical ($CO_3^{-\bullet}$). Secondary by-products of NOM photolysis are ozone and H_2O_2, which explain the presence of these compounds in fresh waters. Their photochemistry leads to reactive species such as hydroxyl radical. Table 2.6 gives the approximate concentrations of these reactive species in waters.

Among the most frequently encountered reactions in the aquatic environment involving short-lived reactive species, those involving the *hydroxyl radical* seem to prevail. This radical is mostly generated through the photolysis of nitrate (Mack and Bolton, 1999; Zepp *et al.*, 1987), but may also be created by photolysis of Fe-OH complexes or H_2O_2. The $^{\bullet}OH$ radical reactions with the organic compounds are hydrogen atom abstraction and addition to the unsaturated bonds. Mabury and Crosby (1994) investigated the degradation of some pesticides in filter-sterilised field water irradiated with simulated sunlight, and attributed their fate to the reaction with the $^{\bullet}OH$ radical generated by the humic matter. The order of rate of decay was similar to that of the rate constants of those pesticides with the $^{\bullet}OH$ radical: carbaryl > carbofuran > MCPA. Torrents *et al.* (1997) studied the nitrate-mediated photolysis of ATZ in natural waters under simulated solar irradiation ($\lambda > 290$ nm), and observed a significant increase of the yield of ATZ degradation in the presence of nitrate as compared to the UV direct photolysis. The increase of natural organic matter content reduced the yield of $^{\bullet}OH$ radical-induced process by 85%. The authors attributed this effect to the scavenging of $^{\bullet}OH$ radicals by NOM.

Molecular *singlet oxygen* is mainly generated as a result of quenching of the triplet state of humic substances by the ground state of the dissolved oxygen:

$$^3NOM^* + {}^3O_2 \rightarrow NOM + {}^1O_2 \tag{2.31}$$

The process occurs with a quantum yield of 0.5–2.6% under broadband irradiation (Haag *et al.*, 1984). The major pathway for $^1O_2$ decay in natural waters is physical quenching by water to the ground state of oxygen. Since the difference between the two energetic levels of oxygen is only 22.7 kcal mol^{-1}, singlet oxygen is the major product resulting from triplet sensitizers in natural waters. $^1O_2$ is an electrophile, and a better electron acceptor than $^3O_2$. It reacts with a few very reactive types of organic compounds, which contain electron-rich double bonds, or are readily oxidised such as mercaptans, anilines, and phenolic structures.

Haag and Mill (1987) studied the direct and indirect photolysis of water-soluble azo-dyes in the presence of humic acids (5 mg L^{-1} as DOC) under irradiation with a xenon (Xe) lamp. The indirect photolysis was induced by the singlet oxygen and the peroxyl radicals generated from the dissolved organic matter.

Gerecke *et al.* (2001) investigated the effect of NOM on the degradation of phenylurea herbicides, that is, isoproturon, diuron, 2,4,6-trimethylphenol, chlorotoluron, fenuron, metoxuron, and fluometuron, under simulated solar irradiation. The authors concluded that the singlet oxygen was not an important oxidant in these systems, and attributed the increased quantum yields of some pesticide degradation to the reaction with $^3NOM^*$.

The *peroxy radicals* ($RO_2^\bullet$) are formed by the reaction of the excited NOM chromophores with dissolved molecular oxygen. These species are not significantly scavenged by NOM (at [NOM] $<5\,mg\,L^{-1}$), have half-lives generally long, and react mostly by H-abstraction with alkyl phenols, aromatic amines, thiophenols and imines (Schwarzenbach *et al.*, 1993).

Hydrated electron and *superoxide anion* generated in irradiated natural waters are not as significant as the other reactive species with respect to the degradation reactions induced to the organic pollutants (Cooper *et al.*, 1989).

Another aspect that should be briefly addressed in the UV light-driven AOP applications is the effect of radiation on the natural organic matter composition, particularly with respect to the by-products formed and their concentrations relative to drinking water regulations. At large fluences (UV doses), the natural organic matter undergoes decomposition, to lower molecular weight organic compounds, such as aldehydes and carboxylic acids (Corin *et al.* (1996)). Liu *et al.* (2002) investigated the formation of disinfection by-products (THMs, HAAs, aldehydes and carboxylic acids) from secondary chlorination of simulated surface water ($3\,mg\,L^{-1}$ total organic carbon as Suwannee River NOM) irradiated with low- and medium-pressure Hg lamps, and pulsed UV technology in the fluence range 0–64,000 $J\,m^{-2}$. Low fluences did not affect chlorinated DBP formation during subsequent chlorination, and fewer THMs and HAAs were formed at high fluences (UV doses), followed by chlorination. At large fluences, UV irradiation resulted in the formation of aldehydes and carboxylic acids that are known to have an adverse effect on drinking water biostability.

Gallard and von Gunten (2002) examined the kinetics of the THMs' formation and chlorine consumption during the chlorination of surface natural waters, and identified the potential precursors of THMs. The study indicated that UV/visible irradiation does not alter THM formation but leads to a higher chlorine demand.

The literature studies on the potential by-product formation provide useful information that increases both general public awareness and regulators' stringency with regard to the maximum admissible levels of such compounds in drinking water. Both critical evaluation of literature information and practical experience in the UV applications, indicate that the extent of formation of what are known as disinfection by-products (DBPs) is highly water quality- and treatment process-dependent. In addition, the water processing involves more and more integrated technologies, where UV application is used in conjunction with other technologies, such as membranes, reverse-osmosis and biologically activated carbon. Such options minimise considerably the levels of DBPs in drinking water, by removal of either their precursors or the compounds themselves.

The irradiation of nitrate-rich waters leads to nitrite as a nitrate photolysis product with a reaction quantum yield dependent on the excitation wavelength, temperature, pH and nitrate concentration (Mack and Bolton 1999, and references therein). Both nitrate and nitrite ion concentrations fall under the regulatory standards for drinking water, with maximum permissible levels for nitrite of 0.1 and 3.3 mg L^{-1}, in European Community countries and North America, respectively. Nitrate and nitrite ions absorb the UV radiation within the 200–400 nm range, with the lower energy-absorption band being rather weak, but still important provided the high transmittance of water above 260 nm. Photolysis of nitrate and nitrite generate reactive species such as the $^{\bullet}OH$ radical, but also $NO^{\bullet}$, $NO_2^{\bullet}$, $ONOO^{\bullet}$ (Mack and Bolton, 1999; Mark *et al.*, 1996). In a laboratory study, Sharpless and Linden (2001) observed that nitrite production in simulated surface waters irradiated with polychromatic light is pH- and light intensity-dependent and increases with the concentration of dissolved organic matter. Mechanistically, such a trend can be explained by the consumption of $^{\bullet}OH$ radicals and oxygen by the dissolved organic matter, therefore limiting the probability of oxidation of nitrite ions back to nitrate. von Sonntag and Schuchmann (1992) addressed the same issue, under the conditions of drinking water disinfection with monochromatic light. The authors concluded that at low fluences required in UV disinfection, formation of nitrite as a nitrate photolysis by-product is of a minor importance.

While recognising the importance of rigorous monitoring of nitrite, from practical application of UV-driven technologies in water processing, it has been observed that nitrite is not always an issue, and, as supported by literature studies, is process-, water quality-, and light source-dependent. In surface waters, nitrite is quickly oxidised to nitrate, as is, in drinking water processing involving chlorination. Also, in UV/oxidant processes, usually applied to environmental contaminant treatment, such as UV/H_2O_2, nitrite is easily oxidised to nitrate by highly reactive $^{\bullet}OH$ radicals [$k(^{\bullet}OH + NO_2^-) = 1.0 \times 10^{10}\ M^{-1}s^{-1}$ (Buxton *et al.*, 1988)].

2.3 UV LAMPS USED IN AOPs AND THEIR SPECTRAL DISTRIBUTIONS

2.3.1 Spectral radiant emittance

Light is produced as a result of the radiative deactivation of the excited states of atoms or molecules. The spectral wavelength ranges emitted by the artificial light sources are determined by the nature of the emitting atoms or molecules, the concentration of the excited states, and the probability of the radiative transitions.

The radiation emission can be related to the concept of 'black body radiator'. The 'black body radiator' is defined as an *idealised* source of radiant energy, which absorbs totally the incident radiation at all wavelengths, and neither reflects nor transmits any of them. The total radiant power emitted by a black body M ($W\,m^{-2}$) depends only on the temperature T (K) (Stefan-Boltzmann law, 1879), and the maximum of the emittance spectrum (λ_{max}, m) shifts toward lower wavelengths as the temperature increases (Wien displacement law, 1894) (Phillips, 1983):

$$M(T) = 5.67 \times 10^{-8} T^4 \quad \lambda_{\max} = 2.8978 \times 10^{-3} T^{-1} \tag{2.32}$$

Planck's quantum theory (1901) explained the black body radiation, and allowed the derivation of the equation of the energy density distribution in a given wavelength range ($\lambda + d\lambda$). Equation (2.33) gives the Planck's distribution of spectral radiant emittance (Phillips, 1983):

$$M_\lambda(T) = \frac{2\pi hc^2}{\lambda^5 (e^{hc/k_B \lambda T} - 1)} \tag{2.33}$$

where $k_B = 1.38 \times 10^{-23}\,\mathrm{J\,K^{-1}}$ is Boltzmann's constant, and the other symbols are as defined previously. If the wavelength interval is 1 nm, the spectral emittance of a black body radiator becomes:

$$M_\lambda(T) = \frac{3.742 \times 10^{-25}}{\lambda^5 (e^{14.39 \times 10^{-3}/\lambda T} - 1)}\ \mathrm{W\,m^{-2}\,nm^{-1}} \tag{2.34}$$

In the above equation, λ has SI unit, m. The equation is valid for an *idealised* black body radiator. For a real radiator – which is the case of some of light sources – the expression above is multiplied by the *spectral emissivity* factor $\varepsilon_{\lambda T}$ which is less than unity, and depends on both wavelength and temperature.

Thus, in order to provide a significant output within 200–300 nm range, an emitter should operate at very high temperatures. Such temperatures are reached in some gas-discharge lamps, but for very short durations. Figure 2.6 shows black body curves for a few temperatures over the 150–1000 nm wavelength range. It suggests how the UVC output varies with plasma temperature in the light sources.

The lamp selection for a particular AOP application is governed by several factors, among which, the most important are:

a. Absorption and action spectrum of the pollutant (UV photolysis) or added oxidant ($^{\bullet}$OH radical-driven AOPs). The lamp spectral emission domain should cover as much as possible the absorption spectrum of the reactant.
b. Spectrum of the water background. As discussed in the previous section, the dissolved organic matter and inorganic constituents of the water exhibit significant light absorption at $\lambda < 230$ nm, and may exert an inner filter effect, diminishing the fraction of light absorbed by the target pollutant.
c. Lamp power and radiant efficiency of the lamp. The photochemical degradation rate is proportional to the spectral irradiance, therefore, to the rate of photons emitted by the lamp.
d. Lamp geometry. The shape and dimensions of the lamp should be suitable to the reactor design, such that a maximum reactor performance is achieved.
e. Manufacturing and operation costs.

The UV lamps used in the AOPs are those that provide a significant output in the 200–300 nm wavelength range. Depending on the operating conditions, they can be classified as *arc lamps*, where the electrons produced through a continuous

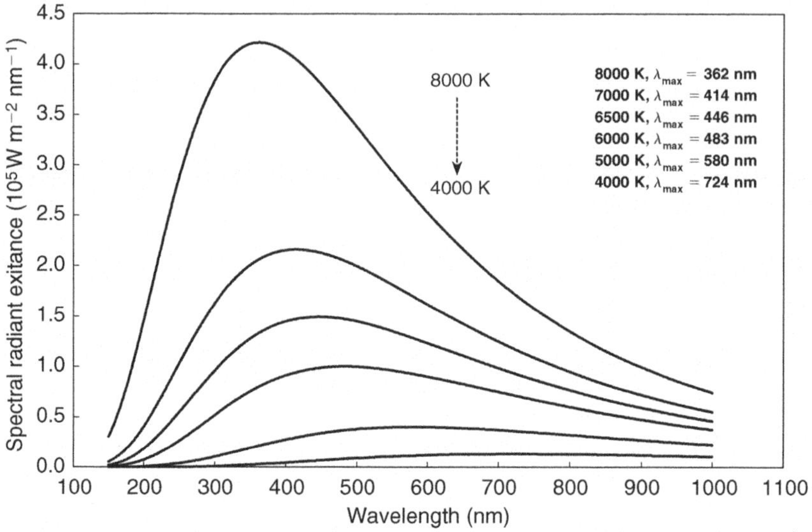

Figure 2.6 Spectral radiant emittance from the black body radiator.

electrical discharge between the two lamp electrodes collide with the gas atoms leading to excited states, and *pulsed lamps*, where the electrical discharge between the electrodes operates in pulsed mode, and the gas is heated to very high temperatures (≥13,000 K). The mercury (Hg)-free *dielectric barrier discharge lamps* (*excimer lamps*) have captured an increasing attention over the past few years, and some of them are now commercially available for laboratory-scale experiments. The Hg and rare gases are commonly used as fill gases in the non-excimer UV lamps. The Hg, which is a liquid in its natural state, has relatively high vapour pressure and low ionisation energy, is relatively chemically inert toward the quartz and electrode materials, and the resonance radiations are in the UV range.

The light sources, either in arc or in pulse discharge mode of operation, emit either quasi-monochromatic or polychromatic radiations. The presence and relative intensity of different lines emitted depend on the pressure of the fill gas.

The theoretical aspects of various light sources as well as their practical applications in environmental treatment are well represented in the published literature (Braun *et al.*, 1991; Calvert and Pitts, 1966; Masschelein, 2000; Phillips, 1983).

2.3.2 Monochromatic light sources and their spectral distributions

2.3.2.1 Low-pressure Hg lamps

The low-pressure Hg lamp operates in either arc ('hot' cathode) or glow discharge mode ('cold' cathode). The fill gas is a mixture of Hg and an inert gas, usually argon (Ar), at a pressure of a few torr, which is much higher than Hg vapour pressure (10^{-3}–10 torr, at the usual operating temperature conditions of ~40°C). The role

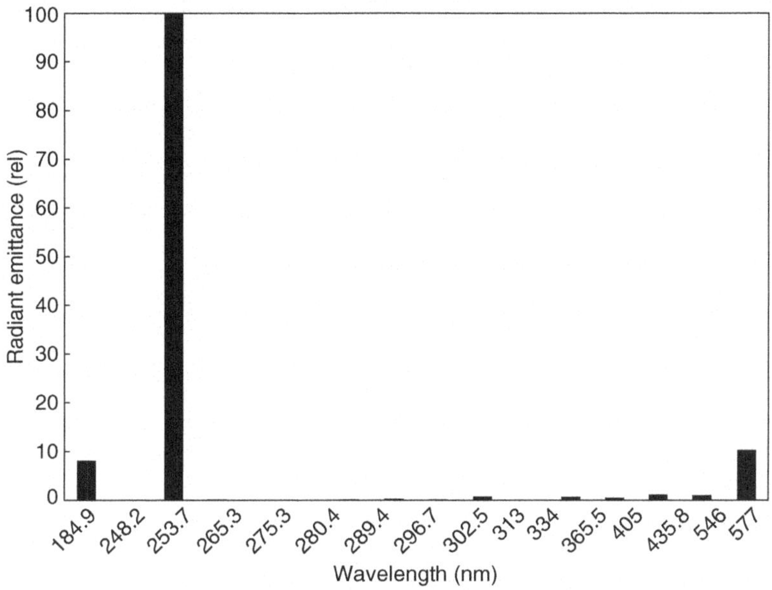

Figure 2.7 Relative spectral distribution of the light emitted from a low-pressure Hg arc lamp [largely taken from Calvert and Pitts (1966)].

of Ar is both to initiate and maintain the discharge, and to amplify the excitation of Hg atoms. The collisions between the highly energetic Ar atoms and the Hg atoms result in energy transfer with the generation of Hg atoms in excited states. Since the Hg vapour pressure is fairly low, the probability of generation of high-energy excited states is very low, therefore, the mostly populated states would be the lowest energy excited states, such as $^3P_1$ and $^1P_1$. The radiative deactivation of these states to the ground state result in the resonance Hg lines 253.7 and 184.9 nm, respectively. Radiative transitions from the upper excited states of the Hg atoms also occur, but their intensities are very low as compared to the 253.7 nm line, given the low populations of those excited states. Figure 2.7 gives the relative radiant emittance from a low-pressure Hg lamp.

As indicated in the figure, the lamp emits mainly (85–90%) at 253.7 nm, and about 7–10% at 184.9 nm. Both radiations are important for the AOP applications, since they are part of the absorption spectra of most of pollutants. The 184.9 nm radiation is filtered out by the ordinary quartz, but is transmitted by the high quality quartz; entering the water, it is absorbed within a few millimetres depth ($\varepsilon = 0.03135\,M^{-1}\,cm^{-1}$), and effects the water photolysis with a quantum yield $\Phi = 0.33$ (Getoff, 1991). The highly reactive $^{\bullet}OH$ radicals generated induce the oxidative degradation of the organic pollutants dissolved in the water. The direct photolysis of the pollutants dissolved in the water occurs as well at 184.9 nm, but water itself is the main absorber of this radiation. There are several commercial applications based on UV photolysis of water, mainly for total organic carbon removal in ultrapure water used in semiconductor, pharmaceutical and food and beverage industries.

The 253.7 nm radiation has proven efficient for remediation of water contaminated with pollutants of various chemical structures, such as unsaturated chlorinated aliphatics and aromatics, nitrosoderivatives, etc.

The energy balance in a low-pressure Hg lamp with 100% power input can be summarised as follows: ~82% is power in arc, ~18% are electrical losses, ~65% is radiation power, and 18% are non-radiative losses (Giller, 2000).

The low-pressure Hg lamps operate near ambient temperature (usually up to 40°C), and the radiant efficiency at 253.7 nm is up to 40%. In general, the temperature of the lamp influences indirectly the lamp efficiency, through the Hg vapour pressure. If the temperature is too low, the Hg condenses reducing the vapour pressure, and thus, the lamp efficiency. Also, if the temperature rises, the Hg vapour pressure increases, reducing the efficiency due to the self-absorption process. The optimum operating temperature of low-pressure Hg lamps is around 40°C. The Hg pressure can be regulated by introducing Hg as amalgams with bismuth, indium and gallium. These are called high intensity low-pressure amalgam lamps. The Hg vapour pressure can be also controlled through the lamp design, particularly with regard to the tube diameter.

Low-pressure Hg lamps have life times up to 25,000 operating hours. The limiting factors of lamp life are the lamp current, electrode failure and solarisation. Due to these effects, the initial radiant efficiency of these lamps decays over the operation time.

2.3.2.2 Excimer lamps

The incoherent excimer lamps are relatively new on the market, and given their narrow emission bands, are regarded as quasi-monochromatic light sources. The radiation is produced in either microwave (Ametepe *et al.*, 1999) or dielectric barrier discharge ('silent' electrical discharge) (Gellert and Kogelschatz, 1991; Ivanov *et al.*, 1996) operation mode, and the emission wavelength depends on the nature of the fill gas. These sources, like low-pressure Hg lamps, operate near room temperature, and emit monochromatic radiation. The electric discharge requires high voltages (up to 20 kV) and high-frequency ranges (kHz to MHz).

A dielectric barrier discharge can be characterised as a transient high-pressure glow discharge (Giller, 2000). One or both lamp electrodes are covered by a dielectric material (such as glass), and on the application of a high voltage at the external electrodes, and a short discharge pulse occurs in the fill gas. Charge accumulation on the dielectricum, quenches the discharge after some nanoseconds. It is also possible to make excimer lamps with regular electrodes inside. The dielectric barrier is not strictly necessary, but greatly enhances efficiency by preventing continuous arc discharge. The generated excimers [dimers consisting from one atom (or molecule) in an excited state and one atom (or molecule) in the ground state, of the same kind] or exciplexes [dimers of the same description as for 'excimers' but from two different kinds] undergo radiative deactivation (vacuum UV or UV radiations) and dissociate to the parent atoms (molecules) in their ground states. Inert gases (Xe, Ar, krypton (Kr)), halogen gases or their mixtures are used as fill gases.

Table 2.7 Excimers and their emission wavelengths (Nohr *et al.*, 1994; Oppenländer, 1994).

Excimer	λ (nm)	Excimer	λ (nm)
Xe_2*	172	Cl_2*	259
ArF*	193	Br_2*	289
KrBr*	207	XeCl*	308
KrCl*	222	I_2*	342

The lamp envelope is often made of high quality quartz, which exhibits high transmittance in the short wavelength range (~70–80% at λ = 172 nm; 92% at λ = 222 nm). Table 2.7 gives some examples of excimers and their emission wavelengths. Some of them are commercially available (Oppenländer, 1994).

There is no self-absorption of the emitted radiation, since there is no ground state of the excimer. The radiant efficiency of the incoherent excimer lamps is about 5–10%, but higher efficiencies up to 40% and 60% were reported only in laboratory research studies (Ivanov *et al.*, 1996; Vollkommer and Hitzschke, 1997). Other features of these lamps, such as almost arbitrary lamp shape design, high power density, no warm-up time (instant radiation), temperature independent, Hg-free, long life time, make them suitable for further development toward the commercial application for water remediation. Competitive light absorption by oxygen and quartz envelopes may limit the usefulness of low-wavelength excimer sources such as 172 nm-Xe excimer in commercial applications.

The incoherent excimer lamps were used in laboratory- and micro-pilot-scale experiments to destroy organic pollutants in water and air, in various reactor geometries, and coupling lamps that emit at different wavelengths (Nohr *et al.*, 1994; Oppenländer, 1998; Oppenländer and Fradl, 1999). The incoherent excimer technology for water treatment has some advantages, such as generation of high local concentrations of reactive species (hydroxyl radicals), no need for addition of oxidants, *in situ* generation of H_2O_2 and O_3, effective photomineralisation of organic contaminants in water by single or consecutive irradiation (e.g. two excimers emitting at different wavelengths, linked in series), no influence of bicarbonate and nitrate ions on the degradation rate, due to the particular geometry of the 172 nm-photolysis of water (Oppenländer, 2002, and references therein).

Since most of the studies using excimer lamps were performed in vacuum UV domain, the degradation of organic contaminants occurred through both direct photolysis, and $^{\bullet}OH$ radical-induced processes, as a result of water photolysis.

2.3.3 Polychromatic light sources and their spectral distributions

2.3.3.1 Medium-pressure Hg lamps

In order to give a high conversion of the electrical energy into radiant energy, a Hg light source should have either high electron temperature or a high Hg atom

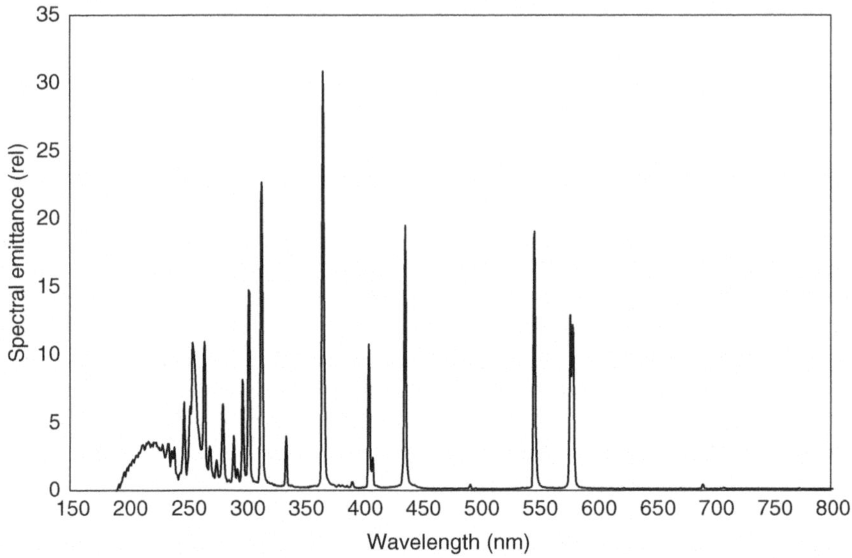

Figure 2.8 Spectral distribution from a medium-pressure Hg lamp.

temperature. In the medium-pressure Hg lamp, the gas pressure reaches $\sim 10^2$–10^4 torr, and a 'local thermal equilibrium' is established, where the temperatures of electrons, atoms and ions become almost identical – this is called a 'plasma temperature' (Phillips, 1983). Plasma temperature during the operation of a medium-pressure lamp is 5000–7000 K. At such temperatures all Hg is in the vapour phase, whereas in the low-pressure lamps liquid Hg is always present. To reach such a state, a 'warm-up' time and a starting gas (usually Ar at a few torr) are required. Medium-pressure lamps operate in arc discharge mode, and are fed usually from an a.c. supply. The temperature at the lamp wall should be maintained at about 600–800°C to ensure a constant spectral distribution of the light during the operation.

Medium-pressure lamps are polychromatic light sources, and the spectral distribution of the radiations emitted covers a wide wavelength range, from VUV to IR. The lamp spectrum (Figure 2.8) is characterised by multiple lines superimposed on a small continuum, which suggests that there is some radiation at all UV wavelengths. The continuum radiation is significant in the UV region, and the spectral lines are fairly broad due to a number of factors, such as Doppler and pressure effects (Phillips, 1983).

The line spectrum in Figure 2.8 can be understood on the basis of a Grotrian diagram, which explains the generation and deactivation of the excited states of Hg atoms (Calvert and Pitts, 1966). The major emission lines from a standard medium-pressure Hg lamp of interest for the AOP applications are at 184.9, 222, 253.7, 265.2, 296.7, 302.8, 313.2 and 365–366 nm in the UV range, and 404.5–407.8, 435.8, 546.1 and 577–570 nm in the visible range of spectrum (Calvert and Pitts, 1966). The 184.9 and 253.7 nm resonance lines are strongly self-absorbed. Altering the concentration of Hg in the arc, the plasma temperature, and the spectral transmission of the lamp envelope can change the spectral distribution of a medium-pressure Hg lamp.

The emission spectrum of the lamp can be modified also by doping the lamps with metal (e.g. thallium, indium and gallium) halides, when the total radiant power remains unchanged, but additional lines appear in the spectrum due to the emission from dopants. The dopants are chosen depending on the photochemical application. Such a practice is suggested in the water UV disinfection using medium-pressure lamps to reduce the output at $\lambda < 240$ nm, which are suspected of changing the natural organic matter composition and convert the nitrate ions naturally present in the water to the undesirable nitrite ions. As discussed in the previous section, nitrite formation in the UV applications may or may not be an issue, provided the role played by the water quality, nitrate concentration in the water, type of UV application. Therefore, the evaluation of nitrite generation in the water processing involving UV light-based technologies should be made on case-by-case basis.

The medium-pressure Hg lamps have become frequently used light sources for various photochemical applications, including the treatment of the environmental pollutants. Their input power varies from 100 W to 30 and 60 kW, and the UVC efficiency per input power is lower than that of low-pressure Hg lamps.

The lifetime of medium-pressure Hg lamps lies in a wide range (1500–10,000) of operating hours. The high temperatures reached by the plasma eventually cause electrode failure, which explains the relatively short lifetime of medium-pressure Hg arcs. To avoid such effects, small amounts of iodine or bromine are added to the filling gas of the lamp; tungsten vapours combine at the wall with halogen vapours and forms a volatile halide which circulates back to the electrodes, regenerating the original amount of tungsten. The UV output declines much faster than the visible output.

2.3.3.2 *Pulsed UV lamps*

The pulsed UV lamp technology has developed over the past two decades and is marketed now for both AOP and UV disinfection applications (Bender, 2000, and references therein; Haag, 1994). These are Hg-free arc lamps, which operate in pulse mode by alternately storing electrical energy in a capacitor and releasing it through an inert gas such as Xe, Ar and Kr (or the mixture of them) by a specifically shaped electrical discharge pulse. The pulse duration is in the microsecond scale range, whereas the interval between pulses is on the order of milliseconds. The electrical discharge heats the fill gas almost instantly to very high temperatures ($\geqslant$13,000 K) that compresses in a thin 'shell' of plasma (ionised gas, electrons and excited states) on the inside surface of the quartz tube. The heated plasma emits blackbody light characteristic to its temperature (see Section 2.3.1). As opposed to the continuous light sources, which emit lines associated with the electronic transitions of the excited states of the fill gas, pulsed lamps emit blackbody radiation (continuum spectrum) on which the electronic transition lines superimpose. Such high temperatures and the shell geometry of the plasma produce UV radiation that spans the 185–3000 nm range, with temperature-specific

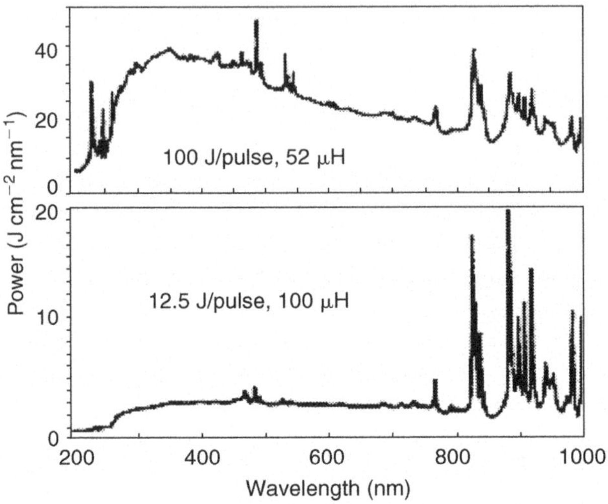

Figure 2.9 Effect of pulse power on the emission spectrum of a pulsed Xe lamp [as taken from Haag (1994)].

maxima in the UV region (usually around 260 nm). Depending how the lamp is operated in a particular application, approx. 38–52% of the output is within 185–400 nm range (Bender, 2000), with an average UVC efficiency of 15–20% (Haag, 1994). Being easily ionised (E^o = 12.1 eV) and with an energy of the lowest excited state E^o = 8.32 eV (Masschelein, 2000), Xe plasmas generally have the greatest efficiency for photon production (Haag, 1994), but Ar plasmas give shorter wavelengths than Xe, at a given power. Since the radiation emitted by the plasma is non-resonant, it is not readily re-absorbed. Pulsed lamp intensities could be in the range of thousands watts per cm (discharge length), delivering high fluences (UV doses), on the order of 10^8 J m^{-2} (Masschelein, 2000). The radiant emittance at the lamp surface ranges from 40,000 to 170,000 W cm^{-2} per pulse with peak power ranging from 2 to 6 MW per pulse, respectively (Bender, 2000). Figure 2.9 shows the effect of peak power on the emission spectrum of a pulsed Xe lamp.

Among the features of pulsed UV lamps are the ability to come to the full power instantly, without any warm-up period, and to shift the emission spectrum by changing the peak pulse power. The latter depends on the peak width, which is controlled by an inductor. The excessively high pulse energies increase thermal stresses, which shorten significantly the lamp lifetime [Purus xenon flashlamp demonstrated a life of 1000 h (Haag, 1994)].

The pulsed UV lamp technology is not as largely applied as the continuous wave (Hg lamp-based systems), but it has been tested both for environmental contaminant removal in water, such as NDMA (Liang *et al.*, 2000; Sharpless *et al.*, 2001), MTBE (Liang *et al.*, 2001), chlorinated aliphatic compounds (trichloroethylene, carbon tetrachloride, dichloromethane, 1,1,1-trichloroethane) (Haag, 1994), and micro-organism inactivation (Marshall *et al.*, 1998; Mofidi *et al.*, 1998).

The UV light sources briefly discussed above are the most representative for the AOP applications. In addition to those, some other monochromatic and polychromatic light sources, such as lasers, continuous wave Xe arc lamps, pulsed Xe white lamps, high-pressure Hg arc lamps, surface discharge Xe lamps, have been used in laboratory studies on the photodegradation of organic and inorganic pollutants. However, since their applicability at the time being is limited at the research level, due to either poor performance or prohibitive cost, they are not covered in this section.

2.4 CONCLUSIONS

This chapter summarises the fundamentals of the UV photolysis process, which is regarded as an environmentally friendly treatment option for water remediation. Assuming that the first requirement – light absorption by the target contaminant – is fulfilled, several other factors should be considered in the evaluation of the opportunity for such an application. Among them, the quantum yield of the contaminant photolysis, pollutant molar absorption coefficients, and water quality characteristics, examined against the spectral power distribution of the light source, are some of the important criteria to consider while approaching feasibility studies regarding the implementation of a UV system.

ACKNOWLEDGEMENTS

Special thanks to Drs. W.L. Cairns, L.M. Gowman, T. Mao, J.A. Robinson, M.R. Sasges, and C.T. Williamson (Trojan Technologies, London, ON, Canada), and Dr. S.R. Stevens (National Research Council, Toronto, ON, Canada), for their valuable comments on the manuscript and kind support throughout this work.

REFERENCES

Handbook of Chemistry and Physics, 72nd edn, 1991, CRC Press, Boca-Raton, FL.

Adam, W. and Oppenländer, T. (1986) 185-nm photochemistry of olefins, strained hydrocarbons, and azoalkanes in solution. *Angew. Chem. Int. Ed.* **25**, 661–672.

Alfassi, Z.B., Huie, R.E. and Neta, P. (1997) Kinetic studies of organic peroxyl radicals in aqueous solutions and mixed solvents. In: *Peroxyl Radicals*, Alfassi, Z.B. (ed.). John Wiley & Sons Ltd., Chichester, England, pp. 235–281.

Ametepe, J.D., Diggs, J., Manos, D.M. and Kelley, M.J. (1999) Characterization and modeling of a microwave driven xenon excimer lamp. *J. Appl. Phys.* **85**(11), 7505–7510.

Anpo, M. and Kubokawa, Y. (1977) Reactivity of excited triplet alkyl ketones in solution. 1. Quenching and hydrogen abstraction of triplet acetone. *Tetrahedron* **50**(8), 1913–1916.

Beltran, F.J., Gonzalez, M., Javier Rivas, J. and Acedo, B. (2000) Determination of kinetic parameters of ozone during oxidations of alachlor in water. *Water Environ. Res.* **72**(6), 689–697.

Bender, J. (2000) Pulsed ultraviolet irradiation for drinking water disinfection. *Water Condit. Purif.* **November**, 48–51.

Bolton, J.R., Bircher, K., Tumas, W. and Tolman, C.A. (2001) Figures-of-merit for the technical development and application of advanced oxidation technologies for both electric- and solar-driven systems (IUPAC technical report). *Pure Appl. Chem.* **73**(4), 627–637.

Bolton, J.R. and Linden, K.G. (2003) Standardization of methods for fluence (UV dose) determination in bench-scale UV experiments. *J. Environ. Eng.* **129**(3), 209–216.

Bolton, J.R. and Stefan, M.I. (2002) Fundamental photochemical approach to the concepts of fluence (UV dose) and electrical energy efficiency in photochemical degradation reactions. *Res. Chem. Intermed.* **28**(7–9), 857–870.

Boule, P., Guyon, C. and Lemaire, J. (1982) Photochemistry and environment: IV-photochemical behaviour of monochlorophenols in dilute aqueous solution. *Chemosphere* **11**(12), 1179–1188.

Braun, A.M., Maurette, M.-T. and Oliveros, E. (1991) *Photochemical Technology*. Wiley, Chichester, UK.

Burgess, E.M. and Lavanish, J.M. (1964) Photochemical decomposition of N-nitrosamines. *Tetrahedron Lett.* **20**, 1221–1226.

Buxton, G.V., Greenstock, C.L., Helman, W.P. and Ross, A.P. (1988) Critical review of rate constants for reactions of hydrated electrons, hydrogen atoms and hydroxyl radicals ($^{\bullet}OH/O^{\bullet-}$) in aqueous solution. *J. Phys. Chem. Ref. Data* **17**(2), 513–886.

Calvert, J.G. and Pitts Jr., J.N. (1966) *Photochemistry*. Wiley, New York.

Canonica, S. and Freiburghaus, M. (2001) Electron-rich phenols for probing the photochemical reactivity of freshwaters. *Environ. Sci. Technol.* **35**(4), 690–695.

Choudhry, G.G., Webster, G.R.B. and Hutzinger, O. (1986) Environmentally significant photochemistry of chlorinated benzenes and their derivatives in aquatic systems. *Toxicol. Environ. Chem.* **13**, 27–83.

Cooper, W.J., Zika, R.G., Petasne, R.G. and Fischer, A.M. (1989) Sunlight-induced photochemistry of humic substances in natural waters: major reactive species. *Adv. Chem. Ser.* **219**, 333–362.

Corin, N., Backlund, P. and Kulovaara, M. (1996) Degradation products formed during UV-irradiation of humic waters. *Chemosphere* **33**, 245–255.

Draper, W.M. (1987) Measurement of quantum yields in polychromatic light: dinitroaniline herbicides. *ACS Symp. Ser.* **327**, 268–281.

Fasnacht, M.P. and Blough, N.V. (2002) Aqueous photodegradation of polycyclic aromatic hydrocarbons. *Environ. Sci. Technol.* **36**, 4364–4369.

Gallard, H. and von Gunten, U. (2002) Chlorination of natural organic matter: kinetics of chlorination and of THM formation. *Water Res.* **36**, 65–74.

Gellert, B. and Kogelschatz, U. (1991) Generation of excimer emission in dielectric barrier discharges. *Appl. Phys. B* **52**, 14–21.

Gerecke, A.C., Canonica, S., Müller, S.R., Scharer, M. and Schwarzenbach, R.P. (2001) Quantification of dissolved natural organic matter (DOM) mediated phototransformation of phenylurea herbicides in lakes. *Environ. Sci. Technol.* **35**(19), 3915–3923.

Getoff, N. (1991) Radiation- and photoinduced degradation of pollutants in water. A comparative study. *Radiat. Phys. Chem.; Int. J. Radiat. Appl. Instrum. Part C* **37**(5/6), 673–680.

Giller, H.F.J.I. (2000) A review of UV lamps. In: *Proceedings of the 'Disinfection 2000: Disinfection of Wastes in the New Millenium' Conference*, New Orleans, LA, March 15–18.

Haag, W.R. (1994) Photooxidation of organic compounds in water and air using low-wavelength pulsed xenon lamps. In: *Aquatic and Surface Photochemistry*, Helz, G.R., Zepp, R.G. and Crosby, D.G. (eds). Lewis Publishers, Boca Raton, pp. 517–530.

Haag, W.R. and Mill, T. (1987) Direct and indirect photolysis of water-soluble azodyes: kinetic measurements and structure–activity relationships. *Environ. Toxicol. Chem.* **6**, 359–369.

Haag, W.R., Hoigné, J., Gassman, E. and Braun, A.M. (1984) Singlet oxygen in surface waters – part II: Quantum yields for its production by some natural humic materials as a function of wavelength. *Chemosphere* **13**, 641–650.

Hatchard, C.G. and Parker, C.A. (1956) A new sensitive chemical actinometer. II. Potassium ferrioxalate as a standard chemical actinometer. *Proc. Royal Soc. London, Ser. A* **235A**, 518–536.

Hessler, D.P., Gorenflo, V. and Frimmel, F.H. (1993a) Degradation of aqueous atrazine and metazachlor solutions by UV/H_2O_2 – influence of pH and herbicide concentration. *Acta Hydrochim. Hydrobiol.* **21**(4), 209–214.

Hessler, D.P., Gorenflo, V. and Frimmel, F.H. (1993b) UV degradation of atrazine and metazachlor in the absence and in the presence of hydrogen peroxide, bicarbonate and humic substances. *J. Water SRT-Aqua* **42**(1), 8–12.

Ho, T.-F.L., Bolton, J.R. and Lipczynska-Kochany, E. (1996) Quantum yields for the photodegradation of pollutants in dilute aqueous solution: phenol, 4-chlorophenol and N-nitrosodimethylamine. *J. Adv. Oxid. Technol.* **1**(2), 170–178.

Hoigné, J., Faust, B.C., Haag, W.R., Scully Jr., F.E. and Zepp, R.G. (1989) Aquatic humic substances as sources and sinks of photochemically produced transient reactants. *Adv. Chem. Ser.* **219**, 363–381.

Ivanov, V.V., Klopovskii, K.S., Mankelevich, Y.A., Rakhimova, T.V., Rulev, G.B., Saenko, V.B. and Rakhimov, A.T. (1996) Experimental and theoretical study of the efficiency of an excimer lamp pumped by a pulse distributed discharge in xenon. *Laser Phys.* **6**(4), 654–659.

Jakob, J., Hashem, T.M., Bürki, S., Guindy, N.M. and Braun, A.M. (1993) Vacuum-ultraviolet (VUV) photolysis of water: oxidative degradation of 4-chlorophenol. *J. Photochem. Photobiol. A: Chem.* **75**, 97–103.

Johns, H.E. (1971) Quantum yields and kinetics of photochemical reactions in solution. In: *Creation and Detection of the Excited States*, Lamola, A.A. (ed.). Marcel Dekker, New York, pp. 123–171.

Koller, L.R. (1965) *Ultraviolet Radiation*, Ballard, S.S. (ed.). John Wiley & Sons, Inc., New York, pp. 1–312.

Kuhn, H.J., Braslavsky, S.E. and Schmidt, R. (1989) Chemical actinometry. *Pure Appl. Chem.* **61**, 187–210.

Larson, R.A. and Weber, E.J. (1994) *Reaction Mechanisms in Environmental Organic Chemistry*. CRC Press, Inc., Boca Raton, FL.

Leifer, A. (1988) *The Kinetics of Environmental Aquatic Photochemistry: Theory and Practice*. ACS Professional Reference Book.

Liang, S., Min, J.H., Davis, M.K., Green, J.F. and Remer, D.S. (2000) Treatment of N-nitrosodimethylamine (NDMA) by pulsed-ultraviolet (UV) irradiation and pulsed-UV/hydrogen peroxide (H_2O_2) processes. In: *Proceedings of the AWWA Water Quality Technology Conference*, Salt Lake City, Utah, November 5–9.

Liang, S., Min, J.H., Yates, R.S., Chou, C.-S. and Kavanaugh, M. (2001) Effects of water quality on destruction of methyl-tert-butyl ether (MTBE) by the pulsed-UV/H_2O_2 process. In: *Proceedings of the AWWA Water Quality Technology Conference*, Nashville, TN, November 11–14.

Liu, W., Andrews, S.A., Bolton, J.R., Linden, K.G., Sharpless, C.M. and Stefan, M.I. (2002) Comparison of disinfection byproduct (DBP) formation from different UV technologies at bench-scale. *Water Sci. Technol.: Water Suppl.* **2**(5–6), 515–521.

Mabury, S.A. and Crosby, D.G. (1994) The relationship of hydroxyl radical reactivity to pesticide persistence. In: *Aquatic and Surface Photochemistry*, Helz, G.R., Zepp, R.G. and Crosby, D.G. (eds). Lewis Publishers, Ann Arbor, MI, pp. 149–161.

Mack, J. and Bolton, J.R. (1999) Photochemistry of nitrite and nitrate in aqueous solution: a review. *J. Photochem. Photobiol. A: Chem.* **128**, 1–13.

Mark, G., Korth, H.-G., Schuchmann, H.-P. and von Sonntag, C. (1996) The photochemistry of aqueous nitrate ion revisited. *J. Photochem. Photobiol. A: Chem.* **101**, 89–103.

Mark, G., Schuchmann, M.N., Schuchmann, H.-P. and von Sonntag, C. (1990) A chemical actinometer for use in connection with UV treatment in drinking-water processing. *J. Water SRT-Aqua* **39**(5), 309–313.

Marshall, T.H., Christy, A.D. and Enochs, M.B. (1998) Emerging electrotechnologies for the disinfection of wastewater. In: *Proceedings of the Conference 'Disinfection '98: The latest Trends in Wastewater Disinfection: Chlorination vs UV Disinfection'*, Baltimore, ML, April 19–22.

Masschelein, W.J. (2000) Utilisation des U.V. dans le traiement des eaux. *Tribune de l'eau* **53**(4–5), 7–107.

Mertens, R. and von Sonntag, C. (1994) Reaction of the OH radical with tetrachloroethene and trichloroacetaldehyde (hydrate) in oxygen-free aqueous solution. *J. Chem. Soc. Perkin Trans.* **2**, 2175–2180.

Mertens, R. and von Sonntag, C. (1995) Photolysis ($\lambda = 254$ nm) of tetrachloroethene in aqueous solutions. *J. Photochem. Photobiol. A: Chem.* **85**, 1–9.

Mill, T. (1990) Detection and characterization of phototransients formed in surface waters. In: *Proceedings of the 199th ACS National Meeting*, Boston, MA, April 22–27.

Mill, T. and Mabey, W. (1985) Photochemical transformations. In: *Environmental Exposure from Chemicals*, Neely, W.B. and Blau, G.E. (eds). CRC Press, Boca Raton, FL.

Mitch, W.A. and Sedlak, D.L. (2002) Formation of N-Nitrosodimethylamine (NDMA) from dimethylamine during chlorination. *Environ. Sci. Technol.* **36**(4), 588–595.

Mofidi, A.A., Baribeau, H., Green, J.F., De Leon, R. and Wolfe, R.L. (1998) Disinfection using high-intensity pulsed-ultraviolet irradiation. In: *Proceedings of the AWWA Water Quality and Technology Conference*, San Diego, CA, November 1–4.

Moore, D. (1996) Photophysical and photochemical aspects of drug stability. In: *Photostability of Drugs and Drug Formulations*, Tonnesen, H.H. (ed.). Taylor & Francis, London.

Murov, S.L., Carmichael, I. and Hug, G.L. (1993) *Handbook of Photochemistry*. Marcel Dekker, Inc., New York, pp. 330–336.

Neumann-Spallart, M. and Getoff, N. (1979) Photolysis and radiolysis of monochloroacetic acid in aqueous solution. *Radiat. Phys. Chem.* **13**, 101–105.

Nick, K. and Schoeler, H.F. (1995) Photochemical degradation of herbicides in water by UV irradiation generated by low pressure mercury lamps. Part I. triazines. *Vom Wasser* **84**, 271–286.

Nick, K., Schoeler, H.F., Mark, G., Söylemez, T., Akhlaq, M.S., Schuchmann, H.-P. and von Sonntag, C. (1992) Degradation of some triazine herbicides by UV radiation such as in the UV disinfection of drinking water. *J. Water SRT-Aqua* **41**(2), 82–87.

Nicole, I., De Laat, J., Dore, M., Duguet, J.P. and Bonnel, C. (1990) Use of UV radiation in water treatment: measurement of photonic flux by hydrogen peroxide actinometry. *Water Res.* **24**(2), 157–168.

Nohr, R.S., MacDonald, J.G., Kogelschatz, U., Mark, G., Schuchmann, H.-P. and von Sonntag, C. (1994) Application of excimer incoherent-UV sources as a new tool in photochemistry: photodegradation of chlorinated dibenzodioxins in solution and adsorbed on aqueous pulp sludge. *J. Photochem. Photobiol. A: Chem.* **79**, 141–149.

Oppenländer, T. (1994) Novel incoherent excimer UV irradiation units for the application in photochemistry, photobiology/medicine and for waste water treatment. *Eur. Photochem. Assoc. Newsl.* **50**, March, 2–8.

Oppenländer, T. (1998) Phootochemical treatment of water: Comparison of incoherent excimer lamps with a medium-pressure mercury lamp. *Chem. Eng. Technol.* **21**(6), 502–505.

Oppenländer, T. and Fradl, M. (1999) TOC destruction of a phenol/water azeotrope by 'photoreactive distillation' through an incoherent vacuum-UV excimer lamp. *Chem. Eng. Technol.* **22**(11), 951–954.

Oppenländer, T. and Gliese, S. (2000) Mineralization of organic micropollutants (homologous alcohols and phenols) in water by vacuum-UV-oxidation (H_2O_2-VUV) with an incoherent xenon-excimer lamp at 172 nm. *Chemosphere* **40**, 15–21.

Oppenländer, T. (2002) *Photochemical Purification of Water and Air. Advanced Oxidation Processes (AOPs): Principles, Reaction Mechanisms, Reactor Concepts*. Wiley-VCH, Weinheim, Germany.

Parker, C.A. (1968) *Photoluminiscence of Solutions: With Applications to Photochemistry and Analytical Chemistry*. Elsevier Publishing Co., New York.

Pfeifer Roland, R., Bolle, M. and Anderson, R.W. (1998) Hydrogen bromide photochemistry: actinometry for determination of absolute power outputs of xenon excimer and other UV/VUV light sources. *J. Phys. D: Appl. Phys.* **31**, 1336–1342.

Phillips, R. (1983) *Sources and Applications of Ultraviolet Radiation*. Academic Press Inc., New York.

Polo, J. and Chow, Y.L. (1976b) Efficient photolytic degradation of nitrosamines. *J. Natl. Cancer Inst.* **56**(5), 997–1001.

Rahn, R.O. (1997) Potassium iodide as a chemical actinometer for 254 nm radiation: use of iodate as an electron scavenger. *Photochem. Photobiol.* **66**(4), 450–455.

Schulz, C.R, Cervantes, P. and Laplace, C. (2001) Development of a flow-through actinometry monitor for dose measurement in UV disinfection reactors. In: *Proceedings of the First International IUVA Congress*, Washington DC, June 14–16.

Schwarzenbach, R.P., Gschwend, P.M. and Imboden, D.M. (1993) Photochemical transformation reactions. In: *Environmental Organic Chemistry*. John Wiley & Sons, Inc., New York, pp. 436–484.

Sharpless, C.M, Chou, C.I., Mofidi, A.A. and Linden, K.G. (2001) N-Nitrosodimethylamine removal from drinking water by UV-treatment: a direct comparison of different UV technologies. In: *Proceedings of the AWWA Water Quality and Technology Conference*, Nashville, TN, November 10–13.

Sharpless, C.M. and Linden, K.G. (2001) UV photolysis of nitrate: effects of natural organic matter and dissolved inorganic carbon and implications for UV water disinfection. *Environ. Sci. Technol.* **35**(14), 2949–2955.

Smith, R.N., Leigthon, P.A. and Leighton, W.G. (1939) The quantum yield of monochloroacetic acid hydrolysis. *J. Am. Chem. Soc.* **61**, 2299–2301.

Sonnemann, G. (1992) Ozon – Natürliche Schwankungen und anthropogene Einflüsse. Akademie Verlag GmbH, Berlin, Germany.

Stefan, M.I. and Bolton, J.R. (1996) Direct photolysis of acetone in aqueous solution. Effect of dissolved oxygen on the photolysis rate. Personal communication.

Stefan, M.I. and Bolton, J.R. (2002) UV direct photolysis of N-nitrosodimethylamine (NDMA): Kinetic and product study. *Helv. Chim. Acta* **85**, 1416–1426.

Stefan, M.I., Bolton, J.R. and Cater, S.R. (1999) UV light-driven degradation of N-nitrosodimethylamine. In: *Proceedings of the 5th International Conference on Advanced Oxidation Technologies for Water Remediation*, Albuquerque, NM, May 24–28, p. 109.

Stefan, M.I., Bolton, J.R. and Rahn, R.O. (2001a) Use of iodide/iodate actinometer together with spherical cells for determination of the fluence rate distribution in UV reactors: validation of the mathematical model. In: *Proceedings of the 1st International IUVA Congress*, Washington DC, May 14–16.

Stefan, M.I., Sharpless, C.M., Linden, K.G. and Bolton, J.R. (2001b) Degradation of atrazine in a bench-scale collimated beam apparatus using low- and medium-pressure UV lamps. In: *Proceedings of the AWWA Water Quality and Technology Conference*, Nashville, TN, November 11–14.

Stefan, M.I., Siddiqui, M., Atasi, K. and Linden, K.G. (2002) Impact of water quality characteristics on the kinetic parameters of NDMA photo-degradation. In: *Proceedings of the AWWA Water Quality and Technology Conference*, Seattle, WA, November 10–14.

Taylor, H.A. (1971) Analytical methods and techniques for actinometry. In: *Analytical Photochemistry and Photochemical Analysis: Solids, Solutions, and Polymers*, Fitzgerald, J.M. (ed.). Marcel Dekker, Inc., New York, pp. 91–115.

Thomas-Smith, T. and Blough, N.V. (2001) Photoproduction of hydrated electron from constituents of natural waters. *Environ. Sci. Technol.* **35**(13), 2721–2726.

Tixier, C., Meunier, L., Bonnemoy, F. and Boule, P. (2000) Phototransformation of three herbicides: chlorbufam, isoproturon, and chlorotoluron. Influence of irradiation on toxicity. *Intl. J. Photoenerg.* **2**, 1–8.

Torrents, A., Anderson, B.G., Bilboulian, S., Johnson, W.E. and Hapeman, C. (1997) Atrazine photolysis: mechanistic investigations of direct and nitrate-mediated hydroxyl radical processes and influence of dissolved organic carbon from the Chesapeake Bay. *Environ. Sci. Technol.* **31**(5), 1476–1482.

Toussaint, G. and Würkert, K. (1978) Nitrat-Nitrit-Nitrosamine in Gewässern. D. Toxikologische Bedeutung von Nitrat-Nitrit-Nitrosaminen: Methämoglobinämie im Säulingsalter. In: *Mitteilung III der Kommission für Wasserforschung in Verbindung mit der Kommission zur Prüfung von Lebensmittelzusatz- und Inhaltsstoffen*, DGF Deutsche Forschungsgemeinschaft, Germany, 136–187.

Tsao, R. and Eto, M. (1994) Effect of some natural photosensitizers on photolysis of some pesticides. In: *Aquatic and Surface Photochemistry*, Helz, G.R., Zepp, R.G. and Crosby, D.G. (eds). Lewis Publishers, Ann Arbor, MI, pp. 163–171.

Vollkommer, F. and Hitzschke, L. (1997) Durchbruch bei der effizienten Erzeugung von Excimer-Strahlung. *Phys. Bl.* **53**(9), 887–889.

von Sonntag, C., Mark, G., Mertens, R., Schuchmann, H.-P. and Schuchmann, M.N. (1993) UV radiation and/or oxidants in water pollution control. *J. Water SRT-Aqua* **42**(4), 201–211.

von Sonntag, C. and Schuchmann, H.-P. (1992) UV disinfection of drinking water and by-product formation – some basic considerations. *J. Water SRT-Aqua* **41**(2), 67–74.

von Sonntag, C. and Schuchmann, H.-P. (1997) Peroxyl radicals in aqueous solutions. In: *Peroxyl Radicals*, Alfassi, Z.B. (ed.). John Wiley & Sons Ltd., Chichester, England, pp. 173–234.

Wan, H.B., Wong, M.K. and Mok, C.Y. (1994) Comparative study on the quantum yields of direct photolysis of organophosphorus pesticides in aqueous solution. *J. Agric. Food Chem.* **42**(11), 2625–2630.

Zellner, R., Exner, M. and Herrmann, H. (1990) Absolute OH quantum yields in the laser photolysis of nitrate, nitrite, and dissolved H_2O_2 at 308 and 351 nm in the temperature range 278–353 K. *J. Atmos. Chem.* **10**, 411–425.

Zepp, R.G. (1978) Quantum yields for reaction of pollutants in dilute aqueous solution. *Environ. Sci. Technol.* **12**(3), 327–329.

Zepp, R.G. (1988) Environmental photoprocesses involving natural organic matter. In: *Humic Substances and Their Role in the Environment*, Frimmel, F.H. and Christman, R.F. (eds). John Wiley & Sons, Inc., New York.

Zepp, R.G. and Baughman, G.L. (1978) Prediction of photochemical transformations of pollutants in the aquatic environment. In: *Aquatic Pollutants. Transformations and Biological Effects*, Hutzinger, O., van Lelyveld, I.M. and Zoeteman, B.C.J. (eds). Pergamon Press, New York, pp. 237–263.

Zepp, R.G., Baughman, G.L. and Sclotzhauer, P.F. (1981a) Comparison of photochemical behavior of various humic substances in water: I. Sunlight induced reactions of aquatic pollutants photosensitized by humic substances. *Chemosphere* **10**, 109.

Zepp, R.G., Baughman, G.L. and Sclotzhauer, P.F. (1981b) Comparison of photochemical behavior of various humic substances in water: photosensitized oxygenations. *Chemosphere* **10**, 119.

Zepp, R.G., Hoigné, J. and Bader, H. (1987) Nitrate-induced photooxidation of trace organic chemicals in water. *Environ. Sci. Technol.* **21**, 443–449.

Zepp, R.G., Schlotzhauer, P.F. and Merritt Sink, R. (1985) Photosensitized transformations involving electronic energy transfer in natural waters: role of humic substances. *Environ. Sci. Technol.* **19**, 74–81.

3

UV light-based applications

Mihaela I. Stefan and Christian T. Williamson

3.1 INTRODUCTION

Advanced oxidation processes (AOPs) have been demonstrated as an effective means of water remediation as compared to the conventional technologies, such as adsorption on activated carbon, air stripping and biodegradation. In general, the light-driven AOPs involve the production of hydroxyl radicals, which react almost non-selectively with the organic pollutants at very high rates. The ultraviolet (UV) direct photolysis process is restricted to those pollutants that exhibit large molar absorption coefficients and quantum yields, and, therefore, has limited applications. Regardless the photochemical process involved in the pollutant removal, the choice of the UV system depends on the cost of equipment, operation and maintenance, which in turn are affected by several factors, such as water quality, system design, lamp maintenance and cost, electrical energy cost, as well as the oxidant dosage (if any).

In this chapter, we shall briefly review the reported laboratory and pilot studies on the UV direct photolysis of organic pollutants, and will present some examples

 Advanced Oxidation Processes for Water and Wastewater Treatment. Edited by Simon Parsons. ISBN: 1 84339 017 5

of current commercial applications of UV photolysis to treat drinking, surface- and groundwaters, emphasizing the critical process efficiency parameters imposed by the reactor, lamp technology and water quality.

3.2 UV DIRECT PHOTOLYSIS OF SPECIFIC COMPOUNDS: LABORATORY- AND PILOT-SCALE EXPERIMENTS

The photochemistry of organic and inorganic pollutants in aqueous phase is well represented in the published literature. The aim of all studies was to develop suitable and economic processes for destruction of water contaminants under UV irradiation from either mono- or polychromatic light sources. The experiments were conducted under various experimental setups using a wide variety of water qualities. The generally low quantum yields and small absorption coefficients over the lamp emission range are the factors that limit the efficiency, therefore, the applicability of direct photolysis process in water remediation. In most cases, comparative studies revealed significantly larger rates of pollutant removal through $^{\bullet}OH$ radical-driven AOPs, such as UV/H_2O_2 (hydrogen peroxide), UV/O_3, $UV/O_3/H_2O_2$, photo-Fenton or photocatalyzed reactions, than those of direct photolysis.

The literature data presented in this section is mostly focused on UV photolysis of selected compounds that are known as environmental aquatic pollutants, most of them falling under regulatory actions. It would be far beyond the scope of this chapter to cover all published work on photodegradation of organic and inorganic pollutants. Such attempts are made by literature reviews and some of the references cited therein (Burrows *et al.*, 2002; Halmann, 1996; Legrini *et al.*, 1993).

3.2.1 Aliphatic compounds

3.2.1.1 Halogenated derivatives

The incidence of these compounds in contaminated waters is caused by either their use as solvents, reagents in industrial organic synthesis, dry-cleaning materials or other applications, or by their formation as disinfection by-products (DBPs) during the water processing, such as chlorination.

The saturated halocarbons cannot be removed efficiently by UV photolysis. Nicole *et al.* (1989) studied the effect of reactor design (annular pathlength, reflecting power and irradiated volume) on the degradation of trihalomethanes ($CHCl_3$, $CHBr_3$, $CHBrCl_2$ and $CHBr_2Cl$) irradiated in dilute (μM range) aqueous solution at 253.7 nm. The photolysis rates of brominated derivatives followed the order, such as $CHBr_3 > CHBr_2Cl > CHBrCl_2$, irrespective of the $\%T$ of the water, therefore, of the amount of light reflected at the reactor inner wall. The rates increased significantly with the reflecting capacity of the reactor wall. No significant photolysis of chloroform (less than 5%), but 95% removal of bromoform was observed over 20-min irradiation, under identical conditions. Sundstrom *et al.* (1986) and Frischhertz *et al.* (1989) observed very low yields of photolysis at 253.7 nm of chlorinated methanes and ethanes, but relatively high conversions in the case of halogenated ethenes, such as trichloroethene (TCE) and tetrachloroethene (PCE). For example, the concentration of TCE decreased by 80% after 40-min irradiation, whereas CH_2Cl_2 concentration was reduced by ~30% within 180 min. However, a strong synergistic effect was observed in the case of both compounds, when exposed to both UV and H_2O_2. Less effect was observed in the case of chloroform and tetrachloroethane, where UV alone was nearly as effective as its combination with H_2O_2. Temperature had a significant effect on the photolysis rate (Sundstrom *et al.*, 1986). In a comparative work on UV and UV/H_2O_2 degradation of chlorinated methanes and ethanes using a low-pressure Hg lamp, Guittonneau *et al.* (1988a) reported very low conversions of these compounds by UV photolysis, and good yields for the UV/H_2O_2 process. Similarly, De Laat *et al.* (1994) concluded that tri- and tetrachloroethanes do not photolyze when irradiated with 253.7-nm radiation, but are efficiently removed in the presence of H_2O_2. The rates of decay for four model compounds were predicted by kinetic models that accounted for both H_2O_2 concentration and water-quality parameters, such as pH, alkalinity and dissolved organic matter content, and were found in very good agreement with the experimental data. Milano and Vernet (1988) measured low yields for the photolysis of 1,2-dibromopropane (50% within 4 h) under broadband irradiation from a 100-W medium-pressure Hg arc.

Chlorinated ethenes, such as PCE, TCE and DCE (1,2-dichloroethene), were identified in groundwater sources, and are currently restricted chemicals in drinking water both in North America and Europe. Mertens and von Sonntag (1995) examined the photolysis of PCE at 253.7 nm from a low-pressure Hg arc, with respect to the product yield, reaction mechanism and the role of oxygen. Overall quantum yields of 0.84 and 0.34 were calculated for PCE photolysis in the presence and the absence of oxygen, respectively. Chloride ion, carbon dioxide, tri- and dichloroacetic acids, hypochlorite, carbon monoxide and trichloroacetaldehyde were among the photodegradation products. Beltran *et al.* (1995) modeled the photolysis of TCE in two surface water qualities in a bench-scale reactor equipped with a 15-W low-pressure Hg lamp. The dissolved organic matter reduced the degradation rate, such that 80% and 54% conversion of TCE was observed within 90-min irradiation in surface water containing 11 and 17 $mg\,L^{-1}$ TOC, respectively. Quantum yields for TCE direct photolysis were determined from three different photochemical kinetic models. The recommended value is $\Phi = 1.712 \pm 0.041$, which is significantly larger than that found by Stefan *et al.* (2003) with polychromatic radiation ($\Phi = 0.46$). The irradiation of 1.1×10^{-3} M TCE in distilled water, in a semi-batch 1-kW medium-pressure reactor (2.18×10^{-4} einstein s^{-1} within 200–300 nm range) resulted in 92% removal of TCE after 10-min irradiation. The reaction mechanism proposed for TCE photodegradation is based on the degradation intermediates observed and quantified during the TCE photolysis, such as chloride ions, formaldehyde, dichloroacetaldehyde, formic, glyoxylic, mono- and dichloroacetic acids (Stefan *et al.*, 2003). The dichloro- and monochloroacetic acids were further photolyzed and removed from the irradiated system during 35- and 55-min irradiation. Haag (1994) investigated the removal of TCE (0.5–500 ppm range) in aqueous phase, both in the absence and in the presence of H_2O_2, using two 1.38-kW pulsed Xe lamps tuned for a peak at 230 nm. The rate of decay and the nature of the intermediates formed depended on the initial concentration of TCE; the amount of the original chlorine concentration (as TCE) was retrieved quantitatively as chloride ions at low initial concentrations of TCE, whereas transient formation of 10–20% dichloroethenes, chlorinated methanes and ethanes were found when starting from 160 ppm TCE. The author concluded that TCE is one of the most rapidly treated compounds studied, and the addition of 250 ppm H_2O_2 increased the rate only by a factor of 2–3, when compared with the direct photolysis rate.

In a comparative study between pulse radiolysis and photolysis (184.9 nm) of DCE, TCE and PCE in aerated aqueous solution, Getoff (1991) reported the reaction quantum yields, mechanistic pathways and degradation products of these pollutants. The large chloride ion yields are explained by both chain reactions carried out by O_2, $O_2^{\bullet -}$ and $^{\bullet}OH$ radical-induced degradation, as the $^{\bullet}OH$ radicals are generated with a $\Phi = 0.33$ through water photolysis. Hirvonen *et al.* (1996) conducted laboratory-scale studies on removal of PCE and TCE in groundwater by H_2O_2-mediated oxidation processes. The irradiations were performed with a 20-W Osram HNS low-pressure lamp (0.8 W/L) on various water qualities, such as distilled water, uncontaminated and contaminated groundwater and drinking water before chlorination. The contaminants were spiked (or adjusted) to a final concentration of 100–200 $\mu g\,L^{-1}$ (ppb), and the efficiencies of various processes, such as UV alone, UV/H_2O_2, O_3 and O_3/H_2O_2, were compared. After 10-min irradiation, 38% and 48% of TCE and PCE, respectively, were removed by direct photolysis, whereas both compounds were below the analytical detection limit in the presence of 135 ppm H_2O_2. Chlorinated acetic acids were observed as by-products; but under prolonged irradiation, in the presence of H_2O_2, they could be further oxidized. In their later work, Hirvonen *et al.* (1998) reported field-scale data on the UV/H_2O_2 process applied to the removal of PCE (76–150 $\mu g\,L^{-1}$) from a groundwater source contaminated by a dry-cleaning facility. The electrical energy per order (E_{EO}) calculated for UV/H_2O_2 process was much lower than that for direct UV photolysis. Di- and trichloroacetic acids were generated as intermediate products; but under the operation conditions, their concentrations were below the drinking water criteria.

3.2.2 Aromatic compounds

3.2.2.1 Aromatic hydrocarbons

The aquatic environment can be contaminated with aromatic or polycyclic aromatic hydrocarbons from industrial and municipal wastewater, runoff or rainwaters. These compounds are found in considerable amounts in surface and marine waters (from Trapido *et al.*, 1994). Wastewaters of chemical, oil shale and oil industry, and urban sewages contain polycyclic aromatic hydrocarbons up to mg L^{-1} (from Trapido *et al.*, 1994). The European Community Guideline for drinking water has set a low standard for polyaromatic hydrocarbons (PAHs) to 200 ng L^{-1}. Maier *et al.*'s studies (2000) showed that the occurrence of PAHs in the distribution system could be linked to stagnation periods, anaerobic conditions and the presence of disinfectants, such as chlorine and chlorine dioxide, which prevent the biofilm growth that exhibits protective effects against PAHs. In a recent study, Bertilsson and Widenfalk (2002) examined the photochemical degradation of PAHs in freshwaters and their impact on bacterial growth.

Miller and Olejnik (2001) examined the photolysis of three polycyclic aromatic compounds, such as benzo[a]pyrene (BAP), chrysene (CHR) and fluorene (FLU), in aqueous solution irradiated with either 253.7 nm (15-W low-pressure Hg lamp) or broadband emission from a 150-W medium-pressure Hg lamp, as function of pH (2.5, 6.8 and 11.7), level of dissolved oxygen, initial concentration of hydrocarbon, and the presence of radical inhibitors. The quantum yields for direct photolysis of BAP, CHR and FLU were 0.014, 0.0031 and 0.0038, respectively, slightly pH dependent in the case of CHR.

Beltran *et al.* (1996b) compared the efficiencies of various AOPs (UV alone, O_3, O_3/H_2O_2, UV/H_2O_2 and UV/O_3) for FLU, phenanthrene and acenaphthene ($\sim 10^{-6}$ M) removal from surface waters with relatively high content of natural organic matter (NOM) (18–30 mg C L^{-1}). A low-pressure Hg lamp was used as a UV light source. About 90% degradation of these compounds was achieved in less than 15 min for phenanthrene, and 30 min for FLU and acenaphthene. In organic-free water, 90% removal of FLU was observed after about 7.5-min irradiation. The other AOPs indicated up to 7–8-fold higher removal rates than UV photolysis. Among the PAHs studied, FLU exhibited the lowest oxidation rates, irrespective of type of AOP applied. Fasnacht and Blough (2002) studied the photodegradation of 12 polycyclic aromatic hydrocarbons in aerated pure water, solutions of Suwannee River fulvic acid and natural waters, using polychromatic light ($\lambda > 290$ nm). The quantum yields for PAHs photodegradation in pure water were found very low [from 3.2×10^{-5} (fluoranthene) to 9.2×10^{-3} (acenaphthene)], and the rate constants were unchanged by the water quality.

The 350-nm photolysis of anthracene ($\Phi = 0.003$) in aerated aqueous solutions leads to endoperoxide and 9,10-anthraquinone, as primary products, which undergo further degradation; in anaerobic solutions, the photoproduct has a bianthranyl structure (Sigman and Zingg, 1994).

Tuhkanen and Beltran (1995) reported a product study on naphthalene photolysis at 253.7 nm in the absence and the presence of H_2O_2. The degradation rates increased with the H_2O_2 concentration, reaching an optimum at 10 mM H_2O_2. Benzofuran and benzofuranone derivatives were identified among the UV photolysis products; such products did not accumulate in the reaction mixture when the UV/H_2O_2 process was applied.

Phenanthrene is a hazardous compound listed as a priority pollutant by the US EPA. Trapido *et al.* (1994) investigated the treatment of phenanthrene by UV photolysis, O_3 and H_2O_2, and their combinations, at pH 3, 7 and 9. Direct photolysis rate was small, and the best treatment process was ozonation in neutral medium. In another study, the same authors (Trapido *et al.*, 1995) determined and compared the rates of removal of FLU, anthracene, phenanthrene, pyrene, fluoranthene, BAP and benzo[g]perylene through the same AOPs. Among the seven PAHs, pyrene ($k_1 = 1.0 \times 10^{-2}\,s^{-1}$) was the most prone to UV photolysis, whereas phenanthrene ($k_1 = 3.0 \times 10^{-4}\,s^{-1}$) was the less photosensitive. It is interesting to note that for studied compounds, the addition of H_2O_2 did not increase the rate constants, and the ozone-based processes occurred with similar rates.

One of the environmental concerns is related to the contamination of water with benzene, ethylbenzene, toluene and xylenes, in their mixture (BTEX), from leaking underground gasoline reservoirs. Such compounds exhibit absorption in the UVC spectral region, but the quantum yields for their direct photolysis are very low to allow their treatment by UV alone, such that the $^{\bullet}OH$ radical-based AOPs are suitable in these cases.

Sundstrom *et al.* (1990) investigated the destruction of trichloroethylene and benzene mixture, and inactivation of *Escherichia coli* and *Bacillus subtilis* spores, using the synergistic effect of H_2O_2 and UV light at 253.7 nm. The studies were conducted in a variety of batch and flow reactor systems. UV alone reduced benzene concentration by about 50% in 90 min. In the presence of H_2O_2, over 98% removal of benzene was observed over the same exposure time. Sundstrom *et al.* (1989) compared the photodegradation rate constants of nine hazardous aromatic compounds, such as benzene, chlorobenzene, toluene, phenol, 2-chlorophenol, 2,4-dichlorophenol, 2,4,6-trichlorophenol (TCP), dimethyl- and diethylphthalate, in the presence and the absence of H_2O_2. The experiments were performed in a recirculating mode batch reactor with a low-pressure Hg lamp (5.3 W at 253.7 nm), at 25°C and pH 6.8. The initial ratio of H_2O_2/pollutant concentrations was 7, with pollutant concentration of 2×10^{-4} M.

In all cases, the UV/H_2O_2 process proceeded at faster rates than UV alone. The ratio of the rate constants of the $^{\bullet}OH$ radical-induced process and of the UV photolysis varied from 1.3 to 47 for 2,4,6-trichlorobenzene (TCB) and diethylphthalate, respectively. In the chloro-substituted benzene series, the ratio decreased as the chlorination degree increased.

The rate constants for photodegradation of phenol, benzene and toluene, in the absence and the presence of H_2O_2, were found as 0.0035, 0.015 and 0.023 min^{-1}, and 0.082, 0.085 and 0.094 min^{-1}, respectively.

3.2.2.2 Phenols

Photochemistry of phenols is pH dependent, given the acid–base equilibria in which these compounds are involved. Audureau *et al.* (1976) studied the kinetics and by-products of phenol photolysis in aqueous solution at various initial concentrations and pH values, under irradiation with a low-pressure Hg lamp. The quantum yield of phenol photo-oxidation in aerated solution was found independent on the initial concentration of phenol within 5×10^{-5}–5×10^{-2} M range, at pH 5.6 ($\Phi = \sim 0.12$), but it depends on pH, varying from ~0.12 at pH 1.6–5.6 to 0.07–0.05 at pH 9.5–13.2, with a minimum value of 0.02 at pH 11.1. Among the primary photoproducts are tri- and tetrahydroxybiphenyls, quinones and diphenols. Ho *et al.* (1996) determined the average quantum yield $\Phi = 0.13 \pm 0.02$ and pH 3–8 for phenol photolysis over a broadband irradiation (Xe flashlamp), which is in good agreement with Audureau *et al.*'s data at 253.7 nm, and suggests that it does not depend on the irradiation wavelength. In a recent study on phenol removal by various AOPs, including UV alone, in a low-pressure Hg lamp annular reactor, Esplugas *et al.* (2002) concluded that phenol degraded by UV/H_2O_2 process with the highest rate, as compared to the other processes. Phenol removal by UV alone was found pH dependent, with low rates in alkaline media.

Kawaguchi (1993) addressed the degradation of 2,4,6-trimethylphenol (TMP) in natural waters photo-induced by humic acid, as irradiated at 313 and 371 nm, and found that the quantum yields are very low ($4.3–9.5 \times 10^{-4}$), and the first-order kinetics is obeyed only at very low concentrations of TMP ($<6.0 \times 10^{-8}$ M and 0.008 mg L^{-1}).

The photolysis of triphenols, such as hydroxyhydroquinone, pyrogallol and phloroglucinol, at 253.7 nm in aerated aqueous solution, occurs with quantum yields independent on concentration for pyrogallol ($\Phi = 0.030–0.035$), but concentration dependent for hydroxyhydroquinone ($\Phi \sim 0.030$ at 1×10^{-2} M and $\Phi \sim 0.015$ at 6×10^{-5} M) and phloroglucinol ($\Phi \sim 0.022$ at 1×10^{-2} M and $\Phi \sim 0.010$ at 1×10^{-4} M) (Perbet *et al.*, 1982).

3.2.2.3 *Halogenated aromatic compounds*

Chlorobenzenes The contamination of soil and aquatic environments by chlorinated aromatic pollutants, such as polychlorobenzenes, polychlorophenols, polychlorodiphenylethers and phenoxy acids originates not only from their use in pesticides, fungicides, solvents, chemical and pulp mill industries, but also as by-products during chlorination. Choudhry *et al.* (1986) reviewed extensively the literature concerning the photochemistry of these classes of compounds, particularly that induced by radiations with $\lambda > 280$ nm.

Chlorobenzene is the best-studied chloroaromatic. In water, it photodissociates via a short-lived triplet state [$\tau = 30$ ns (Tissot *et al.*, 1984)] to the phenyl and chlorine radical, which further react with the solvent leading to phenol and chloride ions as major primary degradation products; acetanilide and other very polar products were detected also as intermediates (Dulin *et al.*, 1986). Quantum yields of 0.37 and 0.50 were determined for chlorobenzene photolysis with a filtered 450-W medium-pressure Hg lamp (irradiation range: 230–410 nm) in water and 0.05-M NaOH solution, respectively (Dulin *et al.*, 1986). The photolysis rate of polychlorinated benzenes decreases as the halogenation number increases. 1,2,4-TCB photolyzes with both concentration and oxygen level-dependent quantum yields: $\Phi = 0.043$ and 0.025 at 1.22×10^{-4} and 1.00×10^{-5} M TCB, respectively, in aerated solutions, and $\Phi = 0.013$ at 1.22×10^{-4} in oxygen-free solutions (Dulin *et al.*, 1986). Guittonneau *et al.* (1988b) compared the photo-oxidation efficiencies of UV and UV/H_2O_2 processes during the irradiation at 253.7 nm (40-W low-pressure Hg lamp) of some chloro- and nitrobenzenes at pH 7.5. The UV photolysis of chloro- and 1,2,4-TCB resulted in 71% and 67% chloride ion release after 45- and 125-min irradiation, respectively, which corresponded to 90% destruction of original pollutant. The increase of optical density of solution during irradiation indicated the formation of phenols as reaction intermediates. Hexachlorobenzene photodegraded by only 20% over 70-min irradiation time. In the presence of H_2O_2, 80% of organic chlorine was mineralized in 4- and 6-min irradiation of chloro- and 1,2,4-TCB, respectively, corresponding to 90% removal of parent contaminant. Hexachlorobenzene was removed by only 50% in 50-min irradiation under similar experimental conditions.

Halogenated phenols The photochemistry of chlorophenols has been studied by many researchers, and it was found that both the kinetics and the photolysis products depend on the degree of chlorination and on the irradiation conditions. Some derivatives, such as 2,4-D, 2,4,5-T and pentachlorophenol (PCP), are well-known pesticides. In general, direct photolysis of polychlorinated phenols is accompanied by chloride ion detachment by either heterolytic or homolytic cleavage of C—Cl bond. Skurlatov *et al.* (1997) studied the photolysis of 2,4,5-TCP and PCP in aerobic and anaerobic conditions at different wavelengths (high- and low-pressure Hg lamps, filtered and unfiltered) and pH values, and, also, under sunlight irradiation using dyes as photosensitizers through a singlet oxygen mechanism. The quantum yield of 2,4,5-TCP photolysis ($\Phi = 0.059 \pm 0.005$) was found both pH and wavelength independent. The rates of 2,4,5-TCP decay and of formation of 2,5-dichlorohydroquinone and 3,4-dichlorocyclopentadiene are independent of the presence of oxygen in the solution, whereas the formation of 2,4- and 2,5-dichlorophenols are more than 10-fold increased in anaerobic conditions.

Wong and Crosby (1981) studied the photolysis of PCP under both sunlight and laboratory UV lamp exposure, and observed a rapid degradation at pH 7.3, leading to tetrahydrocatechol, tetrachlororesorcinol and tetrachlorohydroquinone as primary products, which further underwent photo-oxidation to chloranil, hydroxyquinones and 2,3-dichloromaleic acid. Tetra- and trichlorinated phenols were among the by-products as well. The rate of photolysis at pH 3.3 was much slower than at pH 7.3.

There is the concern that UV treatment of PCP may lead to the generation of highly toxic polychlorinated dibenzo-*p*-dioxins (PCDDs) and furans during the irradiation of PCP with low-pressure Hg lamp (Vollmuth *et al.*, 1994). Ho and Bolton's (1998) toxicity studies during the photolysis of $\sim 1.5 \times 10^{-4}$ M PCP in a bench-scale 1-kW medium-pressure lamp reactor at pH 7 indicated that the relative toxicity index decreased as PCP concentration decreased, concluding that the treatment either does not generate significant levels of acutely toxic intermediates, or any toxic intermediates are rapidly degraded under the specific conditions of their study. The rate of PCP degradation increased in the presence of H_2O_2 [k (UV) = 0.115 min^{-1}; k (UV/H_2O_2) = $\sim$0.66 min^{-1}, at $H_2O_2 \sim 6.7$–37.2×10^{-3} M] (Ho and Bolton, 1998).

Boule *et al.* (1982, 1984, 1985) examined the reaction mechanism of various monochlorophenols in aqueous solution under irradiation at 253.7 nm (molecular form) and 280, 296 and 305 nm (anionic form) at various pH values. The studies revealed the importance of chlorine substitution position on the aromatic ring, on the photochemical behavior of chlorophenols, with respect to both degradation products and quantum yields. 2-Chlorophenol degrades primarily to cyclopentadienic acid and pyrocatechol from the anionic and molecular form, respectively, with quantum yields of 0.30 ($pH > 9$) and 0.02–0.04 ($pH \sim 1$), respectively. 3-Chlorophenol photolysis generates resorcinol as a major product from both anionic and molecular forms, with quantum yields of 0.13–0.15 and 0.09, respectively. The product yield is independent on the oxygen concentration. 4-Chlorophenol photodegrades ($\Phi \sim 0.4$–0.5 and pH 5.5) primarily to 5-chloro-2,4′-dihydroxybiphenyl, with low yields of hydroquinone. Ho *et al.* (1996) determined the average quantum yield for the photolysis of 4-chlorophenol as irradiated with a Xe flashlamp, $\Phi = 0.24 \pm 0.04$, in good agreement with that determined by Boule *et al.* (1982), $\Phi = 0.25$ at 296 nm. The photo-induced degradation of chlorophenols was also reported by other studies (Boule *et al.*, 1999; Miller *et al.*, 1988; Shen *et al.*, 1995).

Chlorinated biphenyls The photolysis of 2- and 4-chlorobiphenyls with a 450-W medium-pressure lamp resulted primarily in 2- and 4-hydroxybiphenyls with quantum yields of 0.29 and 0.002, respectively (Dulin *et al.*, 1986). No biphenyls were observed among the by-products. 2-Hydroxybiphenyl degrades nine times faster than 2-chlorobiphenyl.

2- and 4-Chlorobiphenyl ethers (CBPE) photolyze efficiently in water to release near quantitative amounts of chloride ion and form either dibenzofuran and hydroxybiphenyl ether, respectively. Quantum yields of 0.19 and 0.63 were determined for 2-CBPE and 4-CBPE photolysis with a 450-W medium-pressure Hg lamp, respectively (Dulin *et al.*, 1986).

Chlorinated dioxins and furans Polyhalogenated aromatic compounds are environmental contaminants, with some of the most potent toxins being PCDDs and polychlorinated dibenzofurans (PCDFs). They are classified as endocrine disrupting chemicals, and fall under the US EPA drinking water regulations, being set at a maximum contaminant level (MCL) of 0 $mg\,L^{-1}$. Photochemistry of dioxins was investigated especially after the accidental release of 2,3,7,8-tetrachlorodibenzo-*p*-dioxin (2,3,7,8-TCDD) over a wide area near Seveso, Italy, in 1976. In 1982, Choudhry and Hutzinger (1982) published a review on the photochemical studies on chlorinated dioxins. Dehalogenation was indicated as a major mechanistic pattern through photoreduction processes involving electron donors. Practical implementation of photolytic processes to the waste streams treatment requires product studies to ensure that the technology will not lead to the formation of harmful contaminants. The photochemistry of chlorinated dioxins and furans is still controversial, and a few product studies were published in the literature over the past decade. It is known that they undergo slow degradation in the aquatic environment exposed to sunlight (Dulin *et al.*, 1986), but can be removed faster in the heterogeneous photolcatalysis (Barbeni *et al.*, 1986). Freeman and Hatlevig (1994) investigated the photochemical degradation of 1,2,3,4-TCDD (3×10^{-4}M in acetonitrile) under irradiation at 253.7 nm in the presence of triethanolamine (TEA) as an electron donor in order to elucidate the reaction mechanism. The 1,2,4- and 1,2,3-trichlorodibenzo-*p*-dioxins were the identified reaction products, suggesting a photodehalogenation mechanism. A complete degradation of 1,2,3,4-TCDD was observed after 8 h and 10 min in the absence and the presence of TEA, respectively. Photodegradation of 2-chloro- and 1,2,4-trichlorodibenzo-*p*-dioxins conducted in a mixture of water-organic solvent (methanol and acetonitrile) under monochromatic irradiation at 222 nm (KrCl excimer) and 308 nm (from either an XeCl excimer or a Hg–Xe lamp with monochromator) led to numerous products; among them, chlorohydroxybiphenyls, indicate dioxin ring-opening mechanistic patterns (Nohr *et al.*, 1994). The two selected wavelengths are close to the first and second absorption maxima of chlorodibenzodioxins, and induce the photolysis of the studied compounds with quantum yields in the 0.1–0.2 range. These quantum yields are similar to that determined by Dulin *et al.* (1986) for photolysis of 2,3,7-tridibenzo-*p*-dioxins in hexane at 313 nm ($\Phi = 0.2$), but much lower than those for photolysis of 2,3,7,8-TCDD either at 313 nm or in sunlight, 0.0022 and 0.0007, respectively. Konstantinov *et al.* (2000) performed a product study on the photolysis of 2,3,7,8-TCDD at 300 nm, and measured low yields of dechlorinated by-products at 99.9% conversion of 2,3,7,8-TCDD. Identification of highly chlorinated hydroxybiphenyls as primary photolysis products indicated that C—O rather than C—Cl bond homolysis occurs. High photodegradation yields were observed for four other highly chlorinated dibenzo-*p*-dioxins (99% in 60–100 min) irradiated at 253.7 nm. The authors monitored the hepatotoxicity and estrogenic-binding affinities of 2,3,7,8-TCDD and its degradation by-products in the irradiated samples at different exposure times, and concluded that is unlikely that the products from UV treatment of PCDDs and PCDFs will show either hepato- or estrogen receptor-mediated toxicological effects, and the degradation products should be treated as regular organic waste. In a recent product and mass balance study on 2,3,7,8-TCCD photolysis in water at 302 nm, Rayne *et al.* (2002) identified dihydroxybiphenyls, hydroxyphenoxyphenols and dechlorinated dibenzo-*p*-dioxin. Based on toxicological studies, the authors assessed that UV photolysis of 2,3,7,8-TCCD results in a solution of lower acute toxicity, but one with potential chronic endocrine disrupting and mutagenic effects.

The photolysis of PCDFs in water under either monochromatic ($\lambda = 300$ nm) or sunlight irradiation occurs at low rates, and is photosensitized by dissolved NOM. The rate constants of 2,7-di-, 2,3,7,8-tetra- and 1,2,7,8-tetra-chlorodibenzofurans were 1.3, 0.45 and 0.36 h^{-1}, respectively, and the degradation involved low yields of dechlorinated products (<1%, 17% and 20%, respectively), indicating a different reaction mechanism than in organic solvents where the photoreductive reactions prevail (Dung and O'Keefe, 1994).

Other halogenated aromatics Karpel Vel Leitner *et al.* (1996) investigated the kinetics and mechanisms of photolytic and •OH radical-induced degradation of trifluorobenzene and trifluorotoluene derivatives at 253.7 nm in the presence and the absence of dissolved oxygen. Such compounds are mainly used as intermediates in the production of pharmaceuticals (antibiotics, antidepressants and non-steroidal anti-inflammatories), agrochemicals (herbicides and insecticides) and advanced polymers. The quantum yields determined for the UV photolysis are low ($\Phi = 0.011$–0.015), but the rate constants determined for the reaction with the $^{\bullet}OH$ radical are large ($k \sim 3.7 \times 10^9$ to $4.9 \times 10^9 M^{-1} s^{-1}$), suggesting the UV/$H_2O_2$ process for practical application.

Höfl *et al.* (1997) reviewed the degradation of AOX (defined as the sum of all halogeno-organic compounds, which can be adsorbed by activated carbon) and chemical oxygen demand (COD) by different AOPs (UV, UV/H_2O_2, O_3, UV/O_3 and Fenton's reagent), with two samples from a pharmaceutical wastewater. In the UV-driven processes, a 15-W low-pressure Hg lamp with a batch reactor was employed. UV photolysis was significantly slower than the other AOPs in the removal of AOX from the first water (AOX = $3 mg L^{-1}$ and COD = $670 mg L^{-1}$), but outperformed the O_3 and UV/H_2O_2 processes in the case of the second water (AOX = $5 mg L^{-1}$ and COD = $2700 mg L^{-1}$). It appeared that UV irradiation involved a high selectivity for the degradation of AOX compared to COD. On the other hand, processes involving $^{\bullet}OH$ radicals were less selective, but considerably more effective in COD degradation.

3.2.2.4 Nitroaromatic derivatives

Aromatic nitro-compounds are widely used in the manufacture of dyes, insecticides, pesticides and explosives, and can be generated in the atmospheric environment through photochemical reactions involving nitrogen oxides. They are listed as priority toxic pollutants affecting the central nervous system, liver, kidneys and blood cells, and are severely monitored in the environment. These compounds are fairly resistant to biodegradation, requiring long periods of incubation (O'Connor and Young, 1989), and their aquatic photochemistry was addressed by several studies (Carey and Fox, 1981; Mabey *et al.*, 1983). Nitroaromatic compounds are usually intensely colored, exhibiting absorption bands in both UV and visible range, therefore, they should be susceptible to direct photolysis. However, their quantum yields in pure water are fairly low (10^{-6}–10^{-3}), and undergo photosensitization in natural waters. Photoreductive H-abstraction in the presence of hydrogen donors, photonucleophilic substitution in the presence of electron donors, intra- and intermolecular photoaddition to activated double or triple bonds are the main reaction pathways of nitroaromatic derivatives (Larson and Weber, 1994).

Guittonneau *et al.* (1988b) studied the kinetics of nitrobenzene, chloro-4-nitrobenzene and 4-nitrophenol ($10^{-4} M$) photo-oxidation in a closed batch reactor equipped with a 40-W low-pressure Hg lamp, at 16°C and pH 7.5, in the absence and the presence of H_2O_2. Only 10% removal was observed in all cases after 60-min irradiation without H_2O_2, whereas 90% decrease from the initial concentrations was measured after 36-, 30- and 18-min irradiation of nitrobenzene, chloro-4-nitrobenzene and 4-nitrophenol, respectively, in the presence of $10^{-3} M$ H_2O_2. No change in the organic carbon or nitrogen was observed under UV photolysis over 60-min irradiation, but ~8% and 2% NO_3^- were quantified at 90% reduction of nitrobenzene and 4-nitrophenol in the presence of H_2O_2. Even lower conversions of nitrobenzene, 2-nitrophenol, 4-nitrophenol and 2,4-dinitrophenol exposed to direct broadband irradiation emitted by a 150-W Hg–Xe lamp were reported by Lipczynska-Kochany (1992) in comparatives studies with other AOPs, such as UV/H_2O_2 and photo-Fenton reaction. Under the conditions used in this study, the photodegradation of nitrophenols proceeded at the highest rates in the presence of Fenton reagent.

Alif *et al.* (1987, 1990, 1991) studied the photochemical behavior of nitrophenols in aqueous solutions. The irradiations were performed in either a cylindrical quartz cell located on the axis of a cylindrical mirror using six low-pressure Hg lamps ($\lambda = 253.7$ nm) or a water-cooled reactor using three 125-W high-pressure Hg lamp, with filter ($\lambda = 365$ nm). The authors studied the role of pH, dissolved oxygen and wavelength on the photolytic kinetic parameters. Direct UV photolysis of 2-nitrophenol (Alif *et al.*, 1991) occurs with very low, pH and wavelength-dependent quantum yields. At pH 2.2 (molecular form), the quantum yields are 7.8×10^{-4} (air saturated), 6.8×10^{-4} (degassed), 4.7×10^{-6} (air saturated) and 5.3×10^{-6} (degassed), at 253.7 and 365 nm, respectively. At pH 8.2 (anionic form), the quantum yields increase at 253.7 nm: 1.7×10^{-3} (air saturated), 1.3×10^{-3} (degassed), 2.4×10^{-6} (air saturated) and 2.2×10^{-6} (degassed), at 253.7 and 365 nm, respectively. The product distribution depended strongly on pH, irradiation wavelength and the presence of oxygen. Based on the mass balance, the main photoproducts were 2-nitrosophenol, 3-nitrocatechol, catechol, nitrohydroquinone and cyclopentadiene carboxylic acids. At pH 8.2 and 253.7 nm, large amounts of nitrite and cyclopentadiene carboxylic acids were quantified, as a result of both photohydrolysis and heterolytic scission of C—NO_2 bond of 2-nitrophenol in the excited state. Similar studies were conducted on 3-nitrophenol (Alif *et al.*, 1990). The quantum yields determined at 253.7 nm were much lower than those for 2-nitrophenol. Several organic photoproducts were identified, such as resorcinol, nitroresorcinols, 3- and 4-catechol, nitrohydroquinone and 2,3′-dihydroxy-4-nitrobiphenyl. Very low quantum yields ($\Phi \sim 10^{-5}$–10^{-4}) were measured for 4-nitrophenol, irrespective of the irradiation conditions, 10-fold larger at 253.7 nm than at 313–365 nm (Alif *et al.*, 1987). In aerated acidic and neutral solutions, NO_3^- ions accounted for ~80% of 4-nitrophenol converted.

One concludes that neither nitrophenols nor nitrobenzene can be removed through direct UV photolysis. Several studies on heterogeneous photocatalysis indicated the effectiveness of these processes in destruction of nitroaromatics, including 2,4,6-trinitrotoluene (TNT). The latter is a well-known explosive and very toxic compound. The UV photolysis of TNT occurs with a quantum yield of 10^{-3} at 313 nm (Mabey *et al.*, 1983).

The explosives represent a distinctive class of nitro-derivatives that were formerly produced in military munitions facilities and currently contaminate waste and groundwaters, being usually removed by granulated activated carbon technology. Several studies investigated various AOPs as potential cleanup technologies for such contaminants. Himebaugh (1993) reported bench- and pilot-scale studies on the following explosive nitro-derivatives: 1,3-dinitrobenzene (DNB), 2,4- and

2,6-dinitrotoluene (DNT), 1,3,5-trinitrobenzene (TNB), 2,4,6-TNT, 1,3,5,7-tetranitro-1,3,5,7-tetrazacycloctane (HMX) and 1,3,5-trinitro-1,3,5-triazacyclohexane (RDX). The influent concentrations of these compounds in the studied groundwater varied from 6 mg L^{-1} (TNT) to 600 μm L^{-1} (RDX) and 10 $\mu g\,L^{-1}$ (2,6-DNT), with treatment objectives of 2.9, 0.8 and 0.13 $\mu g\,L^{-1}$, respectively. UV alone (low-pressure Hg lamp) was efficient in the destruction of RDX and TNT. More than 99% destruction of TNT was achieved in 30-min irradiation, and 2,4-DNB and 2,6-DNT were destroyed below the treatment objectives' limits after 60 min.

Bose *et al.* (1998) researched the efficiency of RDX removal by UV, ozone and their combinations in the presence and the absence of H_2O_2 using a 150-W Xe short-arc lamp as a UV source, with a monochromator to isolate the 254-nm radiation. The UV degradation rate was proportional to the UV light intensity.

3.2.3 Pesticides

The worldwide intensive use of agrochemicals, including herbicides, insecticides and fungicides, resulted in the contamination of continental and marine natural waters with pesticides of various chemical structures, which have been confirmed as potential threats for both human health and ecosystem equilibrium. Toxicological studies identified the possible chronic effect of these compounds as carcinogenesis, neurotoxicity, effects on reproduction, and cell development effects (Burrows *et al.*, 2002, and references therein). Most pesticides are either resistant or very slowly degraded by chemical and photochemical processes under typical environmental aquatic conditions. An abundant literature has been published over the last few decades on AOP studies, particularly involving short UV radiations, with regard to their potential to destroy the pesticides to or below the drinking water regulatory standards. The drinking water regulations set in the European Union state clearly a maximum permissible concentration of a particular pesticide of 0.1 and 0.5 $\mu g\,L^{-1}$ for the total load of pesticides. No maximum concentration load of pesticides was established in North America, but there were set MCLs for some individual pesticides based on toxicological data available on that particular compound.

Burrows *et al.* (2002) reviewed thoroughly the state-of-the-art knowledge on pesticides photodegradation, and referenced 295 studies published in the literature on pesticide-related topics. Halmann (1996) summarized the literature information on the photochemical behavior of pesticides-model compounds during various AOPs, including mechanistic studies. Based on their chemical structures, the pesticides can be grouped in over 25 classes. The following brief overview covers the photochemical studies reported in the literature on the pesticides most frequently encountered in the aquatic environment. Some of them, such as chlorinated phenols (e.g. PCP), were discussed above.

3.2.3.1 Triazine pesticides

Nick *et al.* (1992) studied the photolysis of atrazine, simazine, propazine and terbutylazine at 253.7 nm from a Sterisol low-pressure Hg lamp in quartz cells.

The quantum yields of these pesticides and of atrazine primary by-products are within 0.018–0.099 range, and hydroxylated triazine derivatives as resulted from dechlorination at the heteroaromatic ring were identified as major photodegradation products. At low fluences, less than 5% of the triazines were transformed. In a later study, Nick and Schoeler (1995) confirmed these results and found that under similar irradiation conditions; but in the presence of H_2O_2, the atrazine decay was six times faster than the UV photolysis, and the reaction by-products were further degraded. Beltran *et al.* (1996a) reported similar observations.

Palm and Zetzsch (1996) studied the photodegradation rates of the above triazines and ametryn and atraton under polychromatic irradiation (Xe lamp, $240\,nm < \lambda < 300\,nm$) in aqueous buffered solutions (pH $\sim$ 7–9). Atraton exhibited a very low quantum yield ($\Phi < 0.002$).

Hessler *et al.* (1993a,b) studied the UV photolysis of atrazine and metazachlor (anilide pesticide) at 253.7 nm from a 16-W low-pressure Hg lamp. The rate constants ($k_p \sim 10^{-4}\,s^{-1}$) were found dependent on the initial concentration of pesticide, but pH independent within 3–11 range. Irradiations in natural water containing $\sim$14 mg L^{-1} dissolved organic carbon resulted in lower photodegradation rates for both pesticides, due to the optical filter effect of naturally occurring organic carbon that absorbs the 253.7-nm radiation. In the presence of H_2O_2, the rates increased by a factor of 10, depending on the initial concentration of H_2O_2. Frimmel and Hessler (1994) reported kinetic parameters for four triazine pesticides (atrazine, simazine, desethylatrazine and terbutylatrazine) and two anilide pesticides (metazachlor and metolachlor) under irradiation with 253.7-nm radiation in distilled and natural waters. Low quantum yields were determined for triazine pesticides ($\sim$0.04), and $\sim$10 times larger for the chloroanilide pesticides ($\sim$0.35–0.40). Even though overall the humic substances exhibited an inhibitory effect on all six pesticides' photolysis due to the competition in the light absorption, a $\sim$10–15% photosensitization effect of NOM on the pesticides' degradation was estimated by comparison with the distilled water photolysis data.

Torrents *et al.* (1997) reported a thorough mechanistic investigation of the UV direct photolysis and nitrate-induced degradation of atrazine, emphasizing the role of DOC on the rate of removal. A Xe lamp and special glass filters ($\lambda > 290\,nm$), with an average irradiance of 450 W m^{-2}, were used in all experiments. Direct photolysis of atrazine was slow (23% removal over 60 h) yielding to hydroxyatrazine (4-ethylamino-2-hydroxy-6-isopropylamino-S-triazine, 14%), chlorodealkylated and chloroalkyloxidized derivatives (9%). The efficiency of the direct photolysis decreased by 15% with increasing DOC. Irradiation of atrazine (6.5 mg L^{-1}) in the presence of nitrate (11 mg L^{-1} as N) generated $^{\bullet}OH$ radicals, which increased significantly the rate of removal for both atrazine and its degradation by-products. After 60 h irradiation, 87% conversion of atrazine was measured and several by-products were detected. Very little hydroxyderivatives were measured in the $^{\bullet}OH$-driven process, and a strong effect of DOC concentration was observed (85% decrease in efficiency at a level of 3.2 mg L^{-1} DOC).

Rehberg *et al.* (1994) studied the degradation of atrazine by UV oxidation in distilled water and groundwater, via a high-pressure Hg lamp, a Xe flashlamp and a KrCl excimer lamp. Several photo-fragments were identified by GC/MS. UV photolysis alone, either monochromatic at 253.7 nm or polychromatic, does not lead to a complete mineralization of atrazine or other pesticides, even though they decompose significantly (Brun *et al.*, 1993). Tace *et al.* (1992) found that 90% removal of atrazine [$(\text{atrazine})_o = 250\,\mu g\,L^{-1}$] can be achieved with a low-pressure Hg lamp, and under the conditions used in this study, H_2O_2 enhanced significantly the rate of removal.

Wimmer *et al.* (1995) examined the degradation by-products of atrazine by UV and UV/H_2O_2 photolysis; the triazine ring remained intact, only the N-alkyl side chains being split. Oxalate and nitrate ions were formed as further degradation products.

Zwiener *et al.* (1995) determined the rate constants for the UV photolysis of atrazine and its primary degradation products in a pilot plant for drinking water treatment. Rate constants of $k_{obs} = 0.31$, 0.13 and $0.02\,s^{-1}$ were determined for atrazine, desehylatrazine and desethyldesisopropylatrazine, respectively.

Kruithof *et al.* (2001) reported both bench- and pilot-scale experimental data for UV treatment of 10 pesticides of various chemical structures, such as atrazine, pyrazon, diuron, bentazon, bromacil, methabenthiazuron, dicamba, 2,4-D, Trichloroethane (TCA), trichlopyr, and desethyldesisopropylatrazine as the atrazine metabolite, at $\mu g\,L^{-1}$ level, in pretreated IJssel Lake water (The Netherlands). The IJssel Lake water is used by the water treatment plant for drinking water production. The study used medium-pressure Hg lamps, and was conducted both in the absence and the presence of H_2O_2. The O_3/H_2O_2 process was also applied for comparison with the UV-driven processes, with respect to bromate formation, if bromide ions were present in the water. Direct photolysis was found as a significant process for pesticides' destruction, both in the presence and in the absence of H_2O_2, over the wide range of electrical energy doses applied. In the presence of H_2O_2, most pesticides were removed by more than 80%.

Bourgine *et al.* (1997) described the performance of a medium-pressure Hg lamp UV reactor, industrial-scale pilot unit, installed in Aldershot, Hampshire (UK) for groundwater treatment contaminated with several pesticides, such as mecoprop, 4-chloro-2-methylphenoxyacetic acid (MCPA), bromoxynil, linuron, ioxynil, atrazine, diuron, *S*-ethyldipropylthiocarbamate (EPTC), 2,4,5-T,trichlopyr, isoproturon, 2,4-D and dicamba. The quartz tube housing the medium-pressure lamp filtered the radiations with $\lambda < 240$ nm, to minimize the nitrite ion formation from nitrate ions [water characteristics: 35–40 $mg\,L^{-1}$, 0.5–1.5 $mg\,L^{-1}$ DOC, 50–60 SO_4^{2-} $mg\,L^{-1}$, pH 7.1–7.2, $\%T = 87$–93 (253.7 nm, $d = 4$ cm)]. The pilot tests were performed at flow rates from 10 to 100 m^3h^{-1}. The yield of pesticide removal varied from 90% for mecoprop to 10–11% for EPTC.

3.2.3.2 Urea pesticides

Millet *et al.* (1998) investigated the photochemical behavior of two urea herbicides, chlorotoluron and isoproturon, in aqueous solution at pH 4, 7 and 9, using a polychromatic light source (Xe lamp). Both pesticides degraded when irradiated at $\lambda > 240$ nm, with pseudo-first-order rate constants of $k \sim 10^{-4}\,\text{min}^{-1}$, but were photostable at $\lambda > 290$ nm, indicating that any sunlight-induced photodegradation in natural systems is mediated by dissolved organic matter.

Tanaka *et al.* (1982) reported a product study on degradation of monuron and its photoproduct (demethylmonuron) under irradiation with either a 450-W high-pressure Hg lamp filtered through Pyrex glass or sunlight lamps in a reactor. Photochemical demethylation of monuron led to demethylmonuron, 3-(4-chlorophenyl)-1-formyl-1-methylurea, formaldehyde, formic acid and carbon dioxide. Dissolved oxygen increased the photodegradation of monuron, and its by-products.

Mansour *et al.* (1997) studied the photodegradation of six pesticides from various classes: isoproturon, pendimethalin, carbofuran, diazinon, metamitron and terbutylazin using an HPK 124-W high-pressure Hg lamp, with and without pyrex filter. The reaction mechanism depended on the irradiation wavelength (below or above 290 nm). Photosensitization effects of humic acids were observed in photolysis of diazinon, isoproturon and terbutylazin. Carbofuran degraded slowly by UV photolysis at $\lambda > 290$ nm, but very fast in the presence of H_2O_2 or ozone.

3.2.3.3 Carbamate pesticides

The oxidation rate of carbofuran by direct UV radiation is slightly dependent of the pesticide concentration, independent on pH, and significantly dependent on the temperature (Benitez *et al.*, 1995). The irradiations were performed with a high-pressure Hg lamp, in a bench-scale reactor, in the absence and the presence of H_2O_2. Average quantum yields of 0.015–0.0245 were determined for direct photolysis, depending on the experimental conditions. In the presence of H_2O_2, the degradation yield of carbofuran increased (after 5-min irradiation, conversions of $\sim$10% and 42% were calculated in the absence and in the presence of 1.23×10^{-4} M H_2O_2, respectively).

Peterson *et al.* (1988) reported the photodegradation of three pesticides from different classes: carbofuran (carbamate insecticide), fenamiphos sulfoxide (organophosphate pesticide) and propazine (S-triazine herbicide) in aqueous solution irradiated in a closed loop high-pressure Hg lamp photoreactor, at a flow rate of $12\,\text{L}\,\text{min}^{-1}$. At pesticide concentrations of 3.47, 9.55 and $12.6\,\text{mg}\,\text{L}^{-1}$, half-lifetimes of 3.47, 1.03 and 3.76 min were determined for the UV photolysis of carbofuran, fenamiphos and propazine, respectively. In the presence of large concentrations of H_2O_2 ($115–150\,\text{mg}\,\text{L}^{-1}$), as used in this study, no significant increase of the degradation rate (10–20%) was observed. It is not only that such large concentrations of H_2O_2 are not practical for UV applications, but also they do not lead to the expected effect on pesticide degradation rate due to the scavenging of $^{\bullet}OH$ radicals by H_2O_2 itself.

Scheurnert *et al.* (1993) studied the fate of carbofuran and diazinon (insecticides) and pendimethalin (herbicide) under abiotic and biotic conditions in both aqueous humic acid solution and river and lake natural waters. A simulated sunlight irradiation was performed using a pyrex-filtered 125-W high-pressure Hg lamp source ($\lambda > 290$ nm). Carbofuran degraded by less than 10% in 120 min, but the rate was considerably enhanced by the addition of oxidants, such as TiO_2, H_2O_2 or ozone. Diazion photodegradation depended strongly on the natural water quality. Since no significant increase of the rate was observed in aqueous solutions of humic acid, one concluded that other water constituents were responsible of the photosensitized degradation of diazinon in natural waters. The photoproduct distribution of pendimethalin (N-(1-ethylpropyl)-3,4-dimethyl-2,6-dinitrobenzeneamine) depended on the irradiation wavelength. At $\lambda > 290$ nm, the photoreduction of a nitro group to an amino group occurs, whereas at $\lambda < 290$ nm, the dealkylation at the amino group is the major photoreaction.

Carbendazim is a widely used fungicide, but also a degradation product of benomyl fungicide in water. Panadés *et al.* (2000) investigated the photochemistry of carbendazim at various pH values (5, 7, 8 and 11) in a photochemical reactor equipped with a Hg lamp with a broadband emission spectrum (250–750 nm). The UV spectra of carbendazim are pH dependent. As a general trend, the photodegradation rate increases with pH, and a positive effect of dissolved oxygen on the rate was observed, especially at lower pH values. The quantum yields are very low and were found both pH and oxygen dependent, varying from 0.0011 (pH 5) to 0.0063 (pH 11) at 6 mg L^{-1} O_2.

Pirisi *et al.* (1996) studied the photodegradation rates of pirimicarb and primary by-products, as well as the mechanistic pathways and degradation products, at various irradiation wavelengths: 50-W low-pressure Hg lamp (253.7 nm), 125-W high-intensity Hg lamp (full emission spectrum and filtered for $\lambda >$ 290 nm). The main primary photoproducts resulted from dealkylation and photohydrolysis of parent molecule, and were the same irrespective of the irradiation wavelength. The rate constants of pirimicarb photodegradation were in the 10^{-5}–$10^{-3}\,s^{-1}$ range, depending on the irradiation conditions used.

Patterson *et al.* (1998) explored the kinetics and the influence of dissolved organic matter on the photodecomposition of the carbamate pesticides (carbofuran, carbaryl and aldicarb) under irradiation with UV light. The first step of reaction consists in the cleavage of carbamate group from the parent molecule, leading to less toxic products. It is shown that the dissolved organic matter slowed the rate of photolysis.

3.2.3.4 *Organophosphorus pesticides*

Diazinon undergoes both thermal hydrolysis and UV photolysis, when irradiated at 253.7 nm (Ku *et al.*, 1998). The rate constant for hydrolysis of 5.5×10^{-5} M diazinon in aqueous solution decreases from 0.006 min^{-1} at pH 2 to ~0.0002 min^{-1} at pH 4 ($pK_a = 2.4$). No hydrolysis was observed at pH >5. The photolytic rate constant increased from 0.0045 min^{-1} at pH 2 to ~0.01 min^{-1} at pH $\geqslant 5$. The photodegradation rate was found dependent on the irradiance. Same degradation product was identified for both hydrolytic and photolytic degradation of diazinon (2-isopropyl-4-methyl-6-pyrimidinol).

Doong and Chang (1998) studied the photo-induced degradation of diazinon in aqueous solution by UV direct photolysis, UV/H_2O_2, UV/Fe(II)/H_2O_2 and UV/Fe(0)/H_2O_2 processes. The experiments were performed in a 1.2-l hollow cylindrical 100-W medium-pressure Hg lamp (253–578 nm range) photoreactor equipped with a water jacket. The initial concentrations of diazinon, H_2O_2 and Fe(0) or Fe(II) were of 10, 20 and 2.8 mg L^{-1}, respectively. The photodegradation followed pseudo-first-order kinetics, with rate constants of 0.0025, 0.011, 0.011 and 0.009 min^{-1} for direct photolysis, UV/H_2O_2, UV/Fe(II)/H_2O_2 and UV/Fe(0)/ H_2O_2, respectively. H_2O_2 itself was found to cause a dark oxidation of diazinon.

Wan *et al.* (1994) determined the quantum yields for the photolysis of 16 organophosphorus pesticides at 254 and 313 nm. A 200-W medium-pressure Hg lamp and interference filters were used. The pesticide solutions (1.5–3 mg L^{-1}) were irradiated until more than 75% of the initial concentrations of pesticides was degraded.

The photochemical behavior of 80 pesticides (diazinon among them) in simulated natural waters and on the surface of solids, such as agricultural leaves and clay particles, was studied by Hirahara *et al.* (1998) by exposure to 365 nm for 4 h. An efficient removal of oraganophosphorus pesticides was reported, and the results showed that the photodegradation on the surface of solids was faster than that in natural waters.

3.2.3.5 *Chloracetamide pesticides*

Metolachlor is one of the most widely used herbicides. Volatilization, hydrolysis and photolysis are the major natural processes leading to the removal of pesticides from the environment. In the photolytic processes, natural organic matter (NOM) may play an important role, either slowing the pesticide photodegradation through a light filter effect, or enhancing it through photosensitization (Miller and Zepp, 1979).

Mathew and Khan (1996) studied the photodegradation of metolachlor (30 mg L^{-1}) in water at 253.7 nm in the presence of soil minerals (kaolinite, montmorrilonite and goethite) and organic constituents (fulvic acid, 0.01% w/v) at different pH values (3.5–7.0). Irradiations, as long as 6 h, were performed. When the photolysis was conducted in water, without any soil mineral or organic constituents, the calculated half-lives and rate constants for the decay of metolachlor were 1.49 h and $k = 7.76 \times 10^{-3}\,min^{-1}$, and 2.58 h and $k = 4.48 \times 10^{-3}\,min^{-1}$, at pH 5.35 and 7.0, respectively. After 6 h, more than 94% removal was achieved. The calculated half-lives and rate constants in the presence of fulvic acid were 1.28 h and $k = 9.05 \times 10^{-3}\,min^{-1}$, and 1.17 h and $k = 9.89 \times 10^{-3}\,min^{-1}$, at pH 3.46 and 7.0, respectively. Higher rate constants and shorter half-lives were measured when the irradiation was conducted in the presence of soil mineral constituents. The enhancing effect is attributed to iron and titanium present on the surfaces of soil minerals, which may generate hydroxyl radical and other active oxygen species that induced fast photo-oxidation processes. Several degradation intermediates were detected as a result of hydroxylation, dechlorination, dealkylation and oxoquinoline formation processes. The authors concluded that the hydroxylation occurring during photodegradation of metolachlor is associated with detoxification.

Zheng and Ye (2001) identified at least 11 photoproducts of the UV photolysis of butachlor with a 100-W medium-pressure Hg lamp. The half-lives at initial concentration of 20 mg L^{-1} butachlor were only 7.54, 10.56 and 12.22 min in deionized water, river water and paddy water, respectively.

3.2.3.6 Dinitroaniline pesticides

Draper (1987) determined the wavelength-averaged quantum yields of three dinitroaniline herbicides (fluchloralin, isopropalin and profluralin) in near UV region (310–410 nm). Their values ranged from 0.0024 (profluralin) to 0.014 (isopropalin).

3.2.3.7 Chlorophenol pesticides

2,4-D (2,4-dichlorophenoxyacetic acid) is a widespread herbicide known for its toxicity, and well studied with regard to the treatment by various AOPs. Cabrera *et al.* (1997) and Martín *et al.* (1997) studied the photodecomposition of 2,4-D at 253.7 nm in a well-stirred batch-reactor.

3.2.4 N-Nitrosamines

Nitrosamines constitute a class of organic compounds (aliphatic and aromatic) of high environmental concern, as many of them were identified as hepatotoxic, genotoxic, mutagenic and carcinogenic. Among them, N-nitrosodimethylamine (NDMA) is the most studied compound, and is currently efficiently removed from contaminated waters by UV photolysis. NDMA is not an industrially or commercially important chemical, but its occurrence in the human environment (air, surface waters and soil) comes from its potential release from various industries, such as tanneries, amine, pesticide, detergent manufacturing plants, rubber and tire manufacturers, fish and food processing industries, malt beverage industries, etc. It is also a decomposition intermediate of unsymmetrical dimethylhydrazine (rocket fuel component), and can be formed in the reaction of widely occurring primary, secondary or tertiary amines with nitrite, as well as a by-product of water treatment and disinfection processes.

NDMA does not currently fall under the US EPA regulations, but it is listed as a priority pollutant on the *US EPA National Priorities List*, and regulatory actions and advisories are enforced in various states. The California Department of Health and Services has set an action level of 10 ng L^{-1} for NDMA. In Ontario, Canada, the interim maximum acceptable concentration for NDMA is 9 ng L^{-1} since 1994.

Nitrosamines and nitramines are thermally stable in aqueous solutions, do not air strip or adsorb on activated carbon efficiently due to their high solubility in water, and are fairly recalcitrant to biodegradation. However, several studies indicated that they are photolabile, particularly in acidic solutions, both under short UV and solar radiations.

The photochemistry of aqueous solutions of nitrosamines in general, or NDMA in particular, has been investigated by Chow and co-workers (Chow, 1967; Chow *et al.*, 1972; Polo and Chow, 1976a,b), Axenrod and Milne (1968), Grilli *et al.* (1975). Polo and Chow (1976a) used two types of UV reactors to photolyze NDMA (10^{-3} M): one with eight fluorescent lamps surrounding the NDMA solution in a pyrex vessel, the other with a 200-W lamp housed by either a pyrex or quartz sleeve, centered in the 18-L UV reactor. The pH (1.5–11) and oxygen effects on NDMA photodestruction rate were investigated. Half-lives of NDMA photolysis of 1, ~17.5 and 7 h were measured at pH 2.0–3.5, 8.0–9.0 and 11.0, respectively, when using broadband irradiation, whereas 23–27 h within pH range 2.0–5.0 were determined for irradiations at λ >280 nm. Different reaction products were identified, depending on the irradiation wavelength range; at long wavelengths, the photoelimination is the major reaction pathway leading to hyponitrous acid (HNO), whereas at short irradiation wavelengths, photohydrolysis generates mainly nitrous acid (HNO_2), along with dimethylamine (DMA). The study also indicated the dark formation of NDMA from DMA and HNO_2.

Challis and Li's studies (1982) on 5×10^{-3} M aqueous solutions of N-nitrosomorpholine, N-nitrosopiperidine and N-methyl-N-nitrosopiperazine irradiated with an unfiltered 125-W medium-pressure Hg lamp indicated that the photodegradation rate depends on pH and delivered UV dose, rather than on the nitrosamine concentration, and is inhibited by the addition of amine and nitrite ions. The last two effects reflect the formation of N-nitrosamines by photolysis, and were confirmed by the same study during the irradiation of 5×10^{-3} M aqueous solutions morpholine, piperazine and piperidine in the presence of 0.1 M $NaNO_2$. Both N-nitroso- and N-nitramines were formed when nitrite was replaced by nitrate.

Ho *et al.* (1996) determined the average quantum yield of NDMA photolysis (Xe flashlamp) at pH 2.5 and 7.5, as 0.27 ± 0.02 and 0.11 ± 0.03, respectively, over a wide range of NDMA concentration (10^{-5}–10^{-3} M).

Stefan *et al.* (2002) examined the photodegradation rate of NDMA under broadband irradiation (medium-pressure Hg lamp in a collimated beam apparatus) in the absence and the presence of H_2O_2, and observed an insignificant contribution of $^{\bullet}OH$ radical-induced process to the overall rate of NDMA decay. The experiments were performed in simulated surface water under various water qualities, in order to investigate the effect of various parameters, such as DOC (2 and 12 mg L^{-1}), alkalinity (40, 140 and 240 mg L^{-1} as $CaCO_3$), H_2O_2 concentration (0, 5 and 10 mg L^{-1}), temperature (3°C, 13°C and 25°C) and pH (6.5, 7.5 and 8.5), on the photolysis rate of NDMA. The content of NOM increased the reaction quantum yield by ~10%, whereas the rate constant for the UV photolysis decreased by ~35%, due to an inner filter effect exerted by the water background. A positive effect on the rate constant was induced by the increase of temperature. In a recent study, Stefan and Bolton (2002) investigated the product distribution and kinetics of NDMA photolysis in a 1-kW medium-pressure semi-batch reactor at pH 3 and 7. DMA, nitrite and nitrate ions, formaldehyde and formic acid were detected as reaction products. No significant change in TOC was observed since DMA does not treat by UV photolysis and persist in the system. The *pseudo*-first-order rate constants were larger at pH 3 than at pH 7, and depended on the initial concentration of NDMA.

Jobb *et al.* (1994) reported the results of laboratory- and pilot-scale studies on NDMA photolytic removal and formation through the post-treatment processes in a drinking water supply. Medium-pressure Hg lamps were used in the studies and further implemented in a full-scale installation. In laboratory investigations, NDMA was reduced from 100 to less than 5 ng L^{-1}. Pilot studies in a 3 × 10-kW reactor overperformed the laboratory batch reactor, reducing the NDMA concentration to undetectable levels. The addition of H_2O_2 did not improve the yield of NDMA photolysis. No harmful by-products were detected during NDMA photolysis, but formation of NDMA was observed as a result of post-treatment chlorination. As discussed previously, the formation of NDMA as a DBP is not always an issue, and should be evaluated on a case-by-case basis.

3.3 UV PHOTOLYSIS: FULL-SCALE APPLICATIONS

3.3.1 UV Reactors

3.3.1.1 Reactor design concepts

The basic objective of UV reactor design is simple: efficiently and uniformly distribute UV light throughout the reactor, and approach plug-flow behavior of the water as it flows through the reactor, such that an uniform exposure to the radiation field is achieved in the whole treated volume. While the objective is simple, optimizing the application is complicated by the difficulty of predicting the hydraulics, optics and photochemical kinetics of the pollutant in the reactor for a given treatment option, and of combining the corresponding parameters in an adequate computational model.

The main components of a reactor are the lamps, quartz sleeves, ballasts and the reactor housing. The reactor housing and its associated geometry are a function of the configuration and positioning of the lamps and other project design factors, such as flow rate, water transmittance and pressurization requirements. In general, UV reactors can be either pressurized or unpressurized (open channel) reactors, and operate with either polychromatic or monochromatic lamps, depending on the application and water quality parameters. Many commercial reactor housings that are closed vessels are constructed of stainless steel and can be pressure tested up to $\sim 1.4 \times 10^5 \, kg \, m^{-2}$ (Figures 3.1 and 3.2). Open channel (unpressurized) UV systems (Figure 3.3) are, in general, gravity flow systems and are constructed of concrete or stainless steel. UV reactor systems use fluid level sensors, pressure sensors, temperature sensors, irradiance sensors and flow meters to ensure that treatment goals are achieved.

Reactor optimization is water quality and project specific. Therefore, reactor configurations are water quality and site specific. Multiple reactor design parameters impact the efficacy of the UV reactor for a given water quality and project

Figure 3.1 Medium-pressure lamp UV reactor (pressurized system).

Figure 3.2 Low-pressure lamp UV reactor (pressurized system).

Figure 3.3 Open-channel medium-pressure lamp UV reactor.

constraints. The base assumption is that the target contaminant is uniformly distributed in the water. For UV light-driven processes in the presence of oxidants, such as H_2O_2, the uniform distribution of oxidant throughout the radiation field is of critical importance. H_2O_2, dosed upstream of the reactor, must be thoroughly mixed in the water stream prior to its entry into the UV reactor; otherwise, sub-optimal results will be achieved.

Lamp spacing is an important design parameter. Considered together with the fluid flow conditions, there exists an optimum spacing for lamps that will maximize the efficiency of UV treatment. Exceeding or reducing the optimum number of lamps in the UV system results in either poor economic or poor treatment performance of the UV reactor, due to the excessive electrical energy consumption or 'short circuiting' of water containing the contaminant.

Lamp orientation relative to the flow of water is also a critical design factor. Lamps can be placed parallel, normal or even angled to the flow of water. Each lamp orientation has its advantages and disadvantages. A parallel lamp configuration is often used in order to minimize head loss and is efficient if the water velocities are high enough in order to get good radial mixing along the length of the lamp. Normal lamp configurations are often used to distribute light uniformly with respect to the direction of fluid flow within the reactor, and to enhance mixing, while angled configurations attempt to achieve the advantages of both parallel and normal configurations.

Inlet and outlet water flow conditions in combination with other 'mixing devices' within a reactor are also design parameters that can affect the reactor efficiency. Certain reactor configurations are more sensitive to flow being uniform prior to its entry. Some reactors are designed with internal flow diversion and mixing devices to reduce the effect of non-uniform flow entering the reactor. These devices are intended to reduce the impact of elbows and other pipe joints either upstream or downstream of the reactor that would cause the water flow to be non-uniform as it enters the reactor. The overall objective of these hydraulic design elements is to promote adequate fluid mixing within the reactor, so that the light energy delivered to each fluid volume is the same.

All reactor design parameters mentioned above are not only a function of the water quality and treatment goals of a given project, but are also lamp specific. The interaction of the project parameters coupled with the type of lamps (i.e. monochromatic or polychromatic) used will lead to different reactor designs. Very sophisticated and computationally intensive computer models are used to couple the hydraulics, optics (the UV radiation field) and photochemistry for a given reactor and a given application. Often, these models are company proprietary. Computational fluid dynamics (CFD) models are used in conjunction with UV radiation models and chemical equations to derive the most efficient reactor design for a given application. Figure 3.4 shows a CFD image, simulating the particle

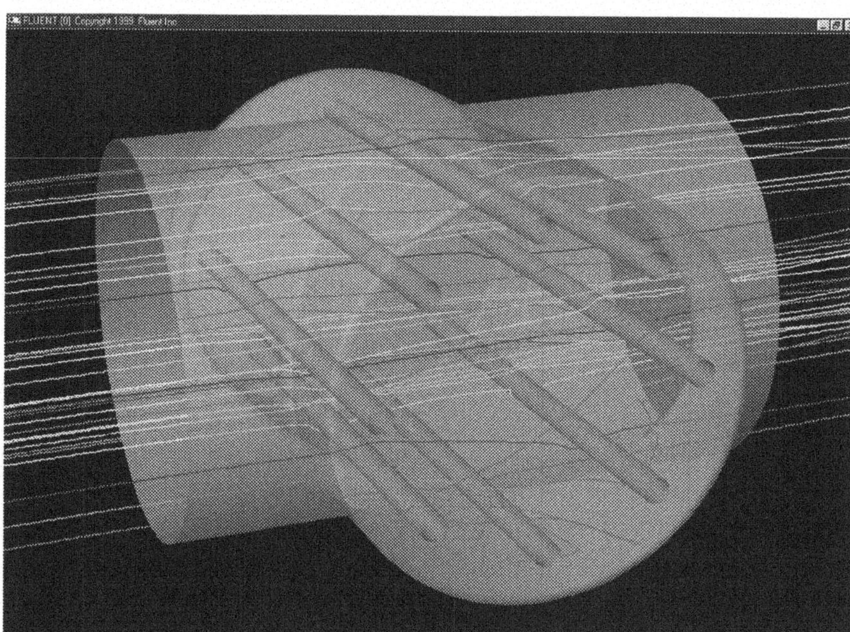

Figure 3.4 CFD image (cross-flow) for a UV reactor.

(micro-organism, chemical compound) paths in a UV reactor. A challenge in UV reactor design is optimization, as each design parameter must be considered in relation to its impact on other reactor design parameters. The manufacturers of UV treatment systems invest millions of dollars per year into optimizing their UV reactors for various applications and water treatment plant conditions.

3.3.1.2 *Using the* E_{EO} *reactor efficiency parameter*

E_{EO} (electrical energy per order, kWh order^{-1} m^{-3}) is an appropriate metric for the description of reactor performance based on the *pseudo*-first-order kinetics reactions undergone by the chemical pollutant, as are the photochemical degradation reactions at low pollutant concentrations. In commercial applications of UV technologies, E_{EO} is used extensively for system comparisons in competitive bid scenarios. As E_{EO} relates the power draw of the reactor to the log reduction of a contaminant in a given water, it is useful in comparing the efficiencies of two reactors treating the same contaminant (regardless of the reactor design and lamp type). A lower E_{EO} leads to lower power costs through the project lifetime. E_{EO} can also be used to compare the performances achieved by certain UV reactor in the treatment of various contaminants under the same water quality conditions.

The practical feature of E_{EO} is that it measures the electrical energy required to achieve a one-log reduction in contaminant concentration in a volume of water. This allows for comparative economic analysis both between various AOPs and between AOPs and conventional treatment technologies, such as carbon adsorption/regeneration, air stripping or incineration.

When adding oxidants, such as H_2O_2, care must be taken to consider the amount of oxidant used in conjunction with the amount of UV used and its resulting impact on E_{EO}. Due to both UV photolysis and UV oxidation simultaneously occurring, the equipment optimization exercise must consider the impact of the oxidant concentration on the treatment process and the engineering and cost implications thereof.

3.3.1.3 *Sizing up to full scale from pilot- or bench-scale testing*

Each site's water is unique and will respond differently to a given AOP process. Factors, such as the composition and concentration of the dissolved constituents, water transmittance (%*T*), suspended solids, etc., affect the treatment of a contaminant. For this reason, testing of the selected AOP on a site's water is required to properly size a full-scale system. This testing is generally performed at bench scale and/or pilot scale. Bench-scale testing consists of experiments designed to determine the kinetic parameters of the pollutant under consideration in the water to be treated. Along with these experiments, it is useful to define the water quality parameters, such as pH, alkalinity, turbidity (NTU), %*T* over the lamp spectral emission range, total suspended solids (TSS), the level of dissolved organic carbon, and, in the case of the hydroxyl radical-based AOPs, the hydroxyl radical demand of the water background. Pilot-scale tests involve a small UV reactor with similar lamp spectral characteristics and similar conceptual design as the full-scale system, set up in the facility, where the full-scale installation will be implemented. However, scaling up to full-size treatment systems from such testing is not trivial. Recalling that the lower the E_{EO} the better, the scale-up process based on pilot E_{EO} data should consider the following:

- Pilot testing often leads to higher E_{EO} values when compared to full-scale results. This is due to low flow rates and the small reactors that are used for pilot tests.
- Hydraulic efficiencies are often gained in full-scale systems from mixing effects due to increased turbulence at higher flow rates. However, there is an optimum correlation between the reactor size and flow rate for which such mixing effects are observed. Outside these optimum conditions, any further increase in the reactor size would result in no additional benefits from hydraulics to the reactor performance.
- As E_{EO} is inversely proportional to the reactor optical pathlength, optical efficiencies are often gained in full-scale systems due to larger reactors and, therefore, longer UV light pathlengths.

Therefore, E_{EO} values determined from small-scale pilot testing can be misleading at face value. If the value is not adjusted for the full-scale system, pilot results can lead to significantly over-sized full-scale systems. Scaling-up factors are *reactor specific* and must be determined based on correlations between contaminant photodegradation rates, water quality parameters, hydraulic/optical/kinetic computer models and full-scale installation experience.

3.3.1.4 Parameters impacting E_{EO}

Several technological parameters affect the E_{EO} determined for a UV reactor.

Reactor design Different reactors (even those using the same lamp type) can have significantly different E_{EO} values for a particular water and contaminant due to reactor characteristics, such as lamp spacing, lamp configuration (e.g. horizontal, vertical, parallel, perpendicular and angled), placement of influent and effluent piping relative to lamps, number of pipe diameters upstream and downstream of the reactor, baffles and other interior mixing devices and/or zones.

Water quality The E_{EO} is impacted by the nature and the concentration of water constituents, which are globally incorporated in the absorption spectrum (or %T) of water. E_{EO} increases as the water transmittance decreases over the spectral UV range of interest. Some common constituents that influence water transmittance are dissolved NOM, nitrate and ferrous/ferric ions. Nitrate will significantly impact the efficiency of polychromatic lamp-based UV technology, and have almost no impact for a monochromatic lamp source (253.7 nm)-based reactor, as nitrate is a strong absorber of UV radiation with $\lambda < 240$ nm. Therefore, the implementation of upstream water treatment processes such as: sedimentation, filtration, reverse osmosis or air stripping, can have a positive effect on the performance of the UV treatment system.

Lamp age E_{EO} increases as a lamp ages, that is, more power is required at the end of the lamp life than at the beginning in order to achieve the same effectiveness, because the UVC output of the lamp decreases.

Contaminant characteristics E_{EO} depends on the molar absorption coefficients of the target contaminant over the spectral range emitted by the lamp, and on the quantum yield of UV photolysis. Since the concentrations of the pollutants are generally low, the E_{EO} does not depend significantly on the pollutant concentration.

Lamp type In general, if a polychromatic lamp is employed for water treatment, the relationship of any parameter with respect to wavelength must be known in order to properly determine theoretically the E_{EO} value. This includes the background water absorption spectrum, the contaminant absorption spectrum and possibly even the contaminant's quantum yield. Conversely, if a monochromatic lamp is used for treatment, the only germane values of the above parameters would be at the wavelength of the lamp's spectral signature, for example 254 nm.

Design flow rate E_{EO} is normalized by flow rate, and one may conclude that E_{EO} should be flow rate independent. Such a generalization is consistent for comparing relatively close flow rates and reactor sizes. However, as was stated previously, efficiencies associated with hydraulic effects are gained as flow rate increases. Comparisons of reactors treating different flow rates should be made very cautiously as empirical evidence and theoretical analyses have shown that the E_{EO} value decreases to an asymptotic value as flow rate increases.

In the UV/oxidant AOP applications, a key constituent that has an effect on water transmittance is the concentration of the oxidant, such as H_2O_2. H_2O_2 undergoes photolysis in a UV reactor to form hydroxyl radicals, which accelerate the degradation of oxidizable compounds in the water. Since H_2O_2 absorbs the UV light, water transmittance decreases as H_2O_2 is added. Thus, an optimum balance between the UV light applied and H_2O_2 dose exists given certain design parameters and economic constraints.

3.3.1.5 Incorporating commercial UV photolysis reactors into a treatment plant

The UV-driven AOPs, such as UV photolysis and UV/H_2O_2, are increasingly being used on a large scale to treat contaminants in water. Often, such systems are used at remediation sites to treat recalcitrant organic compounds, such as NDMA or 1,4-dioxane that are not sufficiently removed by adsorption onto activated carbon. Moreover, the UV technologies applied to contaminant treatment, accomplish disinfection of the water as well. This applies for drinking water and wastewater treatment, wastewater treatment for indirect potable reuse and industrial applications.

As an example, a municipality that draws drinking water from one of the Great Lakes in Central North America must effectively treat the water for both microbiological and chemical constituents. This includes treatment of pathogens, viruses and bacteria as well as low concentrations of pharmaceuticals, industrial chemicals, hydrocarbons, fuel additives, such as MTBE, and taste- and odor-causing chemicals. UV alone or in conjunction with H_2O_2 can provide a barrier to both these types of contaminants.

Processes upstream of the UV reactor that improve the quality of water by removing suspended solids, dissolved organic compounds, dissolved inorganic compounds or turbidity will increase the water transmittance. The increase of transmittance improves the performance of the UV reactor for treating the target compounds. Pre-treatment serves to remove constituents that would compete with the target contaminant for UV light. In general, the UV technologies in real-world applications work very well with other treatment technologies, such as adsorption onto activated carbon, biodegradation, ion exchange and membrane technologies.

Locating the UV reactor in a treatment train for removal of chemical constituents is application specific. For example, at a solvent plume remediation site, where NDMA has been found, the UV reactor would be placed at the end of the train, following air stripping, carbon adsorption, ion exchange or biological process. However, if there were volatile organic compounds (VOCs) in the water stream, it would also be possible to achieve a benefit using the UV in combination with an oxidant, such as H_2O_2, and locating the UV system upstream of the VOC removal process in order to reduce the size of (or eliminate) downstream water treatment processes. In a drinking water treatment plant, a UV system could be placed pre-sedimentation, pre-filtration or post-filtration, depending on the intended purpose of the UV technology. In water reuse applications, UV technology plays an important role in treating compounds, such as NDMA post-membrane filtration. NDMA has been found to form in wastewater treatment facilities with appropriate precursor chemicals and cannot be removed with microfiltration, nanofiltration or even reverse osmosis. It is, however, effectively treated by UV photolysis. For this reason, in indirect potable reuse applications, where high water quality is required, UV technology performs the final treatment step. Regardless of where the UV technology is placed in the treatment train, it acts as a barrier to both contaminants and microorganisms. Each application is unique and, therefore, must be carefully considered.

3.3.2 Full-scale AOPs: case studies

3.3.2.1 NDMA treatment at the Alamitos Barrier Recycled Water project

In Southern California, seawater intrusion into local fresh-water aquifers is a common problem (Figure 3.5). For example, groundwater pumping in the Long Beach, California area over the last 80 years has caused groundwater levels to drop. This has allowed seawater to intrude into the local groundwater aquifer. To prevent additional infiltration by seawater, a coastline barrier of wells injecting

Figure 3.5 Full-scale installation for NDMA treatment at Los Alamitos Barrier, California.

freshwater has been in place since 1965. Since the beginning of the project, imported potable water of drinking water quality has been injected into the aquifer. Recently, however, a treatment plant was installed to facilitate the use of highly treated, recycled water for this application. One obstacle to the use of recycled water, however, was that it contained the contaminant NDMA. As the injected water will mix with potential drinking water in the aquifer, it must be of drinking water quality. In February 2002, the California Department of Health Services (DHS) set a $10\,\mathrm{ng\,L^{-1}}$ (ppt) drinking water action level for NDMA.

Due to its low affinity for carbon and its low volatility, NDMA resists removal by traditional treatment technologies, such as carbon adsorption and air stripping. Due to its low molecular weight, NDMA is not removed by membranes, even reverse osmosis, and is fairly recalcitrant to biodegradation. However, NDMA degrades by photolysis when exposed to UV light within 200–400-nm wavelength range.

The treatment train design at the Alamitos Barrier consists of primary and secondary treatment of wastewater, followed by microfiltration and reverse osmosis that improves considerably the water transmittance by removing many dissolved constituents. Following reverse osmosis treatment, the water flows into a Trojan UVPhox™ reactor system (Figure 3.5) utilizing monochromatic (254 nm) lamps, where the water is treated for NDMA. As an added benefit and requirement, the water is disinfected for pathogenic micro-organisms and bacteria. Ongoing UV treatment system operation and maintenance costs are approximately 0.05 USD per 1000 gallons ($3.785\,\mathrm{m^3}$).

Design parameters:

Flow rate	3 MGD ($\sim$475 $m^3 h^{-1}$)
Influent NDMA concentration	420 $ng L^{-1}$ (ppt)
Effluent NDMA concentration	$<$10 $ng L^{-1}$ (ppt)

3.3.2.2 NDMA and 1,4-dioxane treatment at the La Puente Valley County Water District water treatment facility

Much of the drinking water in the San Gabriel Valley is obtained from groundwater. Along with other contaminants, industrial development has left a significant portion of the Valley's groundwater contaminated with NDMA and 1,4-dioxane. 1,4-Dioxane is primarily used as a stabilizer in industrial solvents, particularly 1,1,1-TCA, where it was added at concentrations of up to 10% by volume. Like NDMA, 1,4-dioxane resists removal by traditional treatment technologies, such as carbon adsorption and air stripping. However, 1,4-dioxane can be treated in a UV reactor by the UV/H_2O_2 process. At the La Puente Valley County Water District, a Trojan Technologies reactor equipped with monochromatic (254 nm) Hg lamps is being used to destroy NDMA as well as generate hydroxyl radicals through H_2O_2 photolysis for the treatment of 1,4-dioxane to produce drinking water (Figure 3.6). The treatment train consists of an air-stripping system (for the removal of certain volatile organic chemicals) followed by an ion exchange system for perchlorate removal, followed by the UV system.

Figure 3.6 Full-scale installation at La Puente Valley County Water District water treatment facility for environmental contaminant removal.

System parameters:

Flow rate	3.6 MGD (~570 $m^3 h^{-1}$)
Influent NDMA concentration	1500 $ng L^{-1}$ (ppt)
Effluent NDMA concentration	<1.0 $ng L^{-1}$ (ppt)
Influent 1,4-dioxane concentration	3.4 $\mu g L^{-1}$ (ppb)
Effluent 1,4-dioxane concentration	<0.5 $\mu g L^{-1}$ (ppb)

Results of performance testing, required by local regulating agencies, show that the UV reactor consistently achieves the above listed treatment targets. Ongoing UV treatment system operation and maintenance costs are less than 0.06 USD per 1000 gallons (3.785 m^3).

3.3.2.3 UV treatment at the Orange County Water District's Groundwater Replenishment System

The Groundwater Replenishment (GWR) System in Orange County, California, will utilize highly treated wastewater to replenish the local groundwater aquifer and prevent seawater intrusion. When completed, it will be the largest water reuse project in the world. To treat the flow of up to 100 MGD, the facility will use lime softening, coagulation, microfiltration, reverse osmosis and UV/H_2O_2. The micro-filters remove any suspended particles still present in water after secondary treatment, the reverse osmosis system removes a significant amount of dissolved compounds, while the UV/H_2O_2 system will treat NDMA. NDMA is present in the water stream at concentrations up to 150 $ng L^{-1}$ and will achieve the disinfection goal as well. A Trojan UVPhox™ reactor system equipped with monochromatic (254 nm) UV lamps will be used.

System parameters:

Flow rate	100 MGD (~15,800 $m^3 h^{-1}$)
Effluent NDMA concentration	<10.0 $ng L^{-1}$ (ppt)

When completed, the system will undergo testing to ensure that NDMA destruction and disinfection is achieved. The anticipated ongoing operation and maintenance cost per 1000 gallons (3785 m^3) is less than 0.06 USD.

3.3.2.4 Treatment of micropollutants at the PWN Water Supply Company North-Holland

The PWN Water Supply Company is located in the North-Holland province of the Netherlands and will utilize the UV/H_2O_2 process with a Trojan UVPhoxHI™ reactor system (polychromatic light sources) to reduce the level of potential contamination of water with micropollutants and meet drinking water guidelines. Such micropollutants could include pesticides (as the facility draws its water from an agriculture area), solvents or other organic compounds. The system will serve water to approximately 500,000 people and will treat a peak flow of ~13 MGD

($2000\,m^3\,h^{-1}$). The system will serve as the primary means of disinfection. The maximum allowable concentrations of the sum of all pesticides in drinking water in Europe is $0.5\,\mu g\,L^{-1}$ (ppb), while any one pesticide may not exceed $0.1\,\mu g\,L^{-1}$ (ppb). The UV/H_2O_2 process will ensure that these limits are consistently achieved.

System parameters:

Maximum flow rate	~13 MGD (~$2000\,m^3\,h^{-1}$)
Reduction of influent micropollutant concentration	80%

3.4 CONCLUSIONS

The use of UV light in drinking, ground- and wastewater remediation is expected to increase as the regulatory standards for the environment and human health protection are more and more enforced, and the UV light-based technologies become more economical and more efficient than the conventional treatment processes. Such a tend will stimulate advanced research in lamp technology and reactor design, as well as toward innovative, environmentally friendly and photochemical processes.

ACKNOWLEDGEMENTS

We gratefully acknowledge the contribution of Mr. Adam D. Festger, M.Sc., Trojan Technologies, London, ON, Canada to the Full-Scale Application Case Studies section, included in this chapter.

We would like to extend our special thanks to Dr. Linda M. Gowman, Trojan Technologies, London, ON, Canada for her valuable comments on the manuscript and kind support throughout this work.

REFERENCES

Alif, A., Boule, P. and Lemaire, J. (1987) Comportement photochimique du nitro-4 phenol en solution aqueuse. *Chemosphere* **16**(10–12), 2213–2223.

Alif, A., Boule, P. and Lemaire, J. (1990) Photochemistry and environment. XII. Phototransformation of 3-nitrophenol in aqueous solution. *J. Photochem. Photobiol. A: Chem.* **50**, 331–342.

Alif, A., Pilichowski, J.F. and Boule, P. (1991) Photochemistry and environment XIII: Phototransformation of 2-nitrophenol in aqueous solution. *J. Photochem. Photobiol. A: Chem.* **59**, 209–219.

Audureau, J., Filiol, C., Boule, P. and Lemaire, J. (1976) Photolyse et photo-oxydation du phénol en solution aqueuse. *J. Chim. Phys.* **73**(6), 513–620.

Axenrod, T. and Milne, G.W.A. (1968) Photochemistry of nitrosamines. Mechanism studies. *Tetrahedron* **24**, 5775–5783.

Barbeni, M., Pramauro, E. and Pelizzetti, E. (1986) Photochemical degradation of chlorinated dioxins, biphenyls, phenols and benzene on semiconductor dispersion. *Chemosphere* **15**(9), 1913–1916.

Beltran, F.J., Gonzalez, M.C., Rivas, J.F. and Jaramillo, J. (1995) Application of photochemical reactor to UV irradiation of trichloroethylene in water. *Chemosphere* **31**(3), 2873–2885.

Beltran, F.J., Gonzalez, M., Rivas, J.F. and Alvarez, P. (1996a) Aqueous UV radiation and UV/H_2O_2 oxidation of atrazine first degradation products: dethylatrazine and deisopropylatrazine. *Environ. Toxicol. Chem.* **15**(6), 868–872.

Beltran, F.J., González, M. and Rivas, J.F. (1996b) Advanced oxidation of polynuclear aromatic hydrocarbons in natural waters. *J. Environ. Sci. Health* **A31**(9), 2193–2210.

Benitez, F.J., Beltran-Heredia, J., Gonzalez, T. and Real, F. (1995) Photooxidation of carbofuran by a polychromatic UV irradiation without and with hydrogen peroxide. *Ind. Eng. Chem. Res.* **34**, 4099–4105.

Bertilsson, S. and Widenfalk, A. (2002) Photochemical degradation of PAHs in freshwaters and their impact on bacterial growth – influence of water chemistry. *Hydrobiologia* **469**, 23–29.

Bose, P., Glaze, W.H. and Maddox, D.S. (1998) Degradation of RDX by various advanced oxidation processes: I. Reaction rates. *Water Res.* **32**(4), 997–1004.

Boule, P., Guyon, C. and Lemaire, J. (1982) Photochemistry and environment: IV-photochemical behaviour of monochlorophenols in dilute aqueous solution. *Chemosphere* **11**(12), 1179–1188.

Boule, P., Guyon, C. and Lemaire, J. (1984) Photochemistry and environment. VI. Direct phototransformation of chlorophenols and interactions with phenol on UV exposure in aqueous solution. *Toxicol. Environ. Chem.* **7**, 97–110.

Boule, P., Guyon, C., Tissot, A. and Lemaire, J. (1985) Phototransformations en milieu aqueux dilué. *J. Chim. Phys.* **82**(5), 513–516.

Boule, P., Othmen, K., Richard, C., Szczepanik, B. and Grabner, G. (1999) Phototransformation of halogenoaromatic derivatives in aqueous solution. *Int. J. Photoenerg.* **1**, 49–54.

Bourgine, F.P., Chapman, J.I. and Martin, S. (1997) Traitement des pesticides par photolyse UV. *Tech., Sci., Methodes: Genie Urbain-Genie Rural* **7–8**, 23–29.

Brun, O.L., Merlet, N., Croue, J.-P. and Dore, M. (1993) Photodegradation of some common herbicides in aqueous media. *Sci. Tech. Eau* **26**(2), 97–101.

Burrows, H.D., Canle, M.L., Santaballa, J.A. and Steenken, S. (2002) Reaction pathways and mechanisms of photodegradation of pesticides. *J. Photochem. Photobiol. B: Biol.* **67**, 71–108.

Cabrera, M.I., Martin, C., Alfano, O.M. and Cassano, A. (1997) Photochemical decomposition of 2,4-dichlorophenoxyacetic acid (2,4-D) in aqueous solution. I. Kinetic study. *Water Sci. Technol.* **35**(4), 31–39.

Carey, J.H. and Fox, M.E. (1981) Photodegradation of lampricide 3-trifluoromethyl-4-nitrophenol (TFM). 1. Pathway of direct photolysis in solution. *J. Great Lakes Res.* **7**, 234–241.

Challis, B.C. and Li, B.F.-L. (1982) Formation of N-nitrosamines and N-nitramines by photolysis. *IARC Sci. Publ.* **41**, 31–40.

Choudhry, G.G. and Hutzinger, O. (1982) Photochemical formation and degradation of polychlorinated dibenzofurans and dibenzo-*p*-dioxins. *Residue Rev.* **84**, 113–132.

Choudhry, G.G., Webster, G.R.B. and Hutzinger, O. (1986) Environmentally significant photochemistry of chlorinated benzenes and their derivatives in aquatic systems. *Toxicol. Environ. Chem.* **13**, 27–83.

Chow, Y.L. (1967) Photochemistry of nitroso compounds in solution. V. Photolysis of N-nitrosodialkylamines. *Can. J. Chem.* **45**, 53–62.

Chow, Y.L., Lau, M.-P., Perry, R.A. and Tam, J.N.S. (1972) Photoreactions of nitroso compounds in solution. XX. Photoreduction, photoelimination, and photoaddition of nitrosamines. *Can. J. Chem.* **50**, 1044–1050.

De Laat, J., Tace, E. and Dore, M. (1994) Degradation of chloroethanes in dilute aqueous solution by H_2O_2/UV. *Water Res.* **28**(12), 2507–2519.

Doong, R. and Chang, W.-H. (1998) Photoassisted iron compound catalytic degradation of organophosphorous pesticides with hydrogen peroxide. *Chemosphere* **37**(13), 2563–2572.

Draper, W.M. (1987) Measurement of quantum yields in polychromatic light: dinitroaniline herbicides. *ACS Symp. Ser.* **327**, 268–281.

Dulin, D., Drossman, H. and Mill, T. (1986) Products and quantum yields for photolysis of chloroaromatics in water. *Environ. Sci. Technol.* **20**(1), 72–77.

Dung, M.H. and O'Keefe, P.W. (1994) Comparative rates of photolysis of polychlorinated dibenzofurans in organic solvents and in aqueous solutions. *Environ. Sci. Technol.* **28**(4), 549–554.

Esplugas, S., Giménez, J., Contreras, S., Pascual, E. and Rodríguez, M. (2002) Comparison of different advanced oxidation processes for phenol degradation. *Water Res.* **36**(4), 1034–1042.

Fasnacht, M.P. and Blough, N.V. (2002) Aqueous photodegradation of polycyclic aromatic hydrocarbons. *Environ. Sci. Technol.* **36**, 4364–4369.

Freeman, P.K. and Hatlevig, S.A. (1994) Photodehalogenation of 1,2,3,4-tetrachlorodibenzo-*p*-dioxin. In: *Aquatic and Surface Photochemistry*, Helz, G.R., Zepp, R.G. and Crosby, D.G. (eds). Lewis Publishers, Boca Raton, FL, pp. 181–186.

Frimmel, F.H. and Hessler, D.P. (1994) Photochemical degradation of triazine and anilide pesticides in natural waters. In: *Aquatic and Surface Photochemistry*, Helz, G.R., Zepp, R.G. and Crosby, D.G. (eds). CRC Press, Boca Raton, pp. 137–147.

Frischhertz, H., Ollram, F., Scholler, F. and Schmidt, E. (1986) Reaction products from halogenated hydrocarbons resulting from UV treatment. *Water Suppl.* **4**, 167–171.

Getoff, N. (1991) Radiation- and photoinduced degradation of pollutants in water. A comparative study. *Radiat. Phys. Chem.; Int. J. Radiat. Appl. Instrum. Part C* **37**(5/6), 673–680.

Grilli, S., Tost, M.R. and Prodi, G. (1975) Degradation of dimethylnitrosoamine catalysed by physical and chemical agents. *Gann* **66**, 481–488.

Guittonneau, S., De Laat, J., Dore, M., Duguet, J.P. and Bonnel, C. (1988a) Photooxidation of volatile chlorinated hydrocarbons in aqueous solution in the presence of hydrogen peroxide. *Rev. Sci. Eau* **1–2**, 35–54.

Guittonneau, S., De Laat, J., Duguet, J.P. and Bonnel, C. (1988b) Comparative study of the photodegradation of aromatic compounds in water by UV and H_2O_2/UV. *Environ. Technol. Lett.* **9**, 1115–1128.

Haag, W.R. (1994) Photooxidation of organic compounds in water and air using low-wavelength pulsed xenon lamps. In: *Aquatic and Surface Photochemistry*, Helz, G.R., Zepp, R.G. and Crosby, D.G. (eds). Lewis Publishers, Boca Raton, pp. 517–530.

Halmann, M.M. (1996) *Photodegradation of Water Pollutants*, CRC Press, Boca Raton, FL.

Hessler, D.P., Gorenflo, V. and Frimmel, F.H. (1993a) Degradation of aqueous atrazine and metazachlor solutions by UV/H_2O_2 – influence of pH and herbicide concentration. *Acta Hydrochim. Hydrobiol.* **21**(4), 209–214.

Hessler, D.P., Gorenflo, V. and Frimmel, F.H. (1993b) UV degradation of atrazine and metazachlor in the absence and in the presence of hydrogen peroxide, bicarbonate and humic substances. *J. Wat. SRT-Aqua* **42**(1), 8–12.

Himebaugh, W.S. (1993) Advanced oxidation process destruction of explosives in water. Paper 93-WP-107.07. In: *The 86th Annual Meeting and Exhibition of the Air and Waste Management Association*, Denver, CO, June 13–18.

Hirahara, Y., Sayato, Y. and Nakamuro, K. (1998) Studies on photochemical behavior of pesticides in environment. *Jpn. J. Toxicol. Environ. Health* **44**(6), 451–461.

Hirvonen, A., Tuhkanen, T. and Kalliokoski, P. (1996) Removal of chlorinated ethylenes in contaminated ground water by hydrogen peroxide mediated oxidation processes. *Environ. Technol.* **17**, 263–272.

Hirvonen, A., Tuhkanen, T., Ettala, M., Korhonen, S. and Kalliokoski, P. (1998) Evaluation of a field-scale UV/H_2O_2-oxidation system for the purification of groundwater contaminated with PCE. *Environ. Technol.* **19**, 821–828.

Ho, T.-F.L. and Bolton, J.R. (1998) Toxicity changes during the UV treatment of pentachlorophenol in dilute aqueous solution. *Water Res.* **32**(2), 489–497.

Ho, T.-F.L., Bolton, J.R. and Lipczynska-Kochany, E. (1996) Quantum yields for the photodegradation of pollutants in dilute aqueous solution: phenol, 4-chlorophenol and N-nitrosodimethylamine. *J. Adv. Oxid. Technol.* **1**(2), 170–178.

Höfl, C., Sigl, G., Specht, O., Wurdack, I. and Wabner, D. (1997) Oxidative degradation of AOX and COD by different advanced oxidation processes: a comparative study with two samples of a pharmaceutical wastewater. *Water Sci. Technol.* **35**(4), 257–264.

Jobb, D.B., Hunsinger, R.B., Meresz, O. and Taguchi, V. (1994) Ultraviolet degradation of N-nitrosodimethylamine (NDMA) in the Ohsweken (Six Nations) water supply. In: *Proceedings of the Annual AWWA Conference*, Denver, CO, June 19–23.

Karpel Vel Leitner, N., Gombert, B., Benabdessalem, R. and Dore, M. (1996) Kinetics and mechanisms of the photolytic and OH radical induced oxidation of fluorinated aromatic compounds in aqueous solutions. *Chemosphere* **32**(5), 893–906.

Kawaguchi, H. (1993) Rates of sensitized photo-oxidation of 2,4,6-trimethylphenol by humic acid. *Chemosphere* **27**(11), 2177–2182.

Konstantinov, A., Johnston, A.M., Cox, B.J., Petrulis, J.R., Orzechowski, M.T., Bunce, N.J., Tashiro, C.H.M. and Chittim, B.G. (2000) Photolytic method for destruction of dioxins in liquid laboratory waste and identification of the photoproducts from 2,3,7,8-TCDD. *Environ. Sci. Technol.* **34**, 143–148.

Kruithof, J.C., Kamp, P.C. and Finch, G.R. (2001) UV/H_2O_2 treatment: the ultimate solution for pesticide control and disinfection. In: *Proceedings of the Annual. AWWA Conference*, Washington DC, June 17–21.

Ku, Y., Chang, J.-L. and Cheng, S.-C. (1998) Effect of solution pH on the hydrolysis and photolysis of diazinon in aqueous solution. *Water Air Soil Poll.* **108**(3–4), 445–456.

Larson, R.A. and Weber, E.J. (1994) *Reaction Mechanisms in Environmental Organic Chemistry*. CRC Press, Inc., Boca Raton, FL.

Legrini, O., Oliveros, E. and Braun, A.M. (1993) Photochemical processes for water treatment. *Chem. Rev.* **93**, 671–698.

Lipczynska-Kochany, E. (1992) Degradation of nitrobenzene and nitrophenols in homogeneous aqueous solution. Direct photolysis versus photolysis in the presence of hydrogen peroxide and the Fenton reacgent. *Water Poll. Res. J. Canada* **27**(1), 97–122.

Mabey, W., Tse, D., Baraze, A. and Mill, T. (1983) Photolysis of nitroaromatics in aquatic systems. 1. 2,4,6-Trinitrotoluene. *Chemosphere* **12**, 1–16.

Maier, M., Maier, D. and LLoyd, B.J. (2000) Factors influencing the mobilisation of polycyclic aromatic hydrocarbons (PAHs) from the coal-tar lining of water mains. *Water Res.* **34**(3), 773–786.

Mansour, M., Feicht, E.A., Behechti, A. and Scheurnert, I. (1997) Experimental approaches to studying the photostability of selected pesticides in water and soil. *Chemosphere* **35**(1/2), 39–50.

Martin, C.A., Cabrera, M.I., Alfano, O.M. and Cassano, A. (1997) Photochemical decomposition of 2,4-dichlorophenoxyacetic acid (2,4-D) in aqueous solution. II. Reactor modeling and verification. *Water Sci. Technol.* **35**(4), 197–205.

Mathew, R. and Khan, S.U. (1996) Photodegradation of metolachlor in waterin the presence of soil mineral and organic constituents. *J. Agric. Food Chem.* **44**(12), 3996–4000.

Mertens, R. and von Sonntag, C. (1995) Photolysis ($\lambda = 254$ nm) of tetrachloroethene in aqueous solutions. *J. Photochem. Photobiol. A: Chem.* **85**, 1–9.

Milano, J.C. and Vernet, J.L. (1988) *Chemosphere* **17**(5), 963–971.

Miller, J.S. and Olejnik, D. (2001) Photolysis of polycyclic aromatic hydrocarbons in water. *Water Res.* **35**(1), 233–243.

Miller, G.C. and Zepp, R.G. (1979) Effects of suspended sediments on photolysis rates of dissolved pollutants. *Water Res.* **13**, 453–459.

Miller, R.M., Singer, G.M., Rosen, J.D. and Bartha, R. (1988) Sequential degradation of chlorophenols by photolytic and microbial treatment. *Environ. Sci. Technol.* **22**, 1215–1219.

Millet, M., Palm, W.U. and Zetzsch, C. (1998) Investigation of the photochemistry of urea herbicides (chlorotoluron and isoproturon) and quantum yields using polychromatic irradiation. *Environ. Toxicol. Chem.* **17**(2), 258–264.

Nick, K. and Schoeler, H.F. (1995) Photochemical degradation of herbicides in water by UV irradiation generated by low pressure mercury lamps. Part I. Triazines. *Vom Wasser* **84**, 271–286.

Nick, K., Schoeler, H.F., Mark, G., Söylemez, T., Akhlaq, M.S., Schuchmann, H.-P. and von Sonntag, C. (1992) Degradation of some triazine herbicides by UV radiation such as in the UV disinfection of drinking water. *J. Wat. SRT-Aqua* **41**(2), 82–87.

Nicole, I., De Laat, J., Dore, M., Duguet, J.P. and Bonnel, C. (1989) The effect of photochemical reactor design on the rate of photolysis of H_2O_2 and of volatile organohalogenated compounds. In: *Proceedings of the Wasser Berlin '89 Conference: Ozone and UV in the Treatment of Water and Other Liquids*, Berlin, Germany, April 10–16.

Nohr, R.S., MacDonald, J.G., Kogelschatz, U., Mark, G., Schuchmann, H.-P. and von Sonntag, C. (1994) Application of excimer incoherent-UV sources as a new tool in photochemistry: photodegradation of chlorinated dibenzodioxins in solution and adsorbed on aqueous pulp sludge. *J. Photochem. Photobiol. A: Chem.* **79**, 141–149.

O'Connor, O.A. and Young, L.L. (1989) Toxicity and anaerobic biodegradability of substituted phenols under methanogenic conditions. *Environ. Toxicol. Chem.* **8**, 853–862.

Palm, W.U. and Zetzsch, C. (1996) Investigation of the photochemistry and quantum yields of triazines using polychromatic irradiation and UV spectroscopy as an analytical tool. *Int. J. Environ. Anal. Chem.* **65**(1–4), 313–329.

Panadés, R., Ibarz, A. and Esplugas, S. (2000) Photodecomposition of carbendazim in aqueous solutions. *Water Res.* **34**(11), 2951–2954.

Patterson, H.H., Bachman, J. and Kanan, S. (1998) Kinetics and the influence of dissolved organic matter on the photodecomposition of the carbamate pesticides carbofuran, carbaryl and aldicarb. In: *Extended Abstracts of the 216th ACS Nationall Meeting, Am. Chem. Soc., Div. Environ. Chem.*, Boston, MA, August 23–27, **38**(2), 105–106.

Perbet, G., Boule, P. and Lemaire, J. (1982) Photochimie et environment. II. Photochimie des triphénols en solution aqueuse diluée. *J. Chim. Phys.* **79**(1), 95–100.

Peterson, D., Watson, D. and Winterlin, W. (1988) The destruction of ground water threatening pesticides using high light UV intensity. *J. Environ. Sci. Health* **B23**(6), 587–603.

Pirisi, F.M., Cabras, P., Garau, V.L., Melis, M. and Secchi, E. (1996) Photodegradation of pesticides. Photolysis rates and half-life of pirimicarb and its metabolites in reactions in water and in solid phase. *J. Agric. Food Chem.* **44**, 2417–2422.

Polo, J. and Chow, Y.L. (1976a) Efficient degradation of nitrosamines by photolysis. *IARC Sci. Publ.* **14**, 473–486.

Polo, J. and Chow, Y.L. (1976b) Efficient photolytic degradation of nitrosamines. *J. Natl. Cancer Inst.* **56**(5), 997–1001.

Rayne, S., Wan, P., Ikonomou, M.G. and Konstantinov, A. (2002) Photochemical mass balance of 2,3,7,8-TeCDD in aqueous solution under UV light shows formation of chlorinated dihydroxybiphenyls, phenoxyphenols, and dechlorination products. *Environ. Sci. Technol.* **36**(9), 1995–2002.

Rehberg, K., Mannschott, P., Erdinger, L., Sonntag, H.G. and Schramm, B. (1994) Degradation of atrazine by UV oxidation and identification of the photodegradation products. *Vom Wasser* **83**, 139–156.

Scheurnert, I., Mansour, M., Dörfler, U. and Schroll, R. (1993) Fate of pendimethalin, carbofuran and diazinon under abiotic and biotic conditions. *Sci. Tot. Environ.* **132**, 361–369.

Shen, Y.-S., Ku, Y. and Lee, K.-C. (1995) The effect of light absorbance on the decomposition of chlorophenols by ultraviolet radiation and UV/H_2O_2 processes. *Water Res.* **29**, 907–914.

Sigman, M.E. and Zingg, S.P. (1994) Anthracene photochemistry in aqueous and heterogeneous media. In: *Aquatic and Surface Photochemistry*, Helz, G.R., Zepp, R.G. and Crosby, D.G. (eds). Lewis Publishers, Boca Raton, FL, pp. 197–206.

Skurlatov, Yu.I., Ernestova, L.S., Vichutinskaya, E.V., Samsonov, D.P., Semenova, I.V., Rod'ko, I.Ya., Shvidky, V.O., Pervunina, R.I. and Kemp, T.J. (1997) Photochemical transformation of polychlorinated phenols. *J. Photochem. Photobiol. A: Chem.* **107**, 207–213.

Stefan, M.I. and Bolton, J.R. (2002) UV direct photolysis of N-nitrosodimethylamine (NDMA): kinetic and product study. *Helv. Chim. Acta* **85**, 1416–1426.

Stefan, M.I., Siddiqui, M., Atasi, K. and Linden, K.G. (2002) Impact of water quality characteristics on the kinetic parameters of NDMA photo-degradation. In: *Proceedings of the AWWA Water Quality and Technology Conference*, Seattle, WA, November 10–14.

Stefan, M.I., Li, K. and Crittenden, J.C. (2003) Photodegradation of trichloroethylene (TCE) in the presence and the absence of hydrogen peroxide. Product study and kinetic modeling (submitted to *Environ. Sci. Technol.*).

Sundstrom, D.W., Klei, H.E., Nalette, T.A., Reidy, D.J. and Weir, B.A. (1986) Destruction of halogenated aliphatics by ultraviolet catalyzed oxidation with hydrogen peroxide. *Hazard. Waste Hazard. Mater.* **3**(1), 101–110.

Sundstrom, D.W., Weir, B.A. and Klei, H.G. (1989) Destruction of aromatic pollutants by UV light catalyzed oxidation with hydrogen peroxide. *Environ. Prog.* **8**(1), 6–10.

Sundstrom, D.W., Weir, B.A., Barber, T.A. and Klei, H.E. (1990) Purification and disinfection of water by ultraviolet light and hydrogen peroxide. In: *Proceedings of the Symposium on Advances in Oxidative Processes and Treatment Contaminated Water and Air*, Toronto, ON, Canada, June 4–5.

Tace, E., De Laat, J. and Dore, M. (1992) Photodegradation of atrazine by UV irradiation in the absence and in the presence of hydrogen peroxide. *J. Fr. Hydrol.* **23**(2), 233–249.

Tanaka, F.S., Wien, R.G. and Mansager, E.R. (1982) Photolytic demethylation of monuron and demethylmonuron in aqueous solution. *Pestic. Sci.* **13**, 287–294.

Tissot, A., Boule, P. and Lemaire, J. (1984) Photochemistry and environment. VII – The photohydrolysis of chlorobenzene: studies of the excited state involved. *Chemosphere* **13**(3), 381–389.

Torrents, A., Anderson, B.G., Bilboulian, S., Johnson, W.E. and Hapeman, C. (1997) Atrazine photolysis: mechanistic investigations of direct and nitrate-mediated hydroxyl radical processes and influence of dissolved organic carbon from the Chesapeake Bay. *Environ. Sci. Technol.* **31**(5), 1476–1482.

Trapido, M., Veressinina, J. and Munter, R. (1994) Ozonation and AOP treatment of phenanthrene in aqueous solutions. *Ozone Sci. Eng.* **16**, 475–485.

Trapido, M., Veressinina, J. and Munter, R. (1995) Ozonation and advanced oxidation processes of polycyclic aromayic hydrocarbons in aqueous solutions. A kinetic study. *Environ. Technol.* **16**, 729–740.

Tuhkanen, T. and Beltran, F.J. (1995) Intermediates of the oxidation of naphthalene in water with the combination of hydrogen peroxide and UV radiation. *Chemosphere* **30**(8), 1463–1475.

Vollmuth, S., Zajc, A. and Niessner, R. (1994) Formation of polychlorinated dibenzo-*p*-dioxins and polychlorinated dibenzofurans during the photolysis of pentachlorophenol-containing water. *Environ. Sci. Technol.* **28**(6), 1145–1149.

Wan, H.B., Wong, M.K. and Mok, C.Y. (1994) Comparative study on the quantum yields of direct photolysis of organophosphorus pesticides in aqueous solution. *J. Agric. Food Chem.* **42**(11), 2625–2630.

Wimmer, B., Bischof, H., Scheuer, C. and Wabner, D. (1995) Oxidative degradation of organic water pollutants. Products and reaction pathways. *Vom Wasser* **85**, 421–432.

Wong, A.S. and Crosby, D.G. (1981) Photodecomposition of pentachlorophenol in water. *J. Agric. Food Chem.* **29**, 125–130.

Zheng, H.-H. and Ye, C.-M. (2001) Identification of UV photoproducts and hydrolysis products of butachlor by mass spectroscopy. *Environ. Sci. Technol.* **35**, 2889–2895.

Zwiener, C., Weil, L. and Niessner, R. (1995) UV and UV/ozone degradation of triazine herbicides in a pilot plant. Estimation of UV photolysis rate constants and quantum yields. *Vom Wasser* **84**, 47–60.

4

UV/H_2O_2 processes

Tuula A. Tuhkanen

4.1 INTRODUCTION

The ultraviolet/hydrogen peroxide (UV/H_2O_2) process is widely studied and applied in water and wastewater purification. Continued advances in the technology have made it a practicable method for both ground- and process-water treatment. Predicting the performance of UV/H_2O_2 technique, water requires several kinds of information or data:

- the identity and concentration of contaminant,
- the organic and inorganic matrix of water and scavengers.

Based on this information, the rate of oxidation can be estimated if the following conditions are known:

a. the identities and concentrations of the oxidants in the system;
b. the rate constant for oxidation by each oxidant at a specific site in a molecule;
c. the kinetic rate law for each process.

UV/H_2O_2/(O_3) technique processes can be applied for several purposes in water and wastewater treatment. These are as follows:

- removal of micro- and macro-pollutants from drinking water;
- removal of the disinfection by-products and their precursors from drinking water (Section 4.7);
- treatment of toxic organics at low concentrations in groundwater;
- treatment of low volume – high-strength wastewaters for detoxification and enhanced degradability (Section 4.8);
- treatment of contaminated soil washing water;
- control of off-gas emissions from volatile organic compound (VOC) stripping.

The technology can be applied *in situ* or as a surface treatment of the soil or the groundwater (AOP handbooks, 1997, 1998; Watts *et al.*, 1990, 1991). During the last year's full-scale applications of UV/H_2O_2, it has become commercially available. Depending on the constituents in the groundwater, chemical oxidation may be used in conjugation with other technologies. These are stripping of volatile organics, granular activated carbon (GAC) treatment (Wang and Latchaw, 1990) or biological oxidation through biologically activated carbon columns (DeWaters *et al.*, 1990) or other processes (Kearnay *et al.*, 1983) depending on the identity and concentration of organics in the groundwater.

Normally, some form of biological treatment will be the most cost-effective technology for organic wastewaters. Some organics, however, are either non-biodegradable or are toxic either to the biological process or aquatic life and must, therefore, be pre- or post-treated. In many cases, pre-treatment by chemical oxidation will affect detoxication as well as enhanced biodegradation (Adams and Scanlan, 1993; Bowers *et al.*, 1991; Habrerel *et al.*, 1993; Wang, 1991). On the other hand, sometimes the effluent is still toxic after biological oxidation due to the presence of non-degradable constituents or biological by-products. Chemical post-oxidation may reduce the toxicity and the amount of organic material in wastewater (Narkis *et al.*, 1980a).

In case of unknown wastewater, toxicity to biological process should first be evaluated using the batch reactor technique or some other evaluation of toxicity. If the wastewater is non-biodegradable and exhibits toxicity, a chemical oxidation study should be conducted (Narkis *et al.*, 1980b).

4.2 THEORY OF UV/H_2O_2

UV/oxidation processes generally involve generation of $OH^{\bullet}$ radical through UV photolysis of H_2O_2 or/and ozone. Most of the information related to these techniques is produced by laboratory-scale experiments. However, full-scale commercial application is already available. Some vendors offer processes, which combines both O_3 and H_2O_2.

The most direct method for generation of hydroxyl radicals is through the cleavage of H_2O_2. Photolysis of H_2O_2 yields hydroxyl radicals by a direct process with a yield of two radicals formed per photon absorbed by 254 nm (Baxendale and Willson, 1957):

$$H_2O_2 + h\nu \rightarrow OH^\bullet + OH^\bullet \tag{4.1}$$

The quantum yield for the photolysis of H_2O_2 by 253.7 nm light in 0.1 N perchloric acid is found to be 1.00 at 25°C and it is independent of concentration and light intensity over the range 2×10^{-5} to 0.1 M peroxide and 4.5×10^{-7} to 5×10^{-4} einstein $L^{-1}\,min^{-1}$. The yield falls to 0.80 at 4°C. The addition of organic acids decreases the quantum yield until it reaches 0.5, when it is independent of acid concentration. This means that the decomposition of peroxide is regenerated in these conditions (Baxendale and Willson, 1957).

The molar extinction coefficient of H_2O_2 at 254 nm is 19.6 $M^{-1}s^{-1}$ (Lay, 1989), which is very low. Comparatively, the molar extinction coefficients of, for example, ozone, naphthalene and pentachlorophenol (PCP) (at pH 7) are 3300, 3300 and 10,000 $M^{-1}cm^{-1}$, respectively. This means that in order to generate a sufficiently high level of OH radicals in a solution, which contains strong photon absorbers, the concentration of H_2O_2 has to be rather high. Low-pressure mercury vapor lamps with a 254-nm peak emission are the most common UV source used in UV/H_2O_2 systems. The maximum absorbance of H_2O_2 occurs at about 220 nm. If low-pressure lamps are used, a high concentration of H_2O_2 is needed to generate sufficient hydroxyl radicals because of low-absorption coefficient. However, high concentration of H_2O_2 scavenges the radicals, making the process less effective:

$$OH^\bullet + H_2O_2 \rightarrow HO_2^\bullet + H_2O \tag{4.2}$$

$$HO_2^\bullet + H_2O_2 \rightarrow OH^\bullet + H_2O + O_2 \tag{4.3}$$

$$HO_2^\bullet + HO_2^\bullet \rightarrow H_2O_2 + O_2 \tag{4.4}$$

To overcome the limitation of low absorbtivity of H_2O_2 by the wavelength of 254 nm, some researchers/vendors use high-intensity, medium-pressure, broadband UV lamps. The third option is high-intensity, xenon flashlamps, whose spectral output can be adjusted to match the absorption characteristics of H_2O_2.

4.2.1 Photolysis

In an advanced oxidation process (AOP), where irradiation is involved, degradation of the organic compound can take place either by direct or indirect photolysis. An organic molecule can undergo photochemical transformation if and only if both of the following conditions are met (Calvert and Pitts, 1966):

- light energy is absorbed by the molecule to produce an electronically excited state molecule,
- chemical transformations of the excited state are competitive with deactivation processes.

The necessity for light absorption is cited as the first law of photochemistry or the Grotthus–Draper law: only the light, which is absorbed by a molecule, can be effective in producing photochemical changes in the molecule. The molecule absorbs light in several regions of electromagnetic spectrum, corresponding to different kinds of molecular transitions.

The rate of direct photolysis of a chemical species at concentration $[C]$ is given by Equation (4.5), which is a combination of Grothus–Draper law and Stark–Einstein law (Leifer, 1988):

$$-\frac{\mathrm{d}[C]}{\mathrm{d}t} = I_o \phi_C f_C (1 - \exp(-A_t)) \tag{4.5}$$

where ϕ_C is the quantum yield of C, that is, the fraction of absorbed radiation that results in photolysis, and I_o is the incident flux of radiation (254 nm in this case). The factor f_C is the ratio of light absorbed by C to that absorbed by other components of the solution, and A_t is the total absorbance of the solution times a factor of 2.3, defined, respectively, as follows:

$$f_C = \frac{\varepsilon_C [C]}{\Sigma \varepsilon_i [C_i]} \tag{4.6}$$

$$A_t = 2.3L \Sigma \varepsilon_i [C_i] \tag{4.7}$$

In Equations (4.6) and (4.7), the subscript i represents any species present in the solution that absorbs light at the specified wavelength, L is the (effective) path length of the photoreactor, and ε_i is the molar absorptivity or molar extinction coefficient of the ith species at the lamp wavelength.

When the concentration of absorbers is so large that $\exp(-A_t)$ is approximately zero, Equation (4.5) simplifies to the following (Leifer, 1988):

$$-\frac{\mathrm{d}[C]}{\mathrm{d}t} = I_o \phi_C f_C \tag{4.8}$$

If C is a minor absorber, then its rate of decay will be first order. If C is the principal absorber, its rate will be zero order. When the concentration of absorbers is small, $[1 - \exp(-A_t)]$ may be expanded in a Taylor series to yield the familiar first-order expression (Leifer, 1988):

$$-\frac{\mathrm{d}[C]}{\mathrm{d}t} = 2.3L \phi_C I_o \varepsilon_C [C] \tag{4.9}$$

Under these conditions, the photolysis of two absorbing substrates will be independent of one another. When the concentration of a single absorber is intermediate in concentration, Equation (4.5) becomes

$$-\frac{\mathrm{d}[C]}{\mathrm{d}t} = \phi_C I_o (1 - \exp(-\alpha [C])) \tag{4.10}$$

where

$$\alpha = 2.3L \varepsilon_C \tag{4.11}$$

4.3 MECHANISMS OF UV/H_2O_2 OXIDATION

The mechanism of oxidation by H_2O_2/UV has been studied extensively and attributed to the photolysis of H_2O_2 (Baxendale and Willson, 1957), which gives rise to the appearance of hydroxyl radicals (Guittonneau, 1989; Glaze and Lay, 1989). Hence, the elimination rate of organic compounds has at least two contributions: direct photolysis and hydroxyl-radical attack. Besides photolysis, the compound C is eliminated by hydroxyl-radical attack:

$$-\frac{d[C]}{dt} = I_o \phi_C f_C (1 - \exp(-A_t)) + k_{OH,C}[OH][C] \tag{4.12}$$

where $k_{OH,C}$ is the second-order reaction rate constant of hydroxyl radical with compound C. Pseudo-first-order rate constant can be used because the concentration of OH radical can be assumed to be constant over the range of reaction and thus it can be incorporated to the rate constant (Stumm and Morgan, 1981):

$$-\ln \frac{C}{C_o} = k't \tag{4.13}$$

where k' = pseudo-first-order rate constant, s^{-1}.

Experimental values of degradation of different compounds are compiled to the Appendix 1. The effect of light source, intensity and dose of H_2O_2 are also listed if provided.

However, there is a limit to the beneficial effects of added peroxide, since H_2O_2 is also an OH-radical scavenger ($k_{OH,H_2O_2} = 2.7 \times 10^7\, M^{-1} s^{-1}$, Christensen *et al.*, 1982):

$$-\frac{d[C]}{dt} = I_o \phi_C f_C (1 - \exp(-A_t)) + k_{OH,C}[OH][C] \tag{4.14}$$

Lot of studies to determine the ratio of H_2O_2 and pollutant has been carried out in order to create a tool for the optimization of treatment process.

4.4 EFFECT OF ALKALINITY

Inorganic carbon species, bicarbonate and carbonate ions are scavengers of hydroxyl radicals (Weeks and Rabani, 1966) and, therefore, are expected to affect the rate of OH-mediated reactions of organic compounds. The reactions with OH radicals are as follows [Equations (4.15) and (4.16)]:

$$OH^\bullet + HCO_3^- \rightarrow {}^\bullet CO_3^- + H_2O \tag{4.15}$$

with $k_{OH,HCO_3} = 1.5 \times 10^7\, M^{-1} s^{-1}$

$$OH^\bullet + CO_3^{2-} \rightarrow {}^\bullet CO_3^- + OH^- \tag{4.16}$$

with $k_{OH,CO_3} = 4.2 \times 10^8\, M^{-1} s^{-1}$

In the presence of bicarbonate–carbonate ions, part of the hydroxyl radicals react to form carbonate ion radicals, $CO_3^{\bullet-}$, via reactions in Equations (4.15) and (4.16).

Although the kinetics and mechanisms of carbonate radicals with organic compounds are poorly understood, there are enough data to suggest that these radicals may react with the organic compounds (Chen and Hoffman, 1974; Chen *et al.*, 1975). These reactions are, however, more selective than those of the hydroxyl radical and the rate constants are lower – how much lower depends on the nature of the organic compounds. Thus, at pH 7, the rate constant of the reaction between carbonate radical and generally quite unreactive benzene is only $3 \times 10^3\,M^{-1}\,s^{-1}$, while that with phenol, which is known to be more reactive than benzene, it is $4.9 \times 10^6\,M^{-1}\,s^{-1}$ (Chen and Hoffman, 1974).

Liao *et al.* (2001b) studied the simultaneous effect of inorganic anions, such as chloride and bicarbonate ions, on the scavenging of hydroxyl radicals in UV/H_2O_2 processes. The model compound of n-chlorobutene (BuCl) was used as the probe of OH radical. By changing the pH conditions (2–9) and the concentration of NaCl (0.25–2500 mM) and $NaHCO_3$ (25 mM), the variation of $OH^\bullet$ concentrations and the rate of H_2O_2 decomposition were compared. Regarding the reaction kinetics in the UV/H_2O_2 process, the $OH^\bullet$ probe compound (BuCl) and H_2O_2 were found to disappear by following the first-order reaction within the initial 10 and 40 min, respectively. In the presence of chloride alone at the pH range of 2–6, the OH-radical concentration in the reaction mixture increased with the increase of OH, and the $OH^\bullet$ concentration at pH 6 was 100 times of that at pH 2. Including bicarbonate species in the solution, the peak $OH^\bullet$ concentration was found at a certain pH, which shifted from 4.5 to 5–7, as the molar ratios of chloride/bicarbonate species increased from 1 to 100. In addition, without bicarbonate species, $OH^\bullet$ concentration decreased significantly with increasing chloride concentration, but remained rather unchanged beyond 1250 mM. In contract, the $OH^\bullet$ scavenging in the presence of bicarbonate species become relatively significant only when the chloride concentration reached beyond 250 mM.

4.5 MODELS

The practical application of UV/H_2O_2 process requires the determination of important design and operational variables, such as the optimum oxidant and irradiation dosages within the reactor. Mathematical models at several different levels have been developed based on known chemical and photochemical principles (Crittenden *et al.*, 1999; Glaze *et al.*, 1992a,b, 1995; Liao and Gurol, 1995; Stefan *et al.*, 1996). Some models also take into account the measured effects of radical scavengers, alkalinity, pH and presence of background organic and inorganic matter. The verification of models is usually carried out against actual laboratory data. Crittnden and Liao used 1-2-dibromo-3-chloropropane, Glaze naphthalene, PCP and Stefan acetone. The models are able to predict the contaminant concentration as a function of the time under different H_2O_2 concentrations.

4.6 LABORATORY-SCALE EXPERIMENTS OF UV/H_2O_2 WITH MODEL COMPOUNDS

4.6.1 General

Most of the organic compounds studied could be decomposed by direct UV radiation. Addition of H_2O_2 accelerated the decomposition except for few pesticides. The comparison of the purification efficiencies and the economical feasibility is difficult due to various kinds of reactors and intensity of lamps used. Especially difficult is to compare results gained with low-, medium- and high-pressure mercury lamps because of their spectral differences. In general, the decomposition of organic compound was considered as single component system – only the disappearance of model compound was followed. Usually, the formed by-products were neglected.

4.6.2 Synthetic solutions

Sundstrom *et al.* (1986) studied the destruction of typical halogenated aliphatics by the combination of UV light and H_2O_2. The chlorinated compounds with carbon–carbon double bonds, such as trichloroethylene, degraded much faster than the other compounds studied. The reacted chlorine was converted quantitatively to chloride ion, indicating that the chlorinated structures were effectively destroyed. Sundstrom *et al.* (1989) also studied the degradation rates of typical aromatic compounds. The pseudo-first-order rate constants for UV and UV/H_2O_2 experiments are given in the parentheses ($\times 10^{-3}\,s^{-1}$): benzene (0.25 and 1.42), chlorobenzene (0.42 and 1.05), toluene (0.38 and 1.57), phenol (0.06 and 1.37), 2-chlorophenol (0.25 and 0.83), 2,4-dichlorophenol (DCP) (0.67 and 1.36), 2,4,6-trichlorophenol (TCP) (1.3 and 1.67), dimethylphthalate (0.02 and 0.78) and diethylphthalate (0.02 and 0.78).

Weir *et al.* (1987) studied the effects of process operating conditions on the destruction of benzene by UV and H_2O_2. The lamp was low-pressure mercury arc lamp with 5.3 W output at 254 nm. The rates of reaction followed pseudo-first-order kinetics and were strongly affected by the contaminant and H_2O_2 molar ratio and by the intensity of UV radiation. Many intermediates were formed, but extending the treatment time could eliminate them. The identified intermediates were phenol, catechol and hydroquinone. No polymerization was observed. They also studied the destruction kinetics of trichloroethene (TCE) (Weir and Sunstrom, 1993). The reaction rate of TCE was first order in UV light intensity and pseudo-first order in TCE concentration. The rate was first order in H_2O_2 concentration at low peroxide levels, but independent of peroxide concentration at high peroxide levels.

Guittonneau *et al.* (1990) compared UV/H_2O_2 and UV/O_3 treatments for the removal of 4-chloronitrobenzene from aqueous solutions by continuous flow reactor. The efficiencies of these two methods increased as the oxidant doses and the irradiation times increased, and significantly decreased in the presence of bicarbonate ions. For a constant oxidant dose, the results showed that the O_3/UV system appeared to be more efficient than the H_2O_2/UV system in removing

4-chloronitrobenzene because of the greater rate of hydroxyl-radical generation by ozone photodecomposition compared to H_2O_2 photolysis.

Ho (1986) studied the synergistic effect of H_2O_2 and UV radiation with medium-pressure mercury arc lamps (450 W) on the decomposition of 2,4-dinitrotoluene (DNT). The degradation pathways of DNT in aqueous solution were side-chain oxidation, which converted DNT to 1,3-dinitrobenzene, hydroxylation of the benzene ring, which converted 1,3-dinitrobenzene to hydroxynitrobenzene derivatives, benzene ring cleavage of the hydroxynitrobenzenes, which produced lower molecular weight carboxylic acids and aldehydes and further photo-oxidation, which eventually converted the lower molecular weight acids and aldehydes to CO_2, H_2O and HNO_3.

Milano *et al.* (1992) studied the intermediate formation from 4-bromodiphenylether degradation by UV/H_2O_2 treatment. The lamp used was 100-W medium-pressure mercury arc lamp. They identified several degradation compounds. The complex degradation mechanism was divided roughly into three steps. Firstly, dehalogenatation produced diphenylether and *p*-hydroxydiphenylether; secondly, decomposition of diphenylether and the formation of phenol and benzene; and thirdly, the opening of the aromatic ring produced carboxylic acids leading to the mineralization. Overall, the mineralization was much more difficult to obtain than that of aliphatic halogen compounds. During degradation of the aromatic ether, at the same time as the formation of the carboxylic acids, macromolecular compounds were formed, which had a humic acid structure and were more resistant to degradation.

Ishikawa *et al.* (1992) studied the photochemical degradation of organic phosphate esters by low-pressure mercury arc irradiation (15 W). The degradation followed first-order kinetics and the reaction rates varied between 0.2 and $11.1 \times 10^{-3}\,s^{-1}$.

According to Yashura *et al.* (1977), the monochlorophenols also followed first-order kinetics in their degradation by UV radiation. The reaction rates were 4-monochlorophenol $1.4 \times 10^{-3}\,s^{-1}$, 2-chlorophenol $2 \times 10^{-4}\,s^{-1}$ and 3-chlorophenol $3 \times 10^{-5}\,s^{-1}$. Addition of H_2O_2 highly accelerated the decomposition rates. Degradation of 2-chlorophenol also followed first-order kinetics, when it was irradiated at a wavelength of 313 nm. The first-order direct and indirect photolysis rate constants were determined in humic solution and in concentrated natural waters. Moza *et al.* (1988) studied the decomposition of 2-chlorophenol, 2,4-DCP and 2,4,6-TCP solutions by UV/H_2O_2 treatment with 125-W medium-pressure lamp. When the initial H_2O_2 concentration was 55 ppm, the following rate constants were obtained for 2-chlorophenol, 2,4-DCP and 2,4,6-TCPs, respectively, in degassed solution: 0.33×10^{-4}, 1.1×10^{-4} and $3.3 \times 10^{-4}\,s^{-1}$. The lower rate constant by UV/H_2O_2 treatment for 2-chlorophenol compared to the direct photolysis can be explained by the lamp used. The medium-pressure lamp produces most of its light at the visible region and less radiation at 254 nm.

2,4-DCP and 2,4,5-TCP were studied by Miller *et al.* (1988). Both compounds were degraded by 300 nm light. Addition of H_2O_2 accelerated the photodecomposition of both chlorophenols. Photodegradation of chlorophenols followed smooth first-order kinetics (pseudo-first order in the presence of H_2O_2). The reaction rate constant of DCP increased from 1.2×10^{-3} to $6.7 \times 10^{-3}\,s^{-1}$, when the H_2O_2 concentration increased from 0 to 0.1 M. The corresponding values for TCP were from 6.7×10^{-3} to $13.4 \times 10^{-3}\,s^{-1}$.

UV/H_2O_2 treatment with 10-W low-pressure vapor mercury arc lamp was shown to be effective in the removal of 2-propanol (Kawaguchi, 1993), phenol, the chemical oxygen demand (COD), TOC contents of the solution (Kawaguchi, 1992a,b), TCE, tetrachloroethene (PCE), 2,4,6-trichloriphenol, PCP (Shulte *et al.*, 1991) and 1,2-dichlorobenzene (Pinto and Rickabauhg).

Peterson *et al.* (1988) studied the efficiency of a high-intensity short-wavelength UV-light system to degrade the pesticides carbofuran, fenamiphos sulfoxide and propazine in aqueous solutions. The first-order rate constants were 3.33×10^{-3}, 11.20×10^{-3} and $3.07 \times 10^{-3}\,s^{-1}$, respectively. Unlike many other substances, addition of H_2O_2 did not accelerate the decomposition. Also captane, chlordane, m-xylene and pentachloronitrobenzene were studied by Peterson *et al.* (1990) with 5000-W high-pressure mercury vapor lamp. Addition of H_2O_2 accelerated only the decomposition of pentachloronitrobenzene.

More data of laboratory-scale UV/H_2O_2 experiments are collected into Table 4.1.

4.6.3 Additional oxidants or treatment steps

Ghaly *et al.* (2001a,b) studied the addition of iron into the UV/H_2O_2 system, and the reaction rate constants of *p*-chlorophenol for UV, UV/H_2O_2, UV/H_2O_2/Fe(II) and UV/H_2O_2/Fe(III). The reaction rate constants increased, respectively, as 1.0×10^{-3}, 2.0×10^{-3}, 1.2×10^{-3} and 1.7×10^{-3}. The energy consumption also decreased significantly from 1105 for UV/H_2O_2 to 184 $kWh\,kg^{-1}$ *p*-chlorophenol for UV/H_2O_2/Fe(III). The disadvantage of introduction of Fenton-type oxidants were the formation of iron precipitation and requirement of pH adjustment to <4.

Trapido *et al.* (2001) studied the combination of ozone with UV radiation and H_2O_2 for the destruction of nitrophenols (NPs). The O_3/UV/H_2O_2 treatment with the admixture of 8 mM ($256\,mg\,L^{-1}$) H_2O_2 at pH 2.5 was the fastest way to degrade NPs. They also noticed almost complete mineralization of organically bound nitrogen and the elimination of toxicity to *Daphnia mangna*.

Akata and Gurol (1992) studied the removal of nitrobenzene and trihalomethane precursors by UV/H_2O_2, system with and without synthetic polymer polyethylene oxide (PEO). The hypothesis was that the polymer could enhance the removal of nitrobenzene by polymerization. The UV/H_2O_2 experiments were conducted in neutral pH (alkalinity 40 and $200\,mg\,L^{-1}$) and the initial concentration of nitrobenzene was 1.5×10^{-4} M. The experiments were carried out in 2-l reactor with the 33-W emission power and in 0.1-mM H_2O_2 solution. The reaction rate of nitrobenzene was $0.76 \times 10^{-6}\,M\,min^{-1}$. The addition of high concentration ($1000\,mg\,L^{-1}$) of polymer significantly retarded the degradation and the reaction rate was only $1.9 \times 10^{-7}\,M\,min^{-1}$ due to the competition of polymer molecules with the nitrobenzene for the hydroxyl radicals. The UV/H_2O_2 treatment was able

Table 4.1 Reaction rate constants for different chemicals in UV/H_2O_2 system.

Compound	Light	$[H_2O_2]$ (mM)	$K \times 10^{-3} s^{-1}$	Reference
Phenol	254 nm + O_3	3.2–12.7	1.74, 1.16	Esplugas *et al.* (2002)
o-Nitrotoluene	Medium pressure 700-W	7.4	2.1	Ghaly *et al.* (2001a,b)
		14.7	3.7	
		29.4	5.7	
		88.2	7.9	
p-Hydroxy-benzoic acid	Medium pressure, 185–436 nm, 3.30×10^{-5} einstein s^{-1}			Beltran-Heredia *et al.* (2001)
Isoprene	254 nm, 2.25 mW m^{-2}	Molar ratio	$k_p + k_{OH}$	Elkanzi *et al.* (2000)
		0	0.40	
		1	0.45	
		4	0.63	
		6	0.70	
Tetrabutyl-azine	125-W high pressure	–	7.40	Sanlaville *et al.* (1996)
		0.2	7.67	
		2.0	9.17	
		2.0	15.83	
Napthalene	1.85×10^{-6} einstein $L^{-1}s^{-1}$	–	0.80	Glaze *et al.* (1992), Tuhkanen (1994)
		1.0	1.6	
		10	6.9	–
		20	5.9	–
		40	4.0	–
		100	2.6	–
	1.05×10^{-6}	–	0.2	–
PCP	1.85×10^{-6}	–	1.0	–
	1.05×10^{-6}	–	0.2	–
	1.85×10^{-6}	1	3.3	–
	1.85×10^{-6}	10	9.2	–
	1.85×10^{-6}	20	7.2	–
	1.85×10^{-6}	100	1.2	–

to remove 40% of the THMFP after 25-min irradiation. The low concentration, 10 mg L^{-1}, of polymer had no effect on removal of THMFP.

The UV/H_2O_2 is commonly used for the treatment of PCE-contaminated groundwater, but the efficiency is usually lowered in the presence of radical scavengers and/or UV-light absorbers. Alibegic *et al.* (2001) studied the improvement of UV/H_2O_2 photolytic system by reducing the $OH^\bullet$-radical scavengers and UV absorbers in the reacting system. Stripped PCE gas was absorbed into a bubble column reactor equipped with UV light, containing distilled water and H_2O_2. Degradation of PCE in the liquid phase was found to follow pseudo-first-order kinetic and the apparent rate constant was in the order of 0.20 s^{-1}.

4.7 DRINKING WATER APPLICATIONS

Removal of the highly potent Ames mutagen, 3-chloro-4-(dichloromethyl)-5-hydroxy-2(5H)-furanone (MX), from an aqueous solution and a decrease in the mutagenic potency of the solution were possible by direct photolysis with the wavelength of 254 nm as well as with thermal, reductive and different kind of oxidative treatments in which generation of hydroxyl radical was involved. The removal of 1 mol of MX required 2 mol of hydroxyl radicals. Photolytic treatment removed 76% of the MX and 79% of the mutagenic potency within 1-h irradiation (Fukui *et al.*, 1991).

Parkinson *et al.* (2001) studied the removal of natural organic matter (NOM) in two Australian waters by UVA, UVB, UVC and UV/H_2O_2. The toxicity of water was monitored by *Vibrio fisheri*-based Microtox test, *Daphna carinata* acute immobilization test and African green monkey kidney cell test. No significant toxicity was observed in any of the tests for the UVA and UVB irradiated samples, also UVC and UVC/H_2O_2. Treated samples did not indicate any toxicity in the Microtox or cytotoxicity analyses. These results differ from that reported by Frimmel (1998), who found that both UVA and UVB irradiated aqueous NOM samples resulted in significant toxicity to *Daphnia magna.* Parkinson observed, on the contrary, that UVC and UV/H_2O_2 treated samples were acutely toxic to the *Daphnia carinata.* The observed toxicity was attributed to photo-oxidative degradation of NOM-metal binding sites, which resulted in the release of bioavailable copper ion. The concentration of free copper was toxic to water fleas, but well below the drinking water guidelines for metals in domestic supplies. Based on the limited data, the authors state that UV/H_2O_2 treatment is unlike to cause taste or health concerns.

Backlund (1992) studied the treatment of natural humic water with UV (254 nm) and H_2O_2 treatment. The dissolved organic carbon content and the UV absorbance of the water decreased substantially during the treatment. The decreases were dependent on the time of the irradiation as well as on the H_2O_2 dose applied. The humus molecules were degraded to smaller fragments during the irradiation. At high UV doses, part of the dissolved organic material was found to precipitate, probably as a result of polymerization. Oxalic acid, acetic acid, malonic acid and n-butanoic acid were the most abundant degradation products detected. No mutagenic activity was generated during the UV or UV/H_2O_2 treatment. Liao *et al.* (2001a) noticed that the complexation of metal into the humic molecule inhibited the humic mineralization and decreased the rate of destruction of the humic matter measured as UV absorption at 254 nm. The experiments were conducted in re-circulated photoreactor with low-pressure mercury lamps (254 nm) with the aqueous solution of humic acid and transition cupric acid. The author mentioned two reasons for the decreased degradation rate: cupric-complexed humic acids are more difficult to be oxidized than the non-complexed humic acids and the strong adsorption of UV radiation by cubric–humic complex.

The degradation of NOM by H_2O_2, solar and high-pressure mercury vapor lamp was studied by Wang *et al.* (2000). The optimum dose of H_2O_2 was found to be 0.01% and the NOM oxidation and H_2O_2 degradation followed first-order and zero-order reaction kinetics, respectively. Carbonate and bicarbonate inhibited the degradation of humic acid. Pseudo-first-order rate constants were determined for the removal of organic carbon measured as non-purgeable dissolved organic carbon (NPDOC). The values varied between 0.0083 and 0.0256 min^{-1} depending on the organic carbon/H_2O_2 ratio and the presence of HCO_3^-/CO_3^- concentration.

Martin *et al.* (2000) studied the decolorization of water for domestic supply by specially designed flat plate reactor. The experiments were carried out in a flat plate photoreactor with circular cross section. It was irradiated from both sides with two tubular, germicidal lamps placed at the focal axis of two cylindrical reflectors of parabolic cross sections. Significant decrease in the concentration of unidentified individual organic compounds was found during the treatment. A simple kinetic model was developed to express the decolorization process using UV and H_2O_2:

$$R^0 = -kC^0_{\mathrm{TOC}}\,\mathrm{e}^{\mathrm{a}}(z)$$

4.8 WASTEWATERS APPLICATIONS

4.8.1 Textile industry

Treatment of textile industry wastewater is frequently studied by AOPs (Galindo and Kalt, 1998; Geargiou *et al.*, 2002; Majcen-Le Marchal *et al.*, 1997; Neamentu *et al.*, 2002; Shen and Wang, 2002). The aims of the study are the decolorization of colorful water, the removal of toxicity and enhancement of biodegradability. The experiments are carried out in the mixture of azo dyes in aqueous solution. Since the dyes are strong absorbers, they act like filters. Due to the competition of radiation, considerably high concentration of H_2O_2 is needed. Ince (1999) has developed a non-linear regression model to determine the most effective level of H_2O_2. Complete degradation of dyes also requires high dose of irradiation, optimal reactor configuration and long treatment times (Cisneros *et al.*, 2002).

Arslan *et al.* (1999, 2000, 2001, 2002a,b) have systematically studied the treatment of textile industry wastewaters by different AOP techniques. The experiments were done with simulated reactive dye bath effluent containing a mixture of monochlorodiazine-type reactive dyes and various auxiliary chemicals at typical concentration encountered in exhausted dye bath liqueurs. In view of electrical energy, UV/H_2O_2 treatment appeared to be a promising alternative for full-scale dye path effluent decolorization. The electrical energy consumption per order (EE/O) for decolorization (525 nm) was 0.633 kWh m^{-3} × order. The additional advantages to iron or ozone-based techniques were that no sludge is formed, significant reduction of COD and relatively short retention times. The degradation via direct photolysis was negligible and the efficiency depended strongly on the H_2O_2 dose and pH. The inherent high pH value of dye bath had more detrimental effect upon the treatment performance than to ozone-based techniques.

El-Dein *et al.* (2002) had studied the combination of UV/H_2O_2 oxidation to biological treatment of synthetic textile industry wastewater (Reactive Black 5). Chemical oxidation alone was not economically feasible; but combined to biological stage, it may offer an economical alternative to conventional treatment technologies.

4.8.2 Paper and pulp industry

Prat *et al.* (1988) studied the use of H_2O_2 in combination with UV radiation from medium-pressure mercury arc lamp in the treatment of wastewater from the chlorine and alkaline extraction stages of paper pulp bleaching. They determined the kinetic constants corresponding to the elimination of color from these effluents at different pH value of the medium. The employment of H_2O_2 treatment together with UV radiation, even under the most favorable, basic conditions, took a very long time and were not economically competitive with the alternative treatment methods.

Shulte *et al.* (1991) determined the influence of parameters, such as the water matrix, OH-radical scavengers, pH, the ratio of H_2O_2 to the contaminant and the UV intensity to the treatment of real wastewater, paper mill wastewater and landfill leachate. The degradation of COD, AOX and formaldehyde was followed in different H_2O_2-based AOPs. Both low- and medium-pressure mercury arc lamps with the output up to 2 kW were used. The activated H_2O_2-based techniques were effective, but the degree of reduction of parameters and the most cost-effective combination of processes should be determined for each individual case.

4.8.3 Olive oil industry

Benitez *et al.* (1996) studied the UV and UV/H_2O_2 treatment of water that contained protocateuic acid (3,4-dihydroxy benzoic acid), *p*-hydroxy phenylacetic acid (Benitez *et al.*, 2001) and phenolic pollutants present in the wastewater from olive oil manufacturing. The degradation was enhanced by addition of H_2O_2. In further studies, Belitez *et al.* (2001) combined AOP treatment with anaerobic treatment of real effluent of olive oil industrial washing water. The chemical pretreatment enhanced the anaerobic degradation. The explanation was suggested to be the elimination of toxic phenolic and polyphenolic compounds.

4.9 PILOT PLANT-SCALE EXPERIMENTS OF UV/H_2O_2

UV/H_2O_2 treatment of process water was studied for power plants by a commercial unit for disinfection, which was comprised of 14 × 30-W low-pressure lamps housed symmetrically around a quartz cell of 10 l capacity. H_2O_2 was injected inline via a metering pump prior to the UV reactor inlet at desired concentrations. The residence time was varied by flow rate. The pilot-scale tests gave clear evidence that significant improvements in TOC and UV absorbance could be obtained after short contact times. In the case where UV/H_2O_2 treatment was a additional stage in the preparation of demineralized water for boiler feed water, the improvements could be gained by converting organic matter to carbon dioxide or to lower molecular weight organic compounds and by degrading the organic chlorides present in the water to ionic chloride, which could then be removed on the anion resins. This treatment could find industrial applications in food and soft drink, pharmaceutical and electronics industries (Clarke and Knowles, 1982).

4.10 DEMONSTRATION-SCALE EXPERIMENTS OF UV/H_2O_2

Technologies to destroy, detoxify or recycle hazardous waste materials, and reduce mobility or volume of hazardous wastes is being demonstrated and evaluated by the United States Environmental Protection Agency (EPA). The Superfund Innovative Technology Evaluation (SITE) Program was created in 1986 to provide information, such as reliable performance and cost data on alternative and innovative technologies.

A UV/oxidation technology developed by Ultrox International was evaluated in treating groundwater contaminated with VOCs in the field at the Lorenz Barrel and Drum site in San Jose, California under the SITE program. The site was used primarily for drum recycling operations from about 1947 to 1987. The drums contained residual organic wastes, solvents, acids, metal oxides and oils (Lewis *et al.*, 1990).

The Ultrox system consists of UV/oxidation reactor module, an air compressor/ ozone generator module, an H_2O_2 feed system and a catalytic ozone decomposition unit. During the operation, contaminated water first comes in contact with H_2O_2, as it flows through the influent line to the reactor. The water then comes in contact with the UV radiation and ozone, as it flows through the reactor at a specified rate to achieve the desired hydraulic retention time.

Removal efficiencies for trichloroethylene and total VOCs were as high as 99% and 90%, respectively. A few compounds were more difficult to treat [1,1-dichloroethane 65% and 1,1,1-trichloroethane (TCA) 85% reduction]. Most VOCs were removed through chemical oxidation but, for example, for 1,1,1-TCA and 1,1-dichloroethane, stripping contributed also toward removal. The treated groundwater met the applicable discharge standards for discharge into a local waterway at the 95% confidence level (Lewis *et al.*, 1990 and EPA/540/5-89/012).

A full-scale UV/H_2O_2 oxidation system that had been designed for 60–80 L per minute flow rate; 200 ppm concentration of H_2O_2 and 180-kW UV power was tested by 13 trial runs. The unit had been in operation for 1 year. Several parameters were varied: residence time, H_2O_2 dose, number of UV bulbs operating and the pH. The parameters were then related to destruction and volatization of the VOCs. The system had been tested previously in a bench scale for acetone, benzene, chlorobenzene, ethylbenzene, tetrahydrofuran, toluene and total xylenes.

At that time, the results had been then excellent, with 99.99% removal even for a worst-case scenario. During the test period in the field, the removal of VOCs varied considerably. The most important factor was the precipitation of many chemicals and minerals on the quartz tubes, which deteriorated dramatically the purification efficiencies (Nyer *et al.*, 1991).

A full-scale treatment system was studied for removal of VOCs from contaminated groundwater by air stripping towers, which were connected to a $UV/H_2O_2/O_3$ reactor in order to remove the VOCs prior to release of the air stripper emissions to the atmosphere. Different combinations of UV radiation and addition of ozone and H_2O_2 were studied. Of the various AOP combinations, the UV radiation alone was the most effective AOP, and the most important operation parameter was the dose of UV radiation. When the UV radiation increased from 0 to 12,000 $W\text{-}s\,L^{-1}$, the removal of carbon tetrachloride also increased. For TCE and PCE, removal increased sharply to 90% with UV dose at 1600 $W\text{-}s\,L^{-1}$ and then gradually to 100% at 12,000 $W\text{-}s\,L^{-1}$. The use of ozone or H_2O_2 did not improve significantly the VOC removal.

4.11 COMMERCIAL APPLICATIONS

The commercial-scale UV/oxidation systems available for contaminated water treatment include Calgon Carbon Corboration Peroxpure™ and Rayox® UV/H_2O_2 systems, Magnum Water Technology, Inc. CAV-OX® UV/H_2O_2 system, WEDECO UV/H_2O_2 and UV/O_3 systems and US Filter/Zimpro, Inc $UV/H_2O_2/O_3$ system (formerly known as Ultrox).

Calgon peroxy-pure™ and Rayox® system use 15-kW medium-pressure mercury arc lamps to generate UV radiation. A typical unit contains portable oxidation units (typically six units in series equipped with a 15-kW lamp each), an H_2O_2 feed module, and acid and base feed module, if needed. In a typical application of the Calgon system, contaminated water is dosed with H_2O_2 before the water enters the first reactor. If the treated water contains high concentration of carbonates and bicarbonates, which are radical scavengers, addition of acid is used to lower the pH and shift the carbonic acid–bicarbonate–carbonate equilibrium to carbonic acid. The base is used for neutralization of the treated water if required.

A typical CAV-OX® system consists of a portable, truck- or skid-mounted module with the following components: a cavitation chamber, an H_2O_2 feed unit and UV reactors. Depending on the application, different kind of UV sources is used. The low-energy CAV-OX® I process uses one UV reactor with six 60-W low-pressure lamps and the reactor is operated at 360 W. The CAV-OX® II uses two UV reactors, each with high-pressure UV lamp operated at 2.5 or 5 kW. In a CAV-OX system, contaminated water is first pumped in cavitation chamber. The H_2O_2 is added between cavitation chamber and UV reactor or sometimes before cavitation unit.

The US Filter $UV/H_2O_2/O_3$ uses UV radiation, ozone and H_2O_2 to oxidize contaminants in water. The major components are the UV/oxidation reactor, ozone generator, H_2O_2 feed tank and the catalytic ozone decomposition unit. The oxidation unit is divided by vertical baffles into chambers and contains low-pressure mercury vapor lamps (65 W each) in quartz sleeves. The contaminated water comes first in contact with H_2O_2 before it is led into the reactor, where it is exposed to UV irradiation and ozone.

The use of combination of different treatment technologies, such as UV/oxidation, GAC filtration, air stripping, chemical precipitation and biological treatment, has proven to be cost effective in many cases. For instance, Calgon Carbon Corporations's Rayox® has coupled UV/oxidation with GAC for the treatment of groundwater contaminated with dichloroethene (DCE), TCE, PCE and DCM. The role of GAC was to easily remove absorbable background organics. UV/oxidation of groundwater contaminated by PCE, TCA, DCM and VC combined to air stripping has reduced the organic load of air stripper and the overall system size and costs. Chemical precipitation has been used to remove the bulk loading of COD and the UV/oxidation has been as a polishing step to destroy the organics to low level. UV/oxidation has been combined also to membrane technologies (Lem, 2002).

4.12 COST ESTIMATION AND PERFORMANCE

EE/O stands for electrical energy consumption per order of magnitude reduction in contaminant concentration in 1 m^3 of treated water. EE/O gives the number of kilowatt-hours of electricity necessary to reduce concentration of a contaminant in 1 m^3 by one order of magnitude. It allows for easy and accurate scale up to a full-scale design and costs. EE/O combines light intensity, residence time and percent destruction into a single measurement. The total energy requirement depends on the initial concentration of contaminant and the required purification efficiency. Once the required UV dose is known, the electrical operating cost can be calculated using the equation:

$$\text{Electrical cost (Euro/m}^3) = \text{UV dose (kWh m}^{-3}) \times \text{Power cost (Euro kWh}^{-1}).$$

Another parameter/variable in the operation costs is the consumption of H_2O_2. The optimal ratio of H_2O_2/contaminant depends on the absorptivity of the contaminant(s) and the matrix. The cost of H_2O_2 can vary considerably. For calculation purposes, a delivered price of 100% H_2O_2 can be used. Lamp replacement costs range from 10% to 20% of the electrical costs (Chemviron, 1997).

According to AOT Handbook, the following rules of thumb can be used when performing a cost comparison analysis:

- Unit operating costs for UV/oxidation treatment increase much more slowly with increasing influent concentrations than activated carbon.
- UV/oxidation treatment costs are almost always less for contaminants that are poor carbon absorber regardless of concentration.
- For average carbon absorbers, the most economical technology must be chosen on a case-by-case basis.

Table 4.2 Cost estimation for UV/H_2O_2 vs. activated carbon (AC) treatment of TCE-contaminated aquifer (Hirvonen *et al.*, 1998).

Cost category	UV/H_2O_2	Activated carbon
Equipment		
Investment × 0.149	12,600	8600
Operation and maintenance		
Maintenance (0.02 × Inv consumables)	1700	1200
AC + disposal (\$0.29 per m^3)		5200
H_2O_2	400	
UV lamps (3 per year)	1500	
Utilities		
Electricity	3100	
Labour	6400	6400
Analytical service	10,000	10,000
Total annual cost	35,700	31,400
Cost \$ per m^3	2.00	1.70

- UV/oxidation capital costs are typically two to three times that of activated carbon.
- For concentrations below 10 ppm, activated carbon is the most cost-effective treatment for good absorbers.

Hirvonen *et al.* (1998) studied the treatment of PCE-contaminated groundwater. The initial concentration was $200\,\mu g\,L^{-1}$ and the target concentration was $2\,\mu g\,L^{-1}$. A H_2O_2 dose of $27\,mg\,L^{-1}$ and electrical energy of $0.5\,kWh\,order^{-1}\,m^{-3}$ was determined experimentally. Costs evaluation of UV/H_2O_2 method was compared with activated carbon adsorption when flow rate was $50\,m^3\,d^{-1}$ and purification period was $360\,d\,a^{-1}$ yielding a total of $18{,}000\,m^3$ in annual level (Table 4.2). The energy cost was \$0.07 per kWh and H_2O_2 cost \$0.87 per kg as 100% H_2O_2. The capital cost of UV/H_2O reactor was calculated as \$85,000. The cost includes purchase of UV reactor, sedimentation and equalization tanks, a shelter for equipments and remotely monitored control system.

Esplugas *et al.* (2002) compared in the laboratory the costs of different kind of UV-based AOTs. The cost per kilogram of degraded phenol was calculated for the 50% (half-lifetime) and 75% of the initial phenol. The cost for UV treatment was 172.2 and 293.1 $\$\,kg^{-1}$, for UV/H_2O_2, 13.1 and 28.7 $\$\,kg^{-1}$, and for $UV/H_2O_2/O_3$, 7.12 and 9.51 $\$\,kg^{-1}$, respectively.

Under some conditions, pre-treatment may be required. Typical forms of pre-treatment that may be used include removal of suspended solids, free phase oil and grease, and iron concentration. It is always recommended that a design test or treatability study be conducted for each application. It allows to characterize the water and to optimize treatment performance.

The effect of water matrix, such as other competing organics and alkalinity (bicarbonate ion), and the effect of the initial concentration and the degradation intermediates are useful to know (Wang, 1990). If we are dealing with known organic chemicals and there are data available on the efficiency of chemical oxidation, studies need be conducted on the toxicity or biodegradability of the oxidation by-products.

Most of the successful applications are related to the treatment of groundwater contaminated by VOCs. The most common constituent has been TCE and PCE. Also benzene, chlorobenzene, di- and trichloro acetic acids, and vinyl chloride have been treated with efficiencies that reach accepted levels [usually maximum contaminant levels (MCLs)]. The cost of treatment depends on the contaminant concentrations: 0.1 ppm (0.13–0.65 $\$\,m^{-3}$), 1 ppm (0.45–1.0 $\$\,m^{-3}$), 10 ppm (0.5–1.5 $\$\,m^{-3}$) (Chemviron, 1997).

The main problem of performance has been scaling of the lamps and formation of precipitations. The quartz sleeves of the lamps are equipped with wipers that periodically clean the tubes. Solids, such as precipitated iron and manganese, can also be filtrated.

The highest costs of the purification of contaminated groundwater are related to the pumping of the water into the soil surface. A pump and treat technique, such as UV/H_2O_2 treatment, is less attractive solution in a case where high volumes of water with quite low contamination level should to be treated.

4.13 FURTHER RESEARCH NEEDS

AOP processes, such as UV/H_2O_2 treatment, are promising environmental engineering techniques. The advantages of this are numerous. $UV/H_2O_2/(O_3)$ systems have been considered as destruction processes, and the final products are carbon dioxide, water and inert salts. Recently, the economical feasibility of total treatment of water stream by chemical oxidation has been set questionable. There are lot of research going on related to the integration of chemical oxidation to other treatment techniques, such as chemical coagulation, membrane techniques, air stripping, activated carbon filtration and biological processes. This kind of approach might solve the limitation of UV/oxidation processes, such as high turbidity, solid particles, heavy metals, high chemical oxygen demand (COD) or biological oxygen demand (BOD) matrix of the treated water stream.

The laboratory-scale experiments with synthetic single compound solution are numerous, but there is still a lack of multicomponent studies. Just the disappearance of one model compound had been studied. In the real environmental engineering applications, the problem is, however, more complicated. First, multiple organic contaminants are routinely expected from materials, such as contaminated groundwaters and industrial wastewaters. Second, in the course of mineralization of any organic contaminant, oxidation will logically involve a series of intermediates on the way to CO_2. Such intermediates make the conversion process multicomponent, even if only a single contaminant existed in the feed.

Usually, there has not been any attempt to separate the role of different mechanism in the overall disappearance of chemical or the role of competing species.

More information is needed of the interplay of various degradation mechanisms, direct photolysis and radical reactions in the case of UV/H_2O_2 treatment.

Due to the high demand of oxidants and energy, the optimization of the conditions of a chemical oxidation is essential for economical feasibility. Both studies with synthetic wastewater, where the parameters can be controlled and feasibility studies with real water and wastewater streams have to be conducted before the new promising techniques can be applied successfully.

REFERENCES

Adams, G.D. and Scanlan, P.A. (1993) Biodegradability enhancement and mineralization of 1,4-dioxane by the hydrogen peroxide advanced oxidation process. In: *Proceeding of the 11th Ozone World Congress, 1933*, Vol. 1, S-3-60-75, San Francisco, USA.

Adams, C.D. and Kuzhikannil, J.J. (2002) Rffect of UV/H_2O_2 on the aerobic bioderradability of quaternary amine surfactants. *Water Res.* **34**, 668–672.

Alibegic, D., Tsuneda, S. and Hirata, A. (2001) Kinetics of tetrachloethyle (PCE) gas degradation and byproducts formation during UV/H_2O_2 treatment in UV-bubble column reactor. *Chem. Eng. Sci.* **56**, 6195–6203.

Arslan, I., Balgioglu, I.A. and Tuhkanen, T. (1999) Oxidative treatment of simulated dyehouse effluent by UV and near-UV light assisted Fenton's reagent. *Chemosphere* **39**, 2767–2783.

Arslan Alaton, I., Balgioglu, I.A. and Bahneman, D. (2002a) Advanced oxidation of a reactive dyebath effluent: comparization of O_3, UV/H_2O_2-C and TiO_2/UV-A processes. *Water Res.* **36**, 1143–1154.

Arslan Alaton, I., Balgioglu, I.A., Tuhkanen, T. and Bahneman, D. (2002b) UV/H_2O_2-C and Fe^{2+}/, UV/H_2O_2-C versus TiO_2/UV-A treatment of reactive dye wastewater. *J. Environ. Eng.* **October**, 903–911.

Backlund, P. (1992) Degradation of aquatic humic material by ultraviolet light. *Chemosphere* **25**, 1869–1878.

Baxendale, J.H. and Willson, J.A. (1957) Photolysis of hydrogen peroxide at high light intensities. *Trans. Faraday Soc.* **53**, 344–356.

Behar, D., Czapski, C. and Duchovny, I. (1979) Carbonate radical in flash photplysis and pulse radiolysis of aqueous carbonate solutions. *J. Phys. Chem.* **74**, 2206–2210.

Beltran, F.J., Encinar, J.M. and Alonso, M.A. (1998) Nitroaromatic hydrocarbon ozonation in water. 2. Combinated ozonation with hydrogen peroxide or UV radiation. *Ind. Eng. Chem. Res.* **37**, 32.

Beltran, F.J., Rivas, J., Alvarez, P.M. and Alonso, M.A. (1999) A kinetic model for advanced oxidation of artomic hydrocarbons in water: application to phenantrene and nitrobenzene. *Ind. Eng. Chem. Res.* **38**, 4189–4199.

Benitez, F.J., Acero, J.L., Real, F.J., Rubio, F.J. and Leal, A.I. (2000) The role of hydroxyl radicals for the decomposition of *p*-hydroxy phenylacetic acid in aqueous solutions. *Water Res.* **35**, 1338–1343.

Benitez, F.J., Acero, J.L., Gonzalez, T. and Garcia, J. (2001) Organic matter removal from wastewater of the black olive industry by chemical and biological procedures. *Process Biochem.* **37**, 257–265.

Bowers, A.R., Cho, S.H. and Singh, A. (1991) Chemical oxidation of aromatic compounds: comparison of H_2O_2, $KMnO_4$ and O_3 for toxicity reduction and improvements in biodegradability. In: *Proceedings of the First International Symposium Chemical Oxidation Technologies for the Nineties*, Eckenfelder, W.W., Bowers, A.R. and Roth, J.A. (eds). Vanderbilt University, Nashville, Tennessee, USA, pp. 11–25.

Bull, R.A. and Zeff, J.D. (1991) Hydrogen peroxide in advanced oxidation processes for treatment of industrial process and contaminated groundwater. In: *Proceedings of the First International Symposium Chemical Oxidation Technologies for the Nineties*, Eckenfelder, W.W., Bowers, A.R. and Roth, J.A. (eds). Vanderbilt University, Nashville, Tennessee, USA, pp. 26–36.

Buxton, G.V., Greenstock, C.L., Helman, W.P. and Ross, A.B. (1988) Critical review of data constants for reactions of hydrated electrons, hydrogen atoms and hydroxyl radicals (OH/O^-) in aqueous solutions. *J. Phys. Chem. Ref. Data* **17**, 513–886.

Calvert, J.G. and Pitts Jr., J.N. (1966) *Photochemistry*. John Wiley & Sons, Inc., New York, USA, 899 pp.

Chemviron Carbon Oxidation Technologies (1997) *AOT Handbook*. Germany.

Chen, S. and Hoffman, M.Z. (1973) Rate constant for the reaction of the carbonate radical with compounds of biochemical interest in neutral aqueous solution. *Rad. Res.* **56**, 40–47.

Chen, S. and Hoffman, M.Z. (1974) Reactivity of the carbonate radical in aqueous solution. Tryptophan and its derivates. *J. Phys. Chem.* **78**, 2099–2102.

Christensen, H.S., Sehested, H. and Corfitzan, H. (1982) Reactions of hydroxyl radicals with hydrogen peroxide at ambient and elevated temperatures. *J. Phys. Chem.* **86**, 15–68.

Clarke, N. and Knowles, G. (1982) High purity water using H_2O_2 and UV radiation. *Effluent Wat. Treat. J.* **September**, 335–341.

Crittenden, J.C., Hu, S., Hand, D.W. and Green, S.A. (1999) A kinetic model for H_2O_2/UV prosess in a completely mixed batch reactor. *Water Res.* **33**, 2315–2328.

Eckenfelder Jr., W.W. (1991) The role of chemical oxidation in wastewater treatment processes, In: *Proceedings of the First International Symposium Chemical Oxidation Technologies for the Nineties*, Eckenfelder, W.W., Bowers, A.R. and Roth, J.A. (eds). Vanderbilt University, Nashville, Tennessee, USA, pp. 1–10.

Eilbeck, W.J. and Mattock, G. (1988) *Chemical Oxidation, in Chemical Processes in Waste Water Treatment*. Ellis Horwood Limited, Chichester, UK, 331 pp.

El-Dein, A.M., Libra, J.A. and Wiesmann, U. (2002) H_2O_2/UV – biological oxidation processes for decolorization and mineralization of Reactive Black 5. In: *The 8th International Conference on Advanced Oxidation Technologies for Water and Air Remediation*, Toronto, Canada, November 17–21.

Esplugas, S., Gimenez, J., Contreras, S., Pascual, E. and Rodrigies, M. (2002) Comparison of different advanced oxidation processes for phenol degradation. *Water Res.* **36**, 1034–1024.

Evens, E.L., Jaysdon, G.G., Robb, I.D. and Swallow, A.J. (1980) Determination by pulse radiolysis of the distribution of solubilities between micellar and nonmicellar phases. *J.C.S. Faraday I* **76**, 528–536.

Frimmel, F.H. (1998) Impact of light on the properities pf aquatic natural organic matter. *Environ. Int.* **24**, 559–571.

Fukui, S., Ogawa, S., Motozuka, T. and Hanasaki, Y. (1991) Removal of 3-chloro-4-(dichloromethyl)-5-hydroxy-2(5)-furanone (MX) in water by oxidative, reductive, thermal and photochemical treatments. *Chemosphere* **23**, 761–775.

Galindo, C. and Kalt, A. (1998) UV-H_2O_2 oxidation of monoazo dyes in aqueous media: a kinetic study. *Dyes Pigments* **40**, 27–35.

Georgiou, D., Melidis, P., Aivasidis, A. and Gimouhopoulos, K. (2002) Degradation of azo-reactive dyes by ultraviolet radiation in the presence of hydrogen peroxide. *Dyes Pigments* **52**, 69–78.

Getoff, N. and Solar, S. (1986) Radiolysis and pulse radiolysis of chlorinated phenols in aqueous solutions. *Radiat. Phys. Chem.* **28**, 443–450.

Ghaly, M.Y., Härtel, G., Mayer, R. and Haseneder, R. (2001a) Photochemical oxidation of *p*-chlorophenol by UV/H_2O_2 and photo-Fenton processes. A comparative study. *Waste Manage.* **21**, 41–47.

Ghaly, M.Y., Härdel, G., Mayer, R. and Haseneder, R. (2001b) Aromatic compounds degradation in water by using ozone and AOPs. A comparative study. *O*-nitrotoluene as a model substrate. *Ozone-Sci. Eng.* **23**, 127–138.

Glaze, W.H. and Kang, J.W. (1989) Advanced oxidation process. Test of a kinetic model for the oxidation of organic compounds with ozone and hydrogen peroxide in a semi-batch reactor. *Ind. Eng. Chem. Res.* **28**, 1580–1587.

Glaze, W.H. and Lay, Y. (1989) Oxidation of 1,2-dibromo-3-chloropropane (DBCP) using advanced oxidation processes. *Ozone in Water Treatment.* New York, USA, IOA, pp. 688–708.

Glaze, W.H., Kang, J.W. and Chapin, D.H. (1987) The chemistry of processes involving ozone, hydrogen peroxide and ultraviolet radiation. *Ozone-Sci. Eng.* **9**, 335–342.

Glaze, W.H., Beltran, F.J., Tuhkanen, T. and Kang, J.W. (1992a) Chemical models of advanced oxidation processes. *Water Poll. Res. J. Canada* **27**, 1580–1587.

Glaze, W.H., Beltran, F., Tuhkanen, T. and Kang, J.-W. (1992b) Chemical models of advanced oxidation processes. *Water Poll. Res. J. Canada* **27**, 23–42.

Glaze,W.H., Lay, Y. and Kang, J.W. (1995) Advanced oxidation processes. A kinetic model for the oxidation of 1,2-dibromo-3-chloropropane in water by combination of hydrogen peroxide and UV radiation. *Ind. Eng. Chem. Res.* **34**, 2314–2323.

Griffiths, J., Chu, K.-Y. and Hawkins, J.C.S. (1976) Photosensitized oxidation of 1-naphtols. *Chem. Commun.* **1976**, 676–677.

Guittonneau, S. (1989) *Contribution a l'Etude de la Photooxidation de Quelques Micropollutants Organochlores en Solution Aqueuse en Presence de Peroxide d'Hydrogen – Comparaison des Systemes Oxydants: H_2O_2/UV, O_3/UV et O_3/H_2O_2.* These de Docteur. Universite de Poitiers, France.

Guittonneau, S., de Laat, J., Dore, M., Duguet, J.P. and Bonnel, C. (1988) Comparative study of the photodegradation of aromatic compounds in water by UV and H_2O_2/UV. *Environ. Technol. Lett.* **9**, 1115–1128.

Hirvonen, A., Tuhkanen, T. and Kalliokoski, P. (1996a) Formation of chlorinated acetic acid during UV/H_2O_2-oxidation of ground water contaminated with chlorinated ethylenes. *Chemosphere* **32**, 1091–1102.

Hirvonen, A., Tuhkanen, T. and Kalliokoski, P. (1996b) Treatment of TCE- and PCE-contaminated groundwater using H_2O_2/UV and O_3/H_2O_2 oxidation processes. *Water Sci. Technol.* **33**, 67–73.

Hirvonen, A., Tuhkanen, T. and Kalliokoski, P. (1996c) Removal of chlorinated ethylenes in contaminated ground water by hydrogen peroxide mediated oxidation processes. *Environ. Technol.* **17**, 263–272.

Hirvonen, A., Tuhkanen, T., Ettala, M. and Kalliokoski, P. (1998) Evaluation of a field-scale UV/H_2O_2-oxidation system for purification of groundwater contaminated with PCE. *Environ. Technol.* **19**, 821–828.

Ho, P.C. (1986) Photooxidation of 2,4-dinitrotoluene in aqueous solution in the presence of hydrogen peroxide. *Environ. Sci. Technol.* **20**, 260–267.

Hodson, J. (1988) The estimation of the photodegradation of organic compounds by hydroxyl radical reaction rate constants obtained from nuclear magnetic resonance spectroscopy. Chemical shift data. *Chemosphere* **17**, 2339–2348.

Ince, N.H. (1999) 'Chritical' effect of hydrogen peroxide in photochemical dye degradation. *Water Res.* **33**, 1080–1084.

Ishikawa, S., Uchimura, Y., Baba, K., Eguchi, Y. and Kido, K. (1992) Photochemical behavior of organic phosphate esters in aqueous solutions irradiated with a mercury lamp. *Bull. Environ. Contam. Toxicol.* **49**, 368–374.

Kawaguchi, H. (1992a) Photooxidation of phenols in aqueous solution in the presence of hydrogen peroxide. *Chemosphere* **24**, 1707–1712.

Kawaguchi, H. (1992b) Determination of direct and indirect photolysis rates of 2-chlorophenol in humic acid solution and natural waters. *Chemosphere* **25**, 635–641.

Kawaguchi, H. (1993) Photo-oxidation of 2-propanol in aqueous solution in the presence of hydrogen peroxide. *Chemosphere* **27**, 577–584.

Klöpffer, W. (1992) Photochemical degradation of pesticides and other chemicals in the environment: a critical assessment of the state of the art. *Sci. Total Environ.* **123/124**, 145–159.

Lay, Y.S. (1989) *Oxidation of 1,2-Dibromo-3-chloropropane in Ground Water Using Advanced Oxidation Processes*. PhD Thesis, University of California at Los Angeles, USA.

Leifer, A. (1988) The kinetics of environmental photochemistry. *Theory and Practice*. ACS Professional Reference Book, York, PA, 304 pp.

Lem, W. (2002) Combination of UV/oxidation with other treatment technologies for the remediation of contaminated waters. In: *The 8th International Conference on Advanced Oxidation Technologies for Air and Water Remediation*, Toronto, Canada, November 17–21.

Lewis, N., Tapudurti, K., Welshans, G. and Foster, R. (1990) A field demonstration of the UV/oxidation technology to treat ground water contaminated with VOCs. *J. Air Waste Manage. Assoc.* **40**, 540–547.

Liao, C.H. and Gurol, M.D. (1995) Chemical oxidation by photolytic decomposion of hydrogen peroxide. *Environ. Sci. Technol.* **29**, 3007–3014.

Liao, C.-H., Lu, M.-C. and Su, S.-H. (2001a) Role of cupric ions in the H_2O_2/UV oxidation of humic acids. *Chemosphere* **44**, 913–919.

Liao, C.-H., Kang, S.-F. and Wu, F.-A. (2001b) Hydroxyl radical scavenging role of chloride and bicarbonate ions in the UV/H_2O_2 process. *Chemosphere* **44**, 1193–1200.

Majcen-Le Marchal, A., Slokar, Y.M. and Taufer, T. (1997) Decoloration of chlorotriazine reactive dyes with UV/H_2O_2. *Dyes Pigments* **33**, 281–298.

Martin, C.A., Alfano, O.M. and Cassano, A.E. (2000) Decolorization of water for domestic supply employing UV radiation and hydrogen peroxide. *Catalysis Today* **60**, 119–127.

Milano, J.C., Yassin-Hussan, S. and Vernet, J.L. (1992) Photochemical degradation of 4-bromodiphenylether: influence of hydrogen peroxide. *Chemosphere* **25**, 353–360.

Mill, T. (1980) Chemical and photooxidation. In: *The Handbook of Environmental Chemistry: Reacton and Processes*, Hutzinger, O. (ed.). Vol. 2, Part A, 307 pp.

Mill, T., Mabey, W.R., Lan, B.Y. and Baraze, A. (1981) Photolysis of polycyclic aromatic hydrocarbons in water. *Chemosphere* **10**, 1281–1290.

Miller, E.M., Singer, G.M., Rosen, J.D. and Bartha, R. (1988) Sequential degradation of chlorophenols by photolytic and microbial treatment. *Environ. Sci. Technol.* **22**, 1215–1219.

Moza, P.N., Fytianos, K., Samanidou, V. and Korte, F. (1988) Photodecomposition of chlorophenols in aqueous medium in presence of hydrogen peroxide. *Bull. Environ. Contam. Toxicol.* **41**, 678–682.

Neamentu, M., Siminieanu, I., Yediler, A. and Kettrup, A. (2002) Kinetic of decolorization and minelaizarion of reactive azo dyes in aqueous solution by the UV/H_2O_2 oxidation. *Dyes Pigments* **53**, 93–99.

Owen, E.D. (1971) Principles of photochemical reactions in aqueous solution. In: *Organic Compounds in Aquatic Environments*, Hunter J.V. (ed.). Marcel Dekker Inc., New York, USA, pp. 387–423.

Parkinson, A., Barry, M.J., Roddick, F.A. and Hobday, M.D. (2001) Preliminary toxicity assessment of water after treatment with UV-irradiation and UV/H_2O_2. *Water Res.* 3656–3664.

Peterson, D., Watson, D. and Winterlin, W. (1988) The destruction ground water threatening pesticides using high intensity UV light. *Environ. Sci. Health* **B23**(6), 587–603.

Peterson, D., Watson, D. and Winterlin, W. (1990) Destruction of pesticides and their formulations in water using short wavelength UV light. *Bull. Environ. Contam. Toxicol.* **44**, 744–750.

Pinto, D. and Rickabauhg, J. Photocatalytic degradation of chlorinated hydrocarbons. *Hazard Ind. Wastes* **23**, 368–373.

Prat, C., Vicente, M. and Esplugas, S. (1988) Treatment of bleaching waters in the paper industry by hydrogen peroxide and ultraviolet radiation. *Water Res.* **22**, 663–668.

Shen, Y.-S. and Wang, D.-K. (2002) Development of photoreactor design equipment for the treatment of dye wastewater by UV/H_2O_2 processes. *J. Hazard. Mater.* **B89**, 267–277.

Shulte, P., Volkmer, M. and Kuhn, F. (1991) Aktiviertes Wasserstoffperoxid zur Beseitigung von Schadstoffen im Wasser (H_2O_2/UV). *Wasser, Luft Boden* **35**, 55–58.

Snoeyink, V.L. and Jenkins, D. (1980) *Water Chemistry.* John Wiley & Sons, New York, USA, 463 pp.

Stumm, W. and Morgan, J.J. (1981) *Aquatic Chemistry*, 2nd edn. John Wiley & Sons, New York, USA, 780 pp.

Sundstrom, D.W., Weir, B.A. and Klei, H.E. Destruction of aromatic pollutants by UV light catalyzed oxidation with hydrogen peroxide. *Environ. Prog.* **8**, 6–11.

Trapido, M., Veressina, Y. and Kallas, J. (2001) Degradation of aqueous nitrophenols by ozone combined with UV-radiation and hydrogen peroxide. *Ozone-Sci. Eng.* **23**, 333–342.

Tuhkanen, T.A. and Beltran, F.J. (1995) Intermediates of the oxidation of naphthalene in water with the combination of hydrogen peroxide and UV radiation. *Chemosphere* **30**, 1463–1475.

USEPA (1998) *Handbook on Advanced Photochemical Oxidation Processes.* EPA/625/ R-98/004, USEPA, December 1998.

von Sontag, C. (1994) Photolytic and radical degradation of chlorinated hydrogarbons and aromatic compounds in aquatic solutions. In: *The First International Conference on Advanced Oxidation Technologies for Water and Air Remediation*, London, Ontario, Canada, June 25–30.

Wang, Y.-T. (1990) Methanogenic degradation of ozonation products of biorefractory or toxic aromatic compounds. *Water Res.* **24**, 185–190.

Wang, Y.-T. (1991) Effects of chemical oxidation on anaerobic treatment of phenols. In: *Proceedings of the First International Symposium Chemical Oxidation Technologies for the Nineties*, Eckenfelder, W.W., Bowers, A.R. and Roth, J.A. (eds). Vanderbilt University, Nashville, Tennessee, USA, pp. 145–156.

Wang, Y.-T. and Latchaw, J.L. (1990) Anaerobic biodegradability and toxicity of hydrogen peroxide oxidation products of phenols. *Res. J. WPCF.* **62**, 234–238.

Wang, G.-S., Hsieh, S.-T. and Hong, C.-S. (2000) Destruction of humic acids in water by UV-light oxidation of hydrogen peroxide. *Water Res.* **34**, 3882–3887.

Watts, R.J., Udell, M.D. and Laung, S.W. (1990) Treatment of pentachloro-contaminated soil using Fenton's reagent. *Hazard. Waste Hazard.* **7**, 335–345.

Watts, R.J., Udell, M.D. and Leung, S.W. (1991) Treatment of contaminated soils using catalyzed hydrogen peroxide. In: *Proceedings of the First International Symposium Chemical Oxidation Technologies for the Nineties*, Eckenfelder, W.W., Bowers, A.R. and Roth, J.A. (eds). Vanderbilt University, Nashville, Tennessee, USA, pp. 37–50.

Weeks, J.L. and Rabani, J. (1966) The pulse radiolysis of deaerated carbonate solutions. I. Transient optical spectrum and mechanism. II. pK for OH radicals. *J. Phys. Chem.* **70**, 2100–2104.

Weir, B.A., Sundstrom, D.W. and Klie, H.E. (1987) Destruction of benzene by ultraviolet light-catalyzed oxidation with hydrogen peroxide. *Hazard. Waste Hazard.* **4**, 167–176.

Weir, B.A. and Sunstrom, D.W. (1993) Destruction of trichloroethylene by UV light-catalyzed oxidation with hydrogen peroxide. *Chemosphere* **27**, 1279–1291.

Wong, A.S. and Crosby, D.G. (1981) Photodecomposition of pentachlorophenol in water. *J. Agric. Food Chem.* **29**, 125–130.

Yasuhara, A., Otsuki, A. and Fuwa, K. (1977) Photodecomposion of odorous phenols in water. *Chemosphere* 659–664.

Zepp, R.G. and Schlotzhauer, P.F. Photoreactivity of selected aromatic hydrocarbons in water. In: *Polynuclear Aromatic Hydrocarbons*, Jones, P.W. and Leber, P. (eds). Ann Arbor Science Publishers, Inc., Ann Arbor, Michigan, USA, pp. 141–158.

Zevos, N. and Sehensted, K. (1978) Pulse raduolysis of aqueous naphtaline solutions. *J. Phys. Chem.* **82**, 138–141.

APPENDIX 1

Radical reactions

Initiation:

(I) $H_2O_2 \rightarrow 2OH^{\bullet}\ k_{UV,H_2O_2}$	$k_1 = 0.5$ (Baxendale and Willson, 1957)

Hydroxyl-radical propagation and termination steps:

(II) $OH^{\bullet} + H_2O_2 \rightarrow O_2^{-} + H_2O + H^{+}$	$k_2 = 2.7 \times 10^7 M^{-1}s^{-1}$ (Christensen *et al.*, 1982)
(III) $OH^{\bullet} + HO_2^{-} \rightarrow O_2^{-} + OH^{-} + H^{+}$	$k_3 = 7.5 \times 10^9 M^{-1}s^{-1}$ (Christensen *et al.*, 1982)
(IV) $OH^{\bullet} + HCO_3^{-} \rightarrow CO_3^{-\bullet} + H_2O$	$k_4 = 1.5 \times 10^7 M^{-1}s^{-1}$ (Weeks and Rabani, 1966)
(V) $OH^{\bullet} + CO_3^{2-} \rightarrow CO_3^{-\bullet} + OH^{-}$	$k_5 = 4.2 \times 10^8 M^{-1}s^{-1}$ (Weeks and Rabani, 1966)
(VI) $OH^{\bullet} + O_2^{-} \rightarrow OH^{-} + O_2$	$k_6 = 8.0 \times 10^9 M^{-1}s^{-1}$
(VII) $OH^{\bullet} + N \rightarrow$ Products	$k_{N,OH} = 1.2 \times 10^{10} M^{-1}s^{-1}$ (Zevos and Sehensted, 1978) $= 5.0 \times 10^9 M^{-1}s^{-1}$ (Evens *et al.*, 1980).

Superoxide- and carbonate-radical propagation and termination steps:

(VIII) $O_2^{-} + H_2O_2 \rightarrow OH^{\bullet} + OH^{-} + O_2$	$k_8 = 16\,kM^{-1}s^{-1}$ (Lay, 1989)
(IX) $O_2^{-} + HCO_3^{-} \rightarrow HO_2^{-} + CO_3^{-\bullet}$	$k_9 = 1.0 \times 10^6 M^{-1}s^{-1}$ (Lay, 1989)
(X) $O_2^{-} + CO_3^{-\bullet} \rightarrow O_2 + CO_3^{-2}$	$k_{10} = 4.0 \times 10^8 M^{-1}s^{-1}$ (Lay, 1989)
(XI) $2O_2^{-} + 2H_2O \rightarrow H_2O_2 + 2OH^{-} + O_2$	$k_{11} = 5.0 \times 10^7 M^{-1}s^{-1}$ (Lay, 1989)
(XII) $CO_3^{-\bullet} + H_2O_2 \rightarrow HCO_3^{-} + O_2^{-} + H^{+}$	$k_{12} = 8.0 \times 10^5 M^{-1}s^{-1}$ (Behar *et al.*, 1970)

(XIII) $CO_3^{-\bullet} + HO_2^- \rightarrow HCO_3^- + O_2^-$ $\quad k_{13} = 3.0 \times 10^7 M^{-1}s^{-1}$ (Behar *et al.*, 1970)

(XIV) $CO_3^{-\bullet} + N \rightarrow \ ?$ $\quad k_{CO_3} = 3000\,M^{-1}s^{-1}$ (assumed)

Equilibria:

(XV) $H_2O_2 \leftrightarrow HO_2^- + H^+$ $\quad pK_p = 11.8$

(XVI) $H_2CO_3 \leftrightarrow HCO_3^- + H^+$ $\quad pK_{CI} = 6.4$

(XVII) $HCO_3^- \leftrightarrow CO_3^{2-} + H^+$ $\quad pK_{C2} = 10.2$

Apparent pseudo-first-order rate constant of direct photolysis of substance i:

(XVIII) Direct photolysis: N $\rightarrow$ Products $k_{UV,N}$

(XIX) $k_{UV,i} = \Phi I_o(1 - \exp(2.3L\Sigma\varepsilon_j[j]))$ $\quad$ if $i = H_2O_2$ then $\Phi_i = 2\Phi$ ($\Phi_H = 0.5$, primary quantum yield) (Baxendale and Willson, 1957)

(XX) $$f_i = \frac{\varepsilon_i[i]}{\sum_{j=1}^{S} \varepsilon_j[j]}$$ S = number of substances that adsorb UV radiation.

Scheme 1 Mechanism of UV/H_2O_2 AOPs, C = Napthalene.

5

Fenton processes

S. Wadley and T.D. Waite

5.1 INTRODUCTION

The Fenton process, in one or other of its various forms, is being increasingly used in the treatment of contaminated water and soil. The conventional 'dark' Fenton process involves the use of one or more oxidising agents [usually hydrogen peroxide (H_2O_2) and/or oxygen] and a catalyst (a metal salt or oxide, usually iron), while the photo-Fenton (or photo-assisted Fenton) process also involves irradiation with sunlight or an artificial light source, which increases the rate of contaminant degradation by stimulating the reduction of Fe(III) to Fe(II). The reactions produce a range of free radicals, which can react with most organic compounds. Reactions involving the highly reactive hydroxyl radical ($^{\bullet}OH$) (or 'hydroxyl radical-like' species) are the most important, and are characteristic of all advanced oxidation processes (AOPs).

The key Fenton reagents, H_2O_2 and iron salts, are relatively inexpensive. Iron is highly abundant (and may be naturally occurring in the system being treated) and is non-toxic (whereas many other potentially suitable metallic catalysts are

 Advanced Oxidation Processes for Water and Wastewater Treatment. Edited by Simon Parsons. ISBN: 1 84339 017 5

toxic to varying degrees). H_2O_2 is relatively easy to transport and handle, as well as being environmentally benign in diluted form.

The Fenton process (in its unmodified form) is efficient only in the pH range 2–4 and is usually most efficient at around pH 2.8 (Pignatello, 1992). Hence the process is inefficient in the pH range of most natural waters (pH 5–9). This is particularly due to the tendency for ferric oxyhydroxide precipitation (which has a low catalytic activity) to occur at pH >3–4 (depending on the iron concentration).

It has been found that addition of certain organic ligands that can complex Fe(III) enables the process to be carried out at higher pH (Balmer and Sultzberger, 1999). This occurs both because complexation limits the loss of Fe(III) by oxyhydroxide precipitation and (in the case of the photo-Fenton process) because organically complexed Fe(III) is more efficiently photolysed than hydrated or inorganically bound Fe(III) (Andreozzi *et al.*, 1999). The higher the concentration of the organic complexant, the greater the extent of Fe(III) complexation. However, since most organic compounds also react with radicals and other oxidants, increasing the concentration of these organic-complexing agents can also lead to a decrease in the rate of oxidation of the contaminant.

The use of organic compounds that can complex Fe(III) is an important modification of the process because the addition of large amounts of acid to achieve the optimum process efficiency, followed by base to neutralise the water after the oxidation is complete, makes the process unsuitable for many applications, due to the accompanying salinity increases. The organic ligands that can be used are usually at least partially consumed in the process and should be biodegradable. Suitable ligands should bind Fe(III) strongly, but allow transformation of Fe(III) to Fe(II) as a result of reductants (such as superoxide) produced in the process. If the ferric complex is photosensitive, light will enhance the formation of Fe(II) as a result of ligand-to-metal charge-transfer induced reduction of the metal centre. Ligands that have been used for this purpose include oxalic and citric acids. It is advantageous if the compounds are naturally occurring in water or may be added from natural sources.

The use of ligands does not solve the problem of iron remaining in solution after treatment, and their use may prevent Fe precipitation by alkali or coagulant addition (if the ligands have not been degraded by the Fenton process). Significant reduction in dissolved Fe can be achieved by the use of heterogeneous Fenton processes, where the iron remains substantially in the solid phase, either as a mineral or as an adsorbed ion. Most of these heterogeneous processes have the added advantage that they can be operated in close to neutral pH range.

Use of Fenton processes can lead to the complete mineralisation of some organic compounds, converting them to CO_2, H_2O and inorganic ions. However, this would usually involve a large excess of chemicals, often preventing the process from being cost effective, hence only partial degradation usually occurs (Huston and Pignatello, 1999). Partial degradation usually reduces the toxicity of the contaminants and hence increases the biodegradability of the residue. However, it is sometimes possible for the process to generate products with the same or higher toxicity than the parent compound (Fernández-Alba *et al.*, 2002). These degradation products are probably not the same as those formed in the environment by the naturally occurring processes.

Fenton-type processes can take many forms, some of which tend to have low efficiencies and can produce large amounts of ferric oxyhydroxide sludge. Hence careful process optimisation and control are required.

5.2 TYPES OF FENTON PROCESSES

A range of Fenton-type processes and their main mechanisms are discussed here. Identification of the most important reactions often depends on the relative concentrations and nature of the chemicals present in the system, the solution and the intensity and wavelength of any incident light.

When dissolved, Fe(II) and Fe(III) are not organically complexed; they may exist in the form of a range of hydrolysis species or other inorganic complexes, depending on the pH, the concentration of iron and the concentrations of inorganic ligands. For solutions in the pH range 2–7, in the absence of any strong inorganic ligands, the main Fe(II) species will be $Fe^{2+}_{(aq)}$. In the case of ferric iron, the main species will be $Fe^{3+}_{(aq)}$ below about pH 3, $FeOH^{2+}_{(aq)}$ close to pH 3 and $Fe(OH)^{+}_{2\,(aq)}$ between about pH 3 and 7. In the following mechanisms, the notation Fe^{II} and Fe^{III} is typically used when referring to the dissolved ferrous and ferric iron, respectively, without reference to the exact species present.

5.2.1 Homogeneous processes

When all the reactants are present in the dissolved phase, the process is said to be homogeneous. These processes may still, however, lead to the precipitation of some insoluble species (either during the Fenton process or in subsequent treatment stages), such as metal hydroxides, but this is not part of the main process. In the case of electrochemically produced Fenton reagents, the reagents are produced at or from solid electrodes, so the process is classified as heterogeneous even though many of the Fenton reactions themselves take place in the bulk aqueous phase. For heterogeneous processes, although many processes occur at the solid–liquid interface, some important reactions also still take place in the bulk aqueous phase, some of which are Fenton-type reactions.

5.2.1.1 Fenton processes

Fe(II)/H₂O₂/dark Reaction between dissolved Fe(II) and H_2O_2 in acidic aqueous solution leads to oxidation of Fe(II) to Fe(III) and is thought to form the highly reactive hydroxyl radical ($^{\bullet}OH$) [reaction (5.1)]. The reaction is spontaneous and can occur without the influence of light:

$$Fe^{II} + H_2O_2 \rightarrow Fe^{III} + {}^{\bullet}OH + OH^{-} \qquad (5.1)$$

This reaction is often known as the Fenton reaction, after Fenton (1894) who first described the process, but is actually the Haber–Weiss reaction, which is a specific example of the Fenton reaction. It was proposed by Haber and Weiss (1934) and further developed by Barb *et al.* (1951a,b) and Walling (1975).

Several reactions can then take place and are given in Table 5.1 (De Laat and Gallard, 1999). This list includes only reactions taking place between the species in reaction (5.1) and excludes other species that may be present in the solution. These reactions show that iron acts as a catalyst; however, since the reduction of Fe(III) is generally much slower than the oxidation of Fe(II), iron exists mainly in the Fe(III) form in these systems (Pignatello, 1992).

Table 5.1 Major reactions involved in the Fenton process.

$Fe^{II} + H_2O_2 \rightarrow Fe^{III} + {}^{\bullet}OH + OH^-$	(5.1a)
$Fe^{III} + H_2O_2 \rightarrow Fe^{II} + HO_2^{\bullet} + H^+$	(5.2)
$Fe^{II} + {}^{\bullet}OH \rightarrow OH^- + Fe^{III}$	(5.3)
$H_2O_2 + {}^{\bullet}OH \rightarrow HO_2^{\bullet} + H_2O$	(5.4)
$Fe^{II} + HO_2^{\bullet} + H^+ \rightarrow Fe^{III} + H_2O_2$	(5.5)
$Fe^{III} + HO_2^{\bullet} \rightarrow Fe^{II} + O_2 + H^+$	(5.6)

The mechanisms of these reactions are still poorly understood and there is much disagreement in the literature on the exact intermediates that form including whether or not the hydroxyl radical itself forms (Dunford, 2002; Kremer, 1999). Other intermediates that have been proposed include a hydrated $Fe^{II}\cdot H_2O_2$ complex and the ferryl ion ($Fe^{IV}O^{2+}$) (Bossmann *et al.*, 1998; Kremer, 1999; Symons and Gutteridge, 1998). More recent studies (Buda *et al.*, 2001; Ensing *et al.*, 2002) have confirmed the existence of the ferryl ion and shown evidence that disfavours the hydroxyl radical as an intermediate.

Reaction (5.1) could be described in terms of an outer sphere mechanism [reaction (5.7)]:

$$Fe^{II} + H_2O_2 \rightarrow Fe^{III} + [H_2O_2^-] \rightarrow Fe^{III} + {}^{\bullet}OH + OH^- \quad (5.7)$$

However, this outer sphere mechanism is thermodynamically unfavourable and unlikely to occur as shown regardless of how the iron is complexed. There is strong evidence in support of an inner sphere mechanism, which involves a transient ferrous peroxide complex (Winterbourn, 1995). Results of studies of the rate of O_2 evolution on reaction of Fe(II) and H_2O_2 confirm the likely involvement of a ferrous peroxide complex ($Fe^{II}\cdot H_2O_2$) with subsequent formation of the ferryl (Fe^{IV}) species, FeO^{2+} (Kremer, 1999). As shown in Figure 5.1, the mechanism proposed by Kremer (1999) also involves formation of the binuclear species $[FeOFe]^{5+}$.

The FeO^{2+} ion is generally considered to be the active oxidising species (rather than the hydroxyl radical) and may react (in a non-free radical process) both with one-equivalent reducing agents of the type HA:

$$FeO^{2+} + HA \rightarrow A + Fe^{3+} + OH^- \quad (5.8)$$

and two-equivalent reducing agents of the type H_2A:

$$FeO^{2+} + H_2A \rightarrow A + Fe^{2+} + H_2O \quad (5.9)$$

There have also been suggestions that the $\{Fe^{II}\cdot H_2O_2\}$ intermediate may be directly involved in oxidation reactions with the particular role of this species

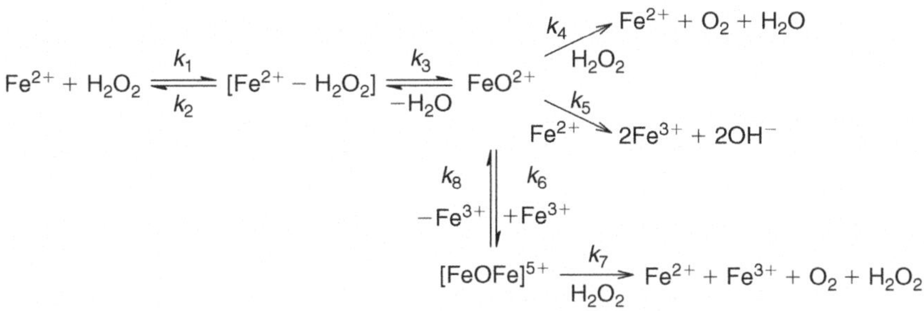

Figure 5.1 Mechanism of Fenton reaction proposed by Kremer (1999).

determined to a large extent by the presence of ligands that mediate transformation of iron between the various states (Wink *et al.*, 1994; Winterbourn, 1995).

Given the uncertainty with regard to the precise mechanism of the Fenton reaction, we will continue to refer to the major reactions as those described in Table 5.1. We will, however, flag where possible that the active species is either the hydroxyl radical or a hydroxyl radical-like species (such as, particularly, the ferryl species FeO^{2+}).

Fe(III)/H_2O_2/dark The iron can initially be in either the Fe(II) or Fe(III) form. The initial degradation rate is slower for Fe(III)/H_2O_2 than for Fe(II)/H_2O_2 because, in the former case, Fe(III) must be reduced to Fe(II) [reaction (5.2)] before hydroxyl radicals (or ferryl species) are produced; however, the overall extent of degradation is independent of the initial iron oxidation state (Safarzadeh-Amiri *et al.*, 1996).

5.2.1.2 Photo-Fenton processes

Fe(II)/H_2O_2/light When irradiated with light of suitable wavelength (180–400 nm), that is, ultraviolet and some visible light, Fe(III) can catalyse the formation of hydroxyl radicals [reaction (5.10)] (or ferryl species):

$$Fe^{III} + H_2O + h\nu \rightarrow Fe^{II} + H^+ + {}^\bullet OH \tag{5.10}$$

This is called the photo-Fenton reaction and is followed by reaction (5.1). Hence, iron becomes cycled between the +II and +III oxidation states. The $^\bullet$OH (or ferryl species) production is determined by the availability of light of suitable wavelength and H_2O_2. In theory, by combination of reactions (5.1) and (5.10), two moles of $^\bullet$OH should be produced per mole of H_2O_2 consumed.

5.2.1.3 Modified photo-Fenton processes

Fe(III)–ligand/H_2O_2/light The quantum yield of Fe(II) production in reaction (5.10) is relatively low and has been estimated by Faust and Hoigné (1990) to be $\Phi_{Fe(II)} = 0.14$ at 313 nm. This quantum yield can increase by an order of magnitude when the Fe(II) is complexed with a carboxylic acid anion, such as oxalate. For example, $\Phi_{Fe(II)} = 1.24$ at 313 nm, with 6 mM ferrioxalate (Murov *et al.*, 1993). The ferrioxalate complex $[Fe(C_2O_4)_3]^{3-}$ is highly photoreactive and the reduction of Fe(III) to Fe(II) can occur at a wavelength further into the visible spectrum (about 550 nm) (Hislop and Bolton, 1999).

The ferrioxalate complex consists of a ferric iron chelated by a varying number ($n = 1$, 2 or 3) of oxalate ions, depending on the concentration of reacting species and the pH. The reaction sequence is initiated by the absorption of light by the ferrioxalate complex, resulting in a ligand-to-metal charge-transfer transition occurring in the complex, which leads to the formation of a charge-transfer state [reaction (5.11)], followed by decomposition of this excited complex, producing Fe(II) and an oxalate radical [reaction (5.12)] (Balmer and Sulzberger, 1999):

$$Fe^{III}(C_2O_4)_n^{(3-2n)} + h\nu \rightarrow Fe^{II}(C_2O_4^{\bullet})_n^{(3-2n)*} \tag{5.11}$$

$$Fe^{II}(C_2O_4^{\bullet})_n^{(3-2n)*} \rightarrow Fe^{II} + (n-1)C_2O_4^{2-} + C_2O_4^{\bullet-} \tag{5.12}$$

The oxalate radical can either just decompose into a carbon dioxide radical and carbon dioxide [reaction (5.13)], or it can react with molecular oxygen to produce the superoxide radical ($O_2^{\bullet-}$), which is in equilibrium with its protonated form ($HO_2^{\bullet}$), and carbon dioxide [reaction (5.14)]:

$$C_2O_4^{\bullet-} \rightarrow CO_2^{\bullet-} + CO_2 \tag{5.13}$$

$$C_2O_4^{\bullet-} + O_2 \rightarrow O_2^{\bullet-} + 2CO_2 \tag{5.14}$$

The superoxide can decompose, forming H_2O_2, by two routes [reactions (5.15) and (5.16)] (Balmer and Sulzberger, 1999):

$$2HO_2^{\bullet}/O_2^{\bullet-} + H^+ \rightarrow H_2O_2 + O_2 \tag{5.15}$$

$$Fe^{II} + HO_2^{\bullet}/O_2^{\bullet-} + H^+ \rightarrow Fe^{III} + H_2O_2 \tag{5.16}$$

$$CO_2^{\bullet-} + Fe^{III}(C_2O_4)_3^{3-} \rightarrow Fe^{II} + CO_2 + 3C_2O_4^{2-} \tag{5.17}$$

When the ferrioxalate is irradiated in the presence of H_2O_2, under ideal conditions, one $^{\bullet}OH$ will be produced for every Fe(II) generated [overall reaction (5.18)]:

$$Fe^{II} + H_2O_2 + 3C_2O_4^{2-} + h\nu \rightarrow Fe^{III}(C_2O_4)_3^{3-} + OH^- + {}^{\bullet}OH \tag{5.18}$$

Here Fe^{II} represents the sum of uncoordinated Fe(II) and $[Fe^{II}C_2O_4]^0$. If a sufficient excess of oxalate is present, Fe(II) will be complexed by either two or three oxalate ions. Since the iron is cycled between the two oxidation states in the same way as in the photo-Fenton reaction, the amount of hydroxyl radicals produced is limited by the availability of light, H_2O_2 and oxalate ions.

For compounds that are resistant to oxidation, such as perchloroalkanes, it is possible to sensitise the removal of the halogen atoms by reduction using carbon dioxide radical anions, which can be generated by irradiation of ferrioxalate in the absence of H_2O_2. The reaction product may then be oxidised by adding H_2O_2 to generate $^{\bullet}OH$ radicals (Huston and Pignatello, 1996).

The initial oxidation of organic compounds usually produces oxygenated intermediates with carboxylic functional groups. These can form complexes with Fe(III) that are photoactive, producing CO_2, organic radicals and Fe(II) on irradiation.

Hence the organic compounds are degraded without the participation of hydroxyl radicals (Safarzadeh-Amiri *et al.*, 1996).

Other ligands that could be used are discussed under Section 5.2.3.5.

Fe(III)–ligand/light A photo-Fenton-type process can be carried out in the absence of added H_2O_2 when dissolved dioxygen is present. Although not strictly a Fenton process, this process has many of the same characteristics and, due to the formation of small amounts of H_2O_2, the Fenton reaction [reaction (5.1)] can proceed in these systems to a significant extent. In natural water exposed to sunlight, low concentrations (up to around 10 μM) of H_2O_2 can be generated (Draper and Crosby, 1983). There is a need for the ligand to be present (in excess over the iron species), otherwise Fe(III) will not remain in solution for sufficiently long periods for the process to take place. Also adequate light irradiation must occur because there is insufficient H_2O_2 present for Fe(II) to be regenerated by reaction (5.2); hence the photolysis reactions [such as reactions (5.10) and (5.11)] are essential. In systems where excess ligand such as oxalate is added, levels of H_2O_2 up to around 80 μM can be attained (Balmer and Sulzberger, 1999).

5.2.1.4 Fenton-like reagents

Numerous transition metal ions and their complexes in their lower oxidation states, such as Cu(I), Cr(II), Co(II) and Ti(III), have the oxidative features of the Fenton reaction. Hence, mixtures of these metal compounds with H_2O_2 are referred to as 'Fenton-like' reagents (Goldstein and Meyerstein, 1999).

5.2.2 Heterogeneous processes

The source of iron used as a catalyst for the Fenton process can be a solid surface, including iron-containing minerals or iron-coated silica particles (sand), as well as iron adsorbed onto zeolites or ion exchange membranes.

A major disadvantage of wastewater treatment by homogeneous Fenton processes is that the iron must usually be removed from the water after treatment is complete. This can usually be achieved by raising the pH of the solution, however this can lead to the production of large volumes of iron-containing sludge. Heterogeneous Fenton processes are of particular interest, since most of the iron remains in the solid phase. The solid phase is usually granular (iron oxides, sand or zeolites) or is a membrane, hence the treated water is easily separated from the iron.

5.2.2.1 Iron oxide catalysts

Heterogeneous Fenton-like processes involving iron oxides can be used for either soil remediation or water treatment. The difference lies only in whether a prepared chemical solution containing the oxidant is added to the soil (usually containing some iron oxides) to be treated or the water to be treated is passed through a prepared catalyst bed containing the iron oxide catalyst in the presence of an oxidant. In either case, high iron content sand should be present; otherwise supplementation of the iron is required, for example, by addition of ferrous sulphate or crushed iron oxide minerals.

Fe(III)oxide/H_2O_2/dark A heterogeneous Fenton process has been patented under the name IROX by Gurol and Lin (1998). The process makes use of a particulate mineral oxide catalyst (size range 10 μm to 5 mm), usually goethite (α-FeOOH) and an oxidant, usually H_2O_2. Alternative Fe(III) oxide catalysts include lepidocrocite (γ-FeOOH), hematite (α-Fe_2O_3), limonite [FeO(OH)·nH_2O], magnetite (Fe_3O_4) as well as other ferric oxides and hydroxy oxides. Goethite is the preferred mineral oxide catalyst mainly because it appears to have the fastest reaction with H_2O_2. Manganese oxides, particularly manganese dioxide (MnO_2), can also be used (Gurol and Lin, 1998). Valentine and Wang (1998) found that goethite had the highest catalytic activity towards quinoline oxidation, while ferrihydrite had the lowest activity of the iron oxides tested. A greatly accelerated decomposition rate of H_2O_2 can take place at neutral pH.

Goethite is a naturally occurring mineral and is widely available. Preparation for use in the process would involve crushing to achieve a suitable particle size range. Goethite may also be prepared synthetically by, for example, reacting a mixture of ferric nitrate and potassium hydroxide and ageing it at an elevated temperature (Schwertmann and Cornell, 1991). The effectiveness of catalysis by the mineral will increase with an increase in surface area. Goethite is almost insoluble in aqueous media over a wide pH range; hence it is able to function as a heterogeneous catalyst without losing a substantial amount of its mass.

According to the patented process (Gurol and Lin, 1998), the catalyst may also be used in combination with another material; for example, granular activated carbon and the catalyst can be prepared by coating a particulate substrate. Suitable substrates include particulate forms of activated carbon, alumina or clay. The particulate catalyst may also be prepared by compacting smaller goethite particles into a larger particulate form.

The oxidising agent can be either H_2O_2 or ozone or combinations of ozone with H_2O_2. H_2O_2 is preferred for treating contaminated water, whereas ozone may be used to treat either contaminated water or gas streams. Further discussion will be limited to the use of H_2O_2. The H_2O_2 is generally used as a 5–35% (by weight) solution to achieve a suitable concentration in the contaminated water. This amount should be slightly in molar excess of the theoretical amount required to achieve the desired extent of oxidisation (Gurol and Lin, 1998).

Oxidation of organic compounds is carried out by the hydroxyl radical (or possibly a ferryl species), which is thought to be produced through the interaction of H_2O_2 with surface sites on the goethite. The oxidation reaction is also thought to take place at these surface sites. This heterogeneous process would appear to have the following advantages:

- The goethite catalyst can be used over extended periods without requiring regeneration or replacement, and can be removed from the treated water by sedimentation or filtration. In the case of the homogeneous Fenton process, the dissolved iron cannot be removed except by precipitation using alkali, producing large volumes of ferric-hydroxide sludge, which is not reusable as a catalyst. Phosphate inhibits the oxidation by preferential adsorption to the surface, hence high phosphate concentrations would limit the life of the catalyst.

- The rate of hydroxyl radical generation increases slightly with increasing pH in the range 5–9, whereas for the homogeneous process the reaction rate decreases sharply in this pH range, with a negligible rate above about pH 4.5, if no complexing agents are used.
- The use of light to regenerate the catalyst is not required.
- The catalytic oxidation efficiency of goethite seems to be relatively insensitive to the inorganic carbonate concentration.
- The catalyst surface has a low affinity for bromide adsorption, hence the production of bromates is minimised.

There are two possibilities for the mechanism of the heterogeneous Fenton-like processes involving the reaction of H_2O_2 with goethite:

1. The Fenton reaction initiated by Fe(II) production due to reductive dissolution of goethite in the presence of H_2O_2, with reprecipitation of Fe(III) as goethite.
2. Catalysis of H_2O_2 decomposition on the goethite surface.

For Case (1), the following steps would be involved (Lu, 2000):

$$\alpha\text{-FeOOH}_{(s)} + 2H^+ + \tfrac{1}{2}H_2O_2 \rightarrow Fe^{II} + \tfrac{1}{2}O_2 + 2H_2O \quad (5.19)$$

$$Fe^{II} + H_2O_2 \rightarrow Fe^{III} + HO^\bullet + OH^- \quad (5.20)$$

$$Fe^{III} + H_2O + OH^- \rightarrow \alpha\text{-FeOOH}_{(s)} + 2H^+ \quad (5.21)$$

For Case (2), a complex series of surface reactions, producing $HO^\bullet$ and $HO_2^\bullet$ radicals, would be involved, of which the most important reactions are (Lin and Gurol, 1998):

$$\equiv Fe^{III}OH + H_2O_2 \rightleftharpoons (H_2O_2)_s \quad (5.22)$$

$$(H_2O_2)_s \rightarrow \equiv Fe^{II} + H_2O + HO_2^\bullet \quad (5.23)$$

$$\equiv Fe^{II} + H_2O_2 \rightarrow \equiv Fe^{III}OH + {}^\bullet OH \quad (5.24)$$

$$HO_2^\bullet \rightleftharpoons H^+ + O_2^{\bullet -} \quad (5.25)$$

$$\equiv Fe^{III}OH + HO_2^\bullet/O_2^{\bullet -} \rightarrow \equiv Fe^{II} + H_2O/OH^- + O_2 \quad (5.26)$$

Case (1) is a rather simplistic approach and would probably require more Fe(II) to be present than could form by the dissolution of goethite in order to obtain the observed rates of reaction. The formation of the hydroxyl radicals occurs by homogeneous catalysis.

Case (2) assumes that no dissolution of goethite occurs, and all reactions occur on the mineral surface. Hence the formation of hydroxyl radicals occurs by heterogeneous catalysis. It is possible that the process occurs by both sets of mechanisms.

Fe(III)oxide/H$_2$O$_2$/light In a study of photo-Fenton degradation of an azo dye using goethite as catalyst, He *et al.* (2002) proposed the following mechanism:

$$\equiv Fe^{III}OH + H_2O_2 \rightarrow \equiv Fe^{III}OOH + H_2O \quad (5.27)$$

$$\equiv Fe^{III}OOH + h\nu \rightarrow \equiv Fe^{IV}{=}O + {}^{\bullet}OH \qquad (5.28)$$

$$\equiv Fe^{IV}{=}O + H_2O \rightarrow \equiv Fe^{III}OH + {}^{\bullet}OH \qquad (5.29)$$

The activity of the surface Fe(III) may be similar to that of a coordinated iron compound in solution (other than hydrolysed Fe(III) itself). UV irradiation of the surface breaks the $\equiv Fe^{III}O{-}OH$ bond producing a surface Fe(IV) species and a hydroxyl radical. The unstable surface Fe(IV) species then reacts with water to form a further hydroxyl radical (He *et al.*, 2002).

5.2.2.2 Supported iron catalysts

Heterogeneous Fenton and photo-Fenton processes using iron ions immobilised on Nafion perfluorinated membranes can be used to degrade organic compounds, including chlorophenols (Maletzky *et al.*, 1999; Sabhi and Kiwi, 2001) and non-biodegradable textile effluents (Balanosky *et al.*, 1999; Fernandez *et al.*, 1998, 1999). Nafion membranes (DuPont 117) are resistant to attack by hydroxyl radicals. These processes can take place in pH neutral solutions with close to equal efficiency as in acidic solutions, which removes the need for costly pH adjustments when subsequent biological degradation is to be used (Balanosky *et al.*, 1999). The reactions of Fe(II) and Fe(III) ions on the membrane surface are equivalent to the reactions of dissolved Fe(II) and Fe(III) ions. The membrane could be reused through many cycles, without leaching out of a significant amount of Fe(III) ions from the membrane, which would cause loss of efficiency (Sabhi and Kiwi, 2001).

Other support media for catalytic ion that have been used include zeolites, such as the Fe-ZSM-5 zeolite (Pulgarin *et al.*, 1995), beads of cross-linked organic material, such as alginate gel beads (Fernandez *et al.*, 2000), modified polyethylene films (Dhananjeyan *et al.*, 2001) and silica/Fe-structured fabrics (Bozzi *et al.*, 2002, 2003).

5.2.2.3 Electrochemical Fenton processes

There are two main types of Fenton process involving the use of *in situ* electrochemically produced reagents. They differ mainly in the form in which the iron enters the system. In cathodic Fenton processes, the iron is added as an Fe(II) or Fe(III) salt, whereas in anodic Fenton processes the source of the iron is a sacrificial iron anode.

Cathodic Fenton processes In cathodic processes, there are several possibilities, in which one or both of the reagents Fe(II) and H_2O_2 are produced *in situ*. The source of Fe(II) may be direct Fe(II) addition or it may be produced by reduction of Fe(III) at the cathode [reaction (5.30)]. The source of H_2O_2 may be either direct H_2O_2 addition or it may be produced by reduction of dioxygen at the cathode [reaction (5.31)] (Ventura *et al.*, 2002):

$$Fe(III) + e^- \rightarrow Fe(II) \qquad (5.30)$$

$$O_2 + 2H^+ + 2e^- \rightarrow H_2O_2 \quad (5.31)$$

Simultaneous Fe(III) reduction and O_2 reduction can take place at the cathode at comparable rates. Since Fe(II) and H_2O_2 are continuously produced at controlled rates, this leads to a more efficient and more complete degradation of the contaminant compared with classic Fenton systems, where all the reagents are typically added at the beginning of the reaction. This is because there are less competitive reactions, which consume the reagents without producing hydroxyl radicals (Ventura *et al.*, 2002).

The electrochemical cell is undivided and has an anode made from inert material, such as platinum, platinised titanium or DSA®. The cathode is made from carbon-containing materials, such as carbon felt (Oturan, 2000) or carbon-polytetrafluoroethylene (Brillas *et al.*, 2000), and is supplied with oxygen.

The most commonly used cathodic Fenton process is the Electro-Fenton® process, developed by Brillas *et al.* (1996). The electro-Fenton systems may also be exposed to irradiation by UV/visible light, leading to photoelectro-Fenton processes (Boye *et al.*, 2002).

Anodic Fenton processes In the anodic Fenton process, an iron electrode is used as the anode and becomes the source of Fe(II). The system consists of two cells, with a salt bridge providing electrical connectivity between the cells and a graphite electrode used as a cathode. Anodic Fenton treatment (AFT) has several advantages over classical Fenton processes (Wang and Lemley, 2001):

1. It occurs under acidic pH conditions (pH 2–3), unlike the cathodic Fenton treatments, which operate under neutral conditions, hence it has a higher efficiency. The pH of the final-treated effluent can be partially neutralised by combining the solutions from the anodic and cathodic half-cells.
2. Since Fe(II) is produced by controlled electrolysis using a sacrificial iron electrode, there is no need to handle Fe(II) salts (which are hydroscopic and easily oxidised) or Fe(III) salts (which are highly corrosive and oxidising).

A modification of the electro-Fenton process, involving the use of a sacrificial iron anode, has been developed by Brillas and Casado (2002). This process (described further in Section 5.4.3) makes use of an undivided electrochemical cell. Since the solution cannot be maintained in the acidic pH range, this process is characterised by precipitation of iron hydroxides.

5.2.2.4 *Coupled Fenton-microbial processes*

One of the concerns with regard to the use of Fenton-type processes is that the water or soil being treated may become sterile, and may be toxic to organisms in subsequent biological-treatment processes due to the low pH conditions and the presence of strong oxidants. According to Howsawkeng *et al.* (2001), simultaneous abiotic–biotic oxidations may exist when H_2O_2 is injected into soils as a microbial oxygen source or as a source of chemical oxidants.

Although H_2O_2 is known to be toxic to microorganisms, bacteria can remain metabolically active in its presence. Howsawkeng *et al.* (2001) showed that while heterotrophic bacterial metabolism can occur in the presence of $^{\bullet}OH$, higher H_2O_2 concentrations (above 4.5 mM) favour $^{\bullet}OH$-mediated abiotic oxidations at the expense of bacterial metabolism.

Although the hydroxyl radicals react with the biomass in these systems, Howsawkeng *et al.* (2001) found that there was a sufficiently high steady-state $^{\bullet}OH$ concentration to react with the probe molecule. During bioremediation, H_2O_2 is injected into the ground as a supply of O_2, which encourages Fenton-like reactions to coexist with the microbial processes. During *in situ* chemical oxidation, where H_2O_2 and catalysts are injected to promote Fenton-like oxidations, microorganisms can survive in the presence of low fluxes of $^{\bullet}OH$ and consume the oxidation products of the contaminants.

Fe(II) and H_2O_2 can be produced by Fe(III)-reducing facultative anaerobic bacteria by alternating between anaerobic and aerobic conditions. This microbially driven Fenton process operates under neutral pH conditions and does not require UV irradiation (McKinzi and DiChristina, 1999).

5.2.3 Effect of system composition and process conditions

Many constituents of the water can affect the reaction rates and reaction stoichiometry due to adsorption to surface active sites on the solid catalyst or through effects on the solution chemistry. Hence a process that has a high efficiency in degrading a contaminant in one type of water may be less efficient in another.

5.2.3.1 pH

The reaction rate of the homogeneous (dark and photo) Fenton process tends to be highest at around pH 3 and decrease with increasing pH. For the case shown in Figure 5.2 (modified photo-Fenton degradation of a dye using oxalate as ligand), the process remained feasible up to pH 6; however at pH 8, negligible degradation took place (Aplin, 2001).

The pH tends to change during the Fenton process, hence pH control is important in most processes.

In a study by Chou *et al.* (2001) in which iron oxyhydroxide and Fenton's reagent were used in a fluidised-bed reactor (FBR), the pH was found to strongly affect the efficiencies of mineralisation and Fe(III) precipitation rates. When treating benzoic acid, they found that the degradation efficiency decreased with increasing pH, while the removal efficiency of total iron increased with pH in the pH range 2.8–4.5.

5.2.3.2 Fe : H_2O_2 ratio

The reaction rate tends to increase with increasing H_2O_2 concentration. However, the ferrous ion and H_2O_2 not only react to form hydroxyl radicals [by reaction (5.1)], but are also scavengers of hydroxyl radicals [by reactions (5.3) and (5.4)]. The ratio of Fe(II) to H_2O_2 should affect the rates of hydroxyl radical production and scavenging. Hence it is important to use the optimum Fe : H_2O_2 ratio. In a study on

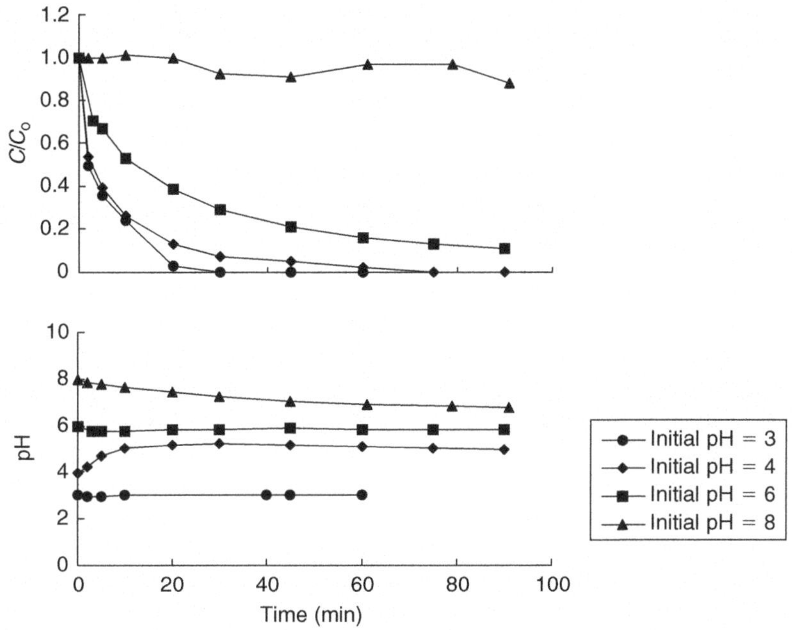

Figure 5.2 Effect of initial pH on degradation of reactive red 235 by a modified photo-Fenton process ($[dye]_0 = 0.1\,g\,L^{-1}$, $[Fe(II)]_0 = 0.1\,mM$, $[oxalate]_0 = 0.3\,mM$, $[H_2O_2]_0 = 1.0\,mM$) (Aplin, 2001).

2,4-D degradation by the anodic Fenton process, Wang and Lemley (2001) found that the optimum Fe(II) : H_2O_2 ratio was 1 : 10. This is close to the theoretical optimum ratio predicted by Tang and Huang (1996) in studies of the dark Fenton process.

5.2.3.3 *Oxygen*

The presence of oxygen reduces the quantum yield of Fe(II) production from Fe(III) oxalates. Balmer and Sulzberger (1999) carried out experiments under nitrogen to remove this effect and also saturated the water with synthetic air to maintain a constant oxygen concentration typical of a natural system. Utset *et al.* (2000) found that some substitution of H_2O_2 by O_2 can occur; however, this limited the extent of degradation of the probe organic.

5.2.3.4 *Inorganic anion concentration*

Degradation rates by a Fenton process can be decreased due to the presence of certain anions that either scavenge hydroxyl radicals or form unreactive complexes with Fe(III). In a study of the degradation of 2,4-dichorophenoxyacetic acid (2,4-D) by Fe(III)/H_2O_2, Pignatello (1992) found that the inhibition of 2,4-D degradation followed the order $SO_4^{2-} \approx Cl^- \gg NO_3^- \approx ClO_4^-$, whereas the inhibition of Fe(III)-catalysed decomposition of H_2O_2 followed the order $SO_4^{2-} \gg Cl^- > NO_3^- \approx ClO_4^-$. It was concluded that sulphate ligands reduced the reactivity of Fe(III), while chloride was responsible for hydroxyl radical scavenging. Scavenging of hydroxyl radicals by chloride was also proposed by Kiwi *et al.* (2000) in a study of the degradation of the azo dye Orange II. Lu *et al.* (1997) found that anions suppressed the degradation of dichlorvos by Fe(II)/H_2O_2 in the order $H_2PO_4^- \gg Cl^- > NO_3^- \approx ClO_4^-$. The inhibition by $H_2PO_4^-$ was attributed to the formation of unreactive Fe(III) complexes.

5.2.3.5 Nature of the ligand

Organic ligands are used as complexing agent for Fe(III) in the modified (photo-) Fenton process. The type of ligand used affects the efficiency of the process. Sun and Pignatello (1992) evaluated a wide range of chelating agents for their chelating ability and catalytic activity in the dark-modified Fenton process and the stability of the chelate. The most catalytically active ligands found in this study included picolinic acid, gallic acid, rhodizonic acid, nitrilotriacetic acid (NTA), hydroxyethyliminodiacetic acid (HEIDA), tetrahydroxy-1,4-quinone and hexaketocyclohexane (HKCH).

When working on degradation of a range of pesticides using picolinic acid, gallic acid and rhodizonic acid as ligands, Sun and Pignatello (1993) found that the efficiency of the process depended on both the type of ligand and type of pesticide being used. These ligands had higher activities at pH 6 (especially for dark Fenton reactions) than oxalate and citrate; however, they tended to be less stable under UV irradiation.

In a study of the degradation of Reactive Red 235 by a modified photo-Fenton process (using a mercury arc lamp, Fe(III), H_2O_2 and a ligand), Aplin (2001) found that the choice of the most effective ligand was dependent on the operating pH. At pH 3, dye degradation was the most rapid for oxalate (and was the same in the absence of ligands). At pH 4.5, degradation was also the most rapid for oxalate, however at pH 6, citrate gave the highest rate. Experiments were also carried out using glucarate, however dye degradation was relatively low due to the low photoactivity of Fe(III)-glucarate complexes. Hence the ligand used must complex Fe(III) sufficiently strongly at the desired operating pH and be strongly photoactive.

In many applications, the oxidation of the chelating ligand may be considered an advantage provided it keeps the iron dissolved and catalytically active long enough for the desired extent of degradation of the contaminant to be achieved. This is because it removes the potential requirement for the ligand to be removed by other processes (Sun and Pignatello, 1992).

5.3 IDENTIFICATION OF DEGRADATION PRODUCTS

Mineralisation of organic compounds by AOPs is seldom complete. If operating conditions exist under which complete mineralisation may occur, the process would be prohibitively expensive. Since hydroxyl radicals (and ferryl species) react non-selectively with organic compounds, a large number of degradation products are formed. It is necessary to identify and quantify these degradation products in order to maximise the process efficiency (Chiron *et al.*, 2000). Since in some cases the degradation products may be more toxic than their parent molecules, it is vital to be able to identify all degradation products.

Suitable methods of simultaneous identification and quantification of degradation products include GC-MS (if the products are sufficiently volatile or if they can be derivativised to form volatile compounds) and HPLC-MS (with an atmospheric pressure ionisation interface) or HPLC-DAD. Many degradation products of a compound have similar structures and are hence difficult to separate by chromatographic techniques, and it is difficult to differentiate between them using their MS or UV–Vis spectra.

5.4 APPLICATIONS

A treatment plant using the Fenton process tends to be less expensive to build and operate than most other AOPs. For example, ozonation requires expensive generators for on-site ozone production, while the titanium dioxide (TiO_2) used in semiconductor-mediated photocatalysis is expensive and difficult to separate from the solution at the end of the treatment (Ormad *et al.*, 2001). The Fenton process requires relatively inexpensive and easily transported chemicals (iron salts and H_2O_2) and, in the case of the photo-Fenton process, a light source which can often be the sun. Hence the Fenton process may be more suitable for on-site treatment processes than some other AOPs.

The operation of Fenton-type processes usually requires strict pH control. In many applications, pH adjustment is required before treatment. Usually an iron sludge forms (after the process is complete and neutralisation is carried out), which requires appropriate disposal (Andreozzi *et al.*, 1999; Venkatadri and Peters, 1993).

Fenton processes have been studied mainly at laboratory scale, with some pilot-scale studies (Venkatadri and Peters, 1993) and a few full-scale applications (Andreozzi *et al.*, 1999).

5.4.1 Possibilities

The Fenton process has a wide range of possible applications. These often involve treatment of a single waste stream, however interesting applications exist which involve coupling of a Fenton process with other water-treatment process (or integration of the process into an existing wastewater-treatment train), and co-treatment of two or more waste streams.

The Fenton reaction also plays an important role in biological systems as it leads to the formation of radicals, which can cause damage in cells. Living systems have protective mechanisms against radicals (and other reactive oxygen species), which are usually referred to as antioxidants. Cells contain ligands to complex any free iron, and hence prevent it from driving oxidative reactions. Iron-mediated reactions in cells can contribute to the cause of many human diseases (Symons and Gutteridge, 1998). Patients with high iron levels are in danger of being affected by these problems. They can be treated using iron-chelating agents that are capable of removing iron from iron stores in the body (Crisponi *et al.*, 1999).

5.4.2 Laboratory-scale studies

In this section, case studies on a range of applications of Fenton-type processes that are still at the experimental stage are discussed. Some of these processes might also have been tested on a pilot scale or be used at industrial-scale plants.

5.4.2.1 Treatment of industrial wastes

Textile dyeing effluent Hydroxyl radicals react with dyes very rapidly; for example, for reactions with the azo dye Orange II, the rate constant is $6.0 \times 10^9\ M^{-1}s^{-1}$ (Kiwi *et al.*, 2000). This reaction destroys the dye chromophore and forms uncoloured hydroxylated products; hence this water can often be reused for dyeing.

Fernandez *et al.* (2000) carried out a study on Fenton-enhanced decolourisation/degradation of the azo dye Orange II using an encapsulated Fe catalyst. The iron was supported on alginate gel beads by complexation with carboxylate groups. The Fe-alginate beads were about 2 mm in diameter and contained highly dispersed Fe species of about 0.5 nm in size. Using energy-dispersive X-ray (EDX) microanalysis, it was shown that most of the iron was on the bead surface. At pH values between 5 and 8, in the presence of H_2O_2 and under visible light irradiation, decolourisation of Orange II took less than 1 h.

Wood pulping effluents De Heredia *et al.* (2001) calculated rate constants for the reaction of a wide range of phenolic compounds with the oxidising species produced in the Fenton system, including 3,4-DHBA (also called protocatechuic acid). They used 0.145 mM Fe(II), 0.5 or 0.25 as much phenolic compound and a 5–40 times stoichiometric excess of H_2O_2. Their rate constant for 3,4-DHBA was $6.7 \times 10^8\ M^{-1}s^{-1}$, the lowest for all the compounds listed. However, the paper does not refer to any complex formation between the phenolic compounds and Fe(III).

Rodriguez *et al.* (1999) stated that 4,5-dimethoxycatechol and 2,5-dimethoxyhydroquinone, which are found in the brown rot fungus, might be ferric chelators and act as electron transport carriers in the Fenton reactions.

Pesticides Pignatello (1992) carried out tests on the dark Fenton process on 2,4-D and 2,4,5-T. It was found that complete mineralisation of 2,4-D and 2,4,5-T could be achieved within 2 h using Fe(III) salts and H_2O_2 in acidic solution, with an optimum pH in the range 2.7–2.8. Oxidation of the herbicides was found to be inhibited by the presence of methanol or chloride ions, since these chemicals scavenge the active oxidant species. Oxidation was also inhibited by sulphate ions, since they complex Fe(III).

Balmer and Sultzberger (1999) carried out tests on the modified Fenton's process (ferrioxalate) in acidic solution (about pH 3.2) on a simulated natural system containing atrazine, without H_2O_2 addition. They found that complete mineralisation of atrazine could be achieved within 125 min in the presence of 0.006 mM total dissolved iron and 0.018 mM oxalate when irradiated with a xenon arc lamp. The time could be reduced to about 20 min by increasing the oxalate concentration by a factor of 10. Complete inhibition of the reaction was obtained when 0.25 mM methanol was present. No degradation was observed when the pH was above 7. In the absence of oxalate, atrazine degradation was considerably slower and did not occur when the pH was above 4.

De Laat *et al.* (1999) compared the rate of oxidation of atrazine (100 $\mu g\,L^{-1}$ at pH 3) using various AOPs [UV/H_2O_2, Fe(III)/UV, Fe(III)/H_2O_2/UV and Fe(II) and Fe(III)/H_2O_2], using UV reactors with low-pressure mercury vapour lamps (emission at 253.7 nm). They found that the rate of degradation of atrazine was more rapid using Fe(III)/UV than using UV/H_2O_2.

In a study on the electro-Fenton and photoelectro-Fenton process (with *in situ* H_2O_2 production) for degrading 2,4-D at pH 3, Brillas *et al.* (2000) found that a 230 ppm solution of 2,4-D with low salt content could be completely mineralised at low current by the photoelectro-Fenton process, while the electro-Fenton process lead to about 90% mineralisation. The rate of loss of 2,4-D itself was the same by both methods, however the higher degradation power of the photo-enhanced form of the process is attributed to the fast photodecomposition of some of the intermediates. These intermediates included chlorinated phenols, such as 2,4-dichlorophenol, 4,6-dichlororesorcinol, chlorohydroquinone and chlorobenzoquinone, as well as short-chain acids, such as glycolic, glyoxylic, maleic, fumaric and oxalic acids.

Chlorophenols Chlorinated aromatic compounds are found in many wastewaters and groundwaters. They tend to be highly toxic and only partially biodegradable. Benitez *et al.* (2001) found that use of the photo-Fenton process enabled partial degradation of a range of chlorophenols. The extent of degradation during the experiments decreased with increasing chlorine substitution of the aromatic ring.

Huang *et al.* (1993) have compared the degradation rates of a large range of chlorophenols by the Fenton process. They have observed that except for tetrachlorophenol and pentachlorophenol (which are more difficult to oxidise), the number of chlorine substitutions does not have a major affect on the degradation rate of the chlorophenol. However, for chlorophenols with the same number of chlorine substitutions the rate of degradation is affected by the position of the chlorine groups on the ring.

Nitrophenols Pulgarin *et al.* (1995) used a heterogeneous photo-Fenton process involving a zeolite-supported iron catalyst (Fe-ZSM-5) to degrade 4-nitrophenol. The 4-nitrophenol (3.6 mM) was photodegraded in a solution containing 1.5% Fe-zeolite and 13 mM H_2O_2 in about 8 h at 35°C, however it required more than 12 h for complete mineralisation to be achieved.

Fuels In a study on the degradation of 1,1-dimethylhydrazine (UDMH), Pestunova *et al.* (2002) compared a range of Cu- and Fe-hydroxide-containing catalysts, supported on oxide carriers and containing less than 2% by weight active metal. UDMH is the main propellant for space rockets and missiles and is a highly toxic substance, with a maximum allowable concentration in water of 0.01 ppm. They found that UDMH could be efficiently oxidised in aqueous solutions by both H_2O_2 and oxygen in the presence of metal-hydroxide catalysts. The pH of the solution affects the mechanism of the process. Under neutral conditions, the main product was found to be the non-toxic 1,1,5,5-tetramethylformazanium cation (at up to 30% yield), which further decomposes to other products with low toxicity. However, in alkaline medium the reaction is not selective, and results in the formation of toxic and stable products, such as DMNA and DMFA (at yields of up to 15%). The Fe-containing catalysts were found to be more resistant to loss of catalytic metal ions than the Cu-containing catalysts, as 0.5% and 30% of the active metal was leached from the catalyst at pH 7, respectively.

5.4.2.2 Soil and groundwater remediation

The use of a Fenton-like process for treatment of organically contaminated soils makes use of the iron oxides naturally present in the soil or may require addition of dissolved iron salts or solid iron oxide particles. In this process, the amount of H_2O_2 used largely determines the operating cost of the process, hence it is important to establish the correct dosage. The efficiency of the process tends to be lower than in aqueous systems due to the effect of the components of the soil matrix. The concentration of the added H_2O_2 depends on the amount of water present, the amount of organic contaminant and the degree of treatment required.

Soil type affects the process in the following ways:

1. *Mineral forms*: Crystalline soils, especially those containing goethite or hematite, are the best catalysts. However, amorphous soils (such as clays) are less effective.
2. *Buffering capacity*: If the carbonate alkalinity of waters is high, the pH will be high and a large amount of mineral acid will be required to decrease the pH sufficiently. If the alkalinity is low, the organic acids produced by the reaction will be sufficient to provide a low pH. In addition, carbonate ions are strong free radical scavengers and hence will retard the Fenton reaction.
3. *Natural organic matter content*: The higher the natural organic matter concentration, the more hydroxyl radicals will be used in degrading this material with less available to degrade the contaminant of concern.

During soil remediation by application of Fenton's reagent, precipitation of the mineral goethite can occur at near neutral pH, which leads to the acceleration of surface catalysed H_2O_2 decomposition as well as blocking of the entry of the reagents into the target zone. Chelating agents, such as citric acid, can be used to control precipitate formation and hence decrease the amount of H_2O_2 required. A study by Rodgers *et al.* (2001) found that, in the pH range 5.3–7.8, the best H_2O_2 utilisation ratio was obtained for Fe : citrate in the ratio 1 : 3.

A process similar to photo-Fenton process has been developed by Hug *et al.* (2001) to treat groundwater contaminated with arsenic. The process oxidises As(III) to As(IV) and then As(V). The process can take place at a pH close to that of the groundwater, usually 6.5–8. In this process, citric acid (in the form of lemon juice) is added to the water, which is then exposed to sunlight. No iron is usually added, because Fe(II) is naturally occurring in As-containing groundwater, however Fe(II) or Fe(III) salts could be added. An Fe(III)-citrate complex is formed which is photoactive. No H_2O_2 is added, however this is assumed to form in small amounts in the system from superoxide radicals, which are generated from a follow-up reaction to Fe(III)-citrate photolysis. Citrate is consumed in the photolysis reactions and once it is sufficiently depleted, Fe(III)-oxyhydroxide precipitates form, induced by the photoproduct of citrate (3-oxoglutaric acid). The As(V) adsorbs onto the precipitates, and can be removed from the system by settling and decantation. The process can be carried out using readily available non-toxic substances (such as lemon juice), sunlight and simple reactors (transparent plastic bottles) and is hence suitable for use on a small scale in developing countries.

5.4.2.3 Pretreatment to biodegradation

Water contaminated with non-biodegradable or toxic organic compounds can be pretreated using the (photo-)Fenton process prior to biological treatment. This can lead to the degradation of these compounds into more biodegradable compounds. Hence biological activity can be increased in water or soil after the Fenton process. However, conditions during the Fenton process are very aggressive due to the presence of excess H_2O_2 and other oxidants and low pH ($pH < 4$) is usually encountered, hence the medium often becomes sterilised. Disinfection is an advantage where pathogens are present, however if subsequent biodegradation is required, the water must be neutralised. The hydroxyl radical (and other radicals) have short half-lives, hence this activity is expected to cease, while the water is still contained.

In many cases, it is necessary to innoculate the soil or water with suitable organisms after treatment. This can be done by mixing in uncontaminated soil or water containing the desired organisms. If conditions are good, the biological population will become as high, or more commonly higher, than before due to the removal of toxic materials.

Miller *et al.* (1996) have found that certain species of bacteria (particularly *Pseudomonas*) will proliferate in the treated soil or water. The presence of these species is desirable, since they enhance the remediation of contaminated soils. The diversity of heterotrophic bacteria is decreased due to sterilisation, however this enables species that are usually overshadowed to proliferate once the oxidant has decayed.

The right balance between treatment times in the Fenton process and the biological process (relatively lower cost) should be determined to optimise the process (Pulgarin *et al.*, 1999). If the Fenton process does not proceed far enough, the water will still be toxic to the biological process, however increasing the time in the Fenton stage increases the cost of treatment more than an equivalent time in the biological stage. Pulgarin *et al.* (1999) found that the problem of high residual H_2O_2 in the Fenton-treated water could be overcome by operating in the semi-continuous mode.

5.4.2.4 Pretreatment to coagulation

The iron used in the Fenton process may be used to replace some of the ferric iron that is dosed during coagulation (Dolejs *et al.*, 1994). Lower sludge production than in conventional water-treatment processes means that Fenton treatment may be feasible for treating water that is contaminated by organic compounds using existing coagulation–flocculation facilities. However, the low pH requirements of Fenton processes may preclude this due to possible corrosion of the plant.

In addition to oxidation of organic compounds and coagulation, use of the Fenton process can cause polymerisation of certain organic compounds (Chang *et al.*, 2002). Coagulation (due to the formation of Fe(III) hydroxides) can assist in the removal of all solutes. If the system is operated as a batch process, formation of Fe(III) hydroxides before oxidation of the contaminant has proceeded far enough must be avoided. Polymerisation only occurs with certain compounds, where reaction with radicals forms intermediates that can combine to form dimers or polymers.

5.4.2.5 Co-treatment of wastes

Some wastes are difficult to treat by Fenton-type processes without addition of other chemicals, such as acids or complexing agents. Co-treatment of waste streams should be considered when one waste contains substances that may assist in the Fenton treatment of another waste. For example, Rivas *et al.* (2002) have recommended that some types of foodstuff wastes containing phenolic substances be combined with wastewater containing pesticides. The phenolic substances form a complex with Fe(III), producing a more effective catalyst.

Other wastes that could be used include those containing aliphatic organic acids (such as citric acid) from food-processing industries. The use of these compounds may make the final effluent after Fenton treatment more amenable to further degradation by microorganisms.

5.4.3 Pilot-plant trials

Chou *et al.* (2001) have carried out pilot-plant studies on the treatment of dying–finishing wastewater and benzoic acid using supported iron oxyhydroxide and Fenton's reagent in a FBR of height 7 m and diameter 1 m. The process involves the simultaneous mineralisation of organic compounds and precipitation of Fe(III). The FBR-Fenton method was shown to be better than the conventional Fenton treatment of benzoic and textile dying–finishing wastewater in terms of mineralisation and Fe(III) precipitation rates. This method can reduce the sludge production compared with the conventional Fenton's method by more than 60%.

Brillas and Casado (2002) have carried out pilot-scale studies on degradation of aniline by the Electro-Fenton® and Peroxi-coagulation processes at constant currents up to 20 A. They used 1000 ppm solutions of aniline in 50 mM Na_2SO_4/H_2SO_4 solution at pH 3 and 40°C. The reactor consisted of a filter-press cell operated in recirculation mode and contained an anode and an oxygen diffusion cathode (made of carbon-PTFE), each of area 0.010 m^2. For the Electro-Fenton Process®, the anode was a platinised titanium mesh or a DSA® plate. For the Peroxi-coagulation process, the anode was an iron plate. For the electro-Fenton process [using 1 mM Fe(II)], 61% degradation of aniline was achieved after 2 h at 20 A. The Peroxi-coagulation process appeared to have a higher degradation power, enabling removal of more 95% of dissolved aniline at 20 A, because some intermediates coagulate with the ferric-hydroxide precipitate that forms. Both these processes have moderate energy costs, which increase with increasing electrolysis time and applied current.

5.4.4 Full-scale applications

There are several full-scale plants in South Africa that use Fenton's reagent to treat wastewater from the textile industry (Vandevivere *et al.*, 1998) and commercial-scale photo-Fenton installations exist in the USA for treatment of water contaminated with volatile organic compounds (VOC) and semivolatile organic compounds (SVOC) (USEPA, 1998). The Calgon Rayox® ENOX water-treatment system is an example of a commercial-scale photo-Fenton system, which has been used for treatment of contaminated groundwater and industrial wastewater.

A variety of patented technologies involving the Fenton reaction are available and used by companies in remediation of contaminated soil and groundwater. The more widely used technologies include:

- The Geo-Cleanse Process® (*in situ* treatment of subsurface environments) by Geo-Cleanse International, Inc. (http://www.geocleanse.com/).
- The CleanOX Process (*in situ* chemical oxidation) by ManTec, Inc. (MECx) (http://www.mecx.net/).
- ISOTECH℠ Modified Fenton's Process by In-Situ Oxidative Technologies, Inc. (http://www.insituoxidation.com/).
- The BIOX® Process (*in situ* coupled chemical oxidation–enhanced biodegradation) by BioManagement Services, Inc. (BMS).
- The On-Contact Remediation Process® by Environmental Business Solutions International, Inc. (EBSI) (http://www.ebsi-inc.com/).

The processes by Geo-Cleanse, CleanOX, ISOTECH and BIOX are *in situ* processes in which H_2O_2 and a catalyst (a proprietary mixture of non-hazardous metallic salts) are injected into the subsurface environment via various techniques. The remediation processes usually take between several hours and several months to complete.

One of the disadvantages of *in situ* chemical oxidation is the violent exothermic reaction that may occur on injection of reagents. In addition, since the Fenton reaction performance is optimal at around pH 3, many applications are precluded due to the presence of alkaline rock, which would neutralise the added acid. Companies appear to have overcome these problems to some extent by using reagents (typically iron-complexing agents of some form) that allow operation in the near to neutral pH range and with minimal temperature change.

The BIOX® Process takes place at neutral pH and ambient groundwater temperature. The process applies a proprietary reagent (containing H_2O_2, iron sulphate, calcium or magnesium oxide and other substances) to the soil. It does not involve the use of injection wells or acid addition. The organic degradation products from the process (which are rendered less toxic) are subsequently consumed by the naturally occurring soil bacteria (Holish *et al.*, 2000).

The ISOTECH^{SM} Modified Fenton's Process uses a proprietary catalytic agent to delay the formation of hydroxyl radicals to allow adequate dispersion of the reagent through the contaminant plume. The catalyst contains an iron-complexing agent that keeps the iron from precipitating and enables operation at circumneutral pH (Greenberg *et al.*, 1998). Kakarla *et al.* (2002) report results full-scale field evaluation of the technology at a former dry-cleaning facility in southern Florida in which a 72% reduction in total chlorinated hydrocarbon contamination occurred following an initial injection with an increase to 90% removal on second injection.

The Geo-Cleanse process operates under slightly acidic conditions (pH 4–6) and is not recommended for groundwater with a pH above 8 or for soil-containing peat. When the chemical oxidation process is finished, naturally occurring microorganisms are used to complete the site remediation (Casey and Bergren, 1999). Numerous examples of the application of Fenton's reagent using the Geo-Cleanse Process are available from Geo-Cleanse International, Inc. and include full-scale treatment of manufactured gas plant hydrocarbons (BTEXM and PAHs), explosives including RDX and TNT and DNAPLs, such as TCE and PCE in soils and groundwaters.

REFERENCES

Andreozzi, R., Caprio, V., Insola, A. and Marotta, R. (1999) Advanced oxidation processes (AOP) for water purification and recovery. *Catal. Today* **53**, 51–59.

Aplin, R. (2001) *Degradation of Reactive Dyes by a Modified Photo-Fenton Process.* PhD Thesis, School of Civil and Environmental Engineering, The University of New South Wales, Australia.

Balanosky E., Fernandez J., Kiwi, J. and Lopez, A. (1999) Degradation of membrane concentrates of the textile industry by Fenton like reactions in iron-free solutions at biocompatible pH values (pH $\approx$ 7–8). *Water Sci. Technol.* **40**(4–5), 417–424.

Balmer, M.E. and Sultzberger, B. (1999) Atrazine degradation in irradiated iron/oxalate systems: effects of pH and oxalate. *Environ. Sci. Technol.* **33**, 2418–2424.

Barb, W.G., Baxendale, J.H., George, P. and Hargrave, K.R. (1951a) Reactions of ferrous and ferric ions with hydrogen peroxide. Part I – The ferrous ion reaction. *Trans. Faraday Soc.* **47**, 462–500.

Barb, W.G., Baxendale, J.H., George, P. and Hargrave, K.R. (1951b) Reactions of ferrous and ferric ions with hydrogen peroxide Part II – The ferric ion reaction. *Trans. Faraday Soc.* **47**, 591–616.

Benitez, F.J., Beltran-Heredia, J., Acero, J.L. and Rubio, F.J. (2001) Oxidation of several chlorophenolic derivatives by UV irradiation and hydroxyl radicals. *J. Chem. Technol. Biotechnol.* **76**, 312–320.

Bossmann, S.H., Oliveros, E., Göb, S., Siegwart, S., Dahlen, E.P., Payawan Jr., L., Straub, M., Wörner, M. and Braum, A.M. (1998) New evidence against hydroxyl radicals as reactive intermediates in the thermal and photochemically enhanced Fenton reactions. *J. Phys. Chem. A* **102**, 5542–5550.

Boye, B., Dieng, M.M. and Brillas, E. (2002) Degradation of herbicide 4-chlorophenoxyacetic acid by advanced electrochemical oxidation methods. *Environ. Sci. Technol.* **36**, 3030–3035.

Bozzi, A., Yuranova, T., Mielczarski, J., Lopez, A. and Kiwi, J. (2002) Abatement of oxalates catalyzed by Fe–silica structured surfaces via cyclic carboxylate intermediates in photo-Fenton reactions. *Chem. Commun.* 2202–2203.

Bozzi, A., Yuranova, T., Mielczarski, E., Mielczarski, J., Buffat, J., Lais, P. and Kiwi, J. (2003) Superior biodegradability of waste waters mediated by immobilized Fe-fabrics compared to Fenton homogeneous reactions. *Appl. Catal. B* **42**, 289–303.

Brillas, E. and Casado, J. (2002) Aniline degradation by Electro-Fenton® and peroxi-coagulation processes using a flow reactor for wastewater treatment. *Chemosphere* **47**(3), 241–246.

Brillas, E., Mur, E. and Casado, J. (1996) Iron(II) catalysis of the mineralization of aniline using a carbon-PTFE O_2-fed cathode. *J. Electrochem. Soc.* **143**(3), L49–L53.

Brillas, E., Calpe, J.C. and Casado, J. (2000) Mineralisation of 2,4-D by advanced electrochemical oxidation processes. *Water Res.* **34**, 2253–2262.

Buda, F., Ensing, B., Gribnau, M.C.M. and Baerends, E.J. (2001) DFT study of the active intermediate in the Fenton reaction. *Chem. Eur. J.* **7**(13), 2775–2783.

Casey, C. and Bergren, C. (1999) Chemical oxidation, natural attenuation drafted in Navy cleanup. *Pollut. Eng.* March 1999 (Suppl).

Chang, C.-Y., Wang, C.-C., Chang, D.-J. and Chang, J.-S. (2002) Combined Fenton-MF process increases acrylonitrile removal. In: *Proceedings of the IWA 3rd World Water Congress*, Melbourne, April 7–12.

Chiron, S., Fernandez-Alba, A., Rodriguez, A. and Garcia-Calvo, E. (2000) Pesticide chemical oxidation: state-of-the-art. *Water Res.* **34**, 366–377.

Chou, S., Huang, Y.-H., Lin, H.-L. and Liao, C.-C. (2001) Treatment of dying–finishing wastewater using iron oxyhydroxide and Fenton's reagent in a fluidised bed reactor. In: *The Seventh International Conference on Advanced Oxidation Technologies for Water and Air Remediation*, Niagara Falls, Ontario, Canada, June 25–29, pp. 89–91.

Crisponi, G., Nurchi, V.M., Silvagni, R. and Faa, G. (1999) Oral iron chelators for clinical use. *Polyhedron* **18**, 3219–3226.

De Heredia, J.B., Torregrosa, J., Dominguez, J.R. and Peres, J.A. (2001) Kinetic model for phenolic compound oxidation by Fenton's reagent. *Chemosphere* **45**, 85–90.

De Laat, J. and Gallard, H. (1999) Catalytic decomposition of hydrogen peroxide by Fe(III) in homogeneous aqueous solution: mechanism and kinetic modelling. *Environ. Sci. Technol.* **33**, 2726–2732.

De Laat, J., Gallard, H., Ancelin, S. and Legube, B. (1999) Comparative study of the oxidation of atrazine and acetone by H_2O_2/UV, Fe(III)/UV, Fe(III)/H_2O_2/UV and Fe(II) or Fe(III)/H_2O_2. *Chemosphere* **39**(15), 2693–2706.

Dhananjeyan, M.R., Mielczarski, E., Thampi, K.R., Buffaat, Ph., Bensimon, M., Kulik, A., Mielczarski, J. and Kiwi, J. (2001) Photodynamics and surface characterization of TiO_2 and Fe_2O_3 photocatalysts immobilized on modified polyethylene films. *J. Phys. Chem. B* **105**, 12046–12055.

Dolejs, P., Kalousková, N., Prados, M. and Legube, B. (1994) Coagulation of humic water with Fenton reagent. In: *Chemical Water and Wastewater Treatment III*, Klute, R. and Hahn, H.H. (eds). Springer-Verlag, Berlin, pp. 117–129.

Draper, W.M. and Crosby, D.G. (1983) The photochemical generation of hydrogen peroxide in natural waters. *Arch. Environ. Contam. Toxicol.* **12**, 121–126.

Dunford, H.B. (2002). Oxidations of iron(II)/(III) by hydrogen peroxide: from aquo to enzyme. *Coord. Chem. Rev.* **233–234**, 311–318.

Ensing, B., Buda, F., Blöchl, P.E. and Baerends, E.J. (2002) A Car-Parrinello study of the formation of oxidising intermediates from Fenton's reagent in aqueous solution. *Phys. Chem. Chem. Phys.* **4**, 3619–3627.

Faust, B.C. and Hoigné, J. (1990) Photolysis of iron(III)-hydroxy complexes as sources of hydroxyl radicals in clouds, fog and rain. *Atmos. Environ.* **24A**(1), 79–89.

Fenton, H.J.H. (1894) Oxidation of tartaric acid in the presence of iron. *Chem. Soc. J. Lond.* **65**, 899–910.

Fernández-Alba, A.R., Hernando, D., Agüera, A., Cáceres, J. and Malato, S. (2002) Toxicity assays: a way for evaluating AOPs efficiency. *Water Res.* **36**, 4255–4262.

Fernandez, J., Bandara, J., Lopez, A., Albers, P. and Kiwi, J. (1998) Efficient photo-Fenton catalysis mediated by Fe ions on Nafion membranes active in the abatement of non-biodegradable azo-dye. *Chem. Commun.* **14**, 1493–1494.

Fernandez, J., Bandara, J., Lopez, A., Buffat, P. and Kiwi, J. (1999) Photoassisted Fenton degradation of nonbiodegradable azo dye (Orange II) in Fe-free solutions mediated by cation transfer membranes. *Langmuir* **15**, 185–192.

Fernandez, J., Dhananjeyan, M.R., Kiwi, J., Senuna, Y. and Hilborn, J. (2000) Evidence for Fenton photoassisted processes mediated by encapsulated Fe ions at biocompatible pH values. *J. Phys. Chem. B* **104**, 5298–5301.

Goldstein, S. and Meyerstein, D. (1999) Comments on the mechanism of the 'Fenton-like' reaction. *Acc. Chem. Res.* **32**(7), 547–550.

Greenberg, R.S., Andrews, T., Kakarla, P.K.C. and Watts, R.J. (1998) In-situ Fenton-like oxidation of volatile organics: laboratory, pilot, and full-scale demonstrations. *Remediation* **8**(2), 29–42.

Gurol, M.D. and Lin, S. (1998) *Continuous Catalytic Oxidation Process*. US Patent No. 5,755,977.

Haber, F. and Weiss, J.J. (1934) The catalytic decomposition of hydrogen peroxide by iron salts. *Proc. Roy. Soc. (London), Ser. A.* **147**, 332–345.

He, J., Tao, X., Ma, W. and Zhao, J. (2002) Heterogeneous photo-Fenton degradation of an azo dye in aqueous H_2O_2/iron oxide dispersions at neutral pHs. *Chem. Lett.* (1), 86–87.

Hislop, K.A. and Bolton, J.B. (1999) The photochemical generation of hydroxyl radicals in the UV-vis/ferrioxalate/H_2O_2 system. *Environ. Sci. Technol.* **33**, 3119–3126.

Holish, L.L., Lundy, W.L. and Nuttall, H.E. (2000) Indiana 'land banned' herbicide release. *Soil Sed. Groundwater* **October/November**, 14–18.

Howsawkeng, J., Watts, R.J., Washington, D.L., Teel, A.L., Hess, T.F. and Crawford, R.L. (2001) Evidence for simultaneous abiotic–biotic oxidations in a microbial-Fentons's system. *Environ. Sci. Technol.* **35**, 2961–2966.

Huang, C.P., Dong, C. and Tang, Z. (1993) Advanced chemical oxidation: its present role and potential future in hazardous waste treatment. *Waste Manage.* **13**, 361–377.

Hug, S.J., Canonica, L., Wegelin, W., Gechter, G. and von Gunten, U. (2001) Solar oxidation and removal of arsenic at circumneutral pH in iron containing waters. *Environ. Sci. Technol.* **35**, 2114–2121.

Huston, P.L. and Pignatello, J.J. (1999) Degradation of selected pesticide active ingredients and commercial formulations in water by the photo-assisted Fenton reaction. *Water Res.* **33**(5), 1238–1246.

Huston, P.L. and Pignatello, J.J. (1996) Reduction of perchloroalkanes by ferrioxalate-generated carboxylate radical preceding mineralization by the photo-Fenton reaction. *Environ. Sci. Technol.* **30**(12), 3457–3463.

Kakarla, P.K., Andrews, T., Greenberg, R.S. and Zervas, D.S. (2002) Modified Fenton's processes for effective in-situ chemical oxidation – laboratory and field evaluation. *Remediation* **12**, 23–36.

Kiwi, J., Lopez, A. and Nadtochenko, V. (2000) Mechanism and kinetics of the OH-radical intervention during Fenton oxidation in the presence of a significant amount of radical scavenger (Cl^-). *Environ. Sci. Technol.* **34**(11), 2162–2168.

Kremer, M.L. (1999) Mechanism of the Fenton reaction. Evidence for a new intermediate. *Phys. Chem. Chem. Phys.* **1**, 3595–3605.

Lin, S. and Gurol, M.D. (1998) Catalytic decomposition of hydrogen peroxide on iron oxide: kinetics, mechanism, and implications. *Environ. Sci. Technol.* **32**(10), 1417–1423.

Lu, M. (2000) Oxidation of chlorophenols with hydrogen peroxide in the presence of goethite. *Chemosphere* **40**, 125–130.

Lu, M.-C., Chen, J.-N. and Chang, C.-P. (1997) Effect of inorganic ions on the oxidation of dichlorvos insecticide with Fenton's reagent. *Chemosphere* **35**(10), 2285–2293.

Maletzky, P., Bauer, R., Lahnsteiner, J. and Pouresmael, B. (1999) Immobilisation of iron ions on Nafion and its applicability to the photo-Fenton method. *Chemosphere* **38**(10), 2315–2325.

McKinzi, A.M. and DiChristina, T.J. (1999) Microbially driven Fenton reaction for transformation of pentachlorophenol. *Environ. Sci. Technol.* **33**(11), 1886–1891.

Miller, C.M., Valentine, R.L., Roehl, M.E. and Alvarez, P.J.J. (1996) Chemical and microbiological assessment of pendimethalin-contaminated soil after treatment with Fenton's reagent. *Water Res.* **30**(11), 2579–2586.

Murov, S.L., Carmichael, I. and Hug, G.L. (1993) *Handbook of Photochemistry*, 2nd edn. Marcel Dekker, New York, pp. 299–305.

Ormad, M.P., Ovelleiro, J.L. and Kiwi, J. (2001) Photocatalytic degradation of concentrated solutions of 2,4-dichlorophenol using low energy light. Identification of intermediates. *Appl. Catal. B: Environ.* **32**(3), 157–166.

Oturan, M.A. (2000) An ecologically effective water treatment technique using electrochemically generated hydroxyl radicals for *in situ* destruction of organic pollutants: application to herbicide 2,4-D. *J. Appl. Electrochem.* **30**, 475–482.

Pestunova, O.P., Elizarova, G.L., Ismagilov, Z.R., Kerzhentsev, M.A. and Parmon, V.N. (2002) Detoxication of water containing 1,1-dimethylhydrazine by catalytic oxidation with dioxygen and hydrogen peroxide over Cu- and Fe-containing catalysts. *Catal. Today* **75**, 219–225.

Pignatello, J.J. (1992) Dark and photoassisted Fe^{3+}-catalysed degradation of chlorophenoxy herbicides by hydrogen peroxide. *Environ. Sci. Technol.* **26**(5), 944–951.

Pulgarin, C., Peringer, P., Albers, P. and Kiwi, J. (1995) Effect of Fe-ZSM-5 zeolite on the photochemical and biochemical degradation of 4-nitrophenol. *J. Mol. Catal. A: Chem.* **95**(1), 61–74.

Pulgarin, C., Invernizzi, M., Parra, S., Sarria, V., Polania, R. and Péringer, P. (1999) Strategy for the coupling of photochemical and biological flow reactors useful in mineralization of biorecalcitrant industrial pollutants. *Catal. Today* **54**, 341–352.

Rodgers, P., Bachas, L. and Bhattacharyya, D. (2001) Oxidation of organic contaminants in groundwater by chelate-modified Fenton's regent. In: *The First International Conference on Oxidation and Reduction Technologies for In-Situ Treatment of Soil and Groundwater*, Niagara Falls, Ontario, Canada, June 25–29, pp. 82–83.

Rodriguez, J., Contreras, D., Parra, C., Freer, J., Baeza, J. and Durán, N. (1999) Pulp mill effluent treatment by Fenton-type reactions catalyzed by iron complexes. *Water Sci. Technol.* **40**, 351–355.

Sabhi, S. and Kiwi, J. (2001) Degradation of 2,4-dichlorophenol by immobilised iron catalysts. *Water Res.* **35**(8), 1994–2002.

Safarzadeh-Amiri, A., Bolton, J.R. and Cater, S.R. (1996) The use of iron in advanced oxidation processes. *J. Adv. Oxid. Technol.* **1**(1), 18–26.

Schwertmann, U. and Cornell, R.M. (1991) *Iron Oxides in the Laboratory: Preparation and Characterization.* VCH Verlagsgessellschaft, Weinheim, p. 64.

Sun, Y. and Pignatello, J.J. (1992) Chemical treatment of pesticide wastes. Evaluation of Fe(III) chelates for catalytic hydrogen peroxide oxidation of 2,4-D at circumneutral pH. *J. Agric. Food. Chem.* **40**(2), 322–327.

Sun, Y. and Pignatello, J.J. (1993) Activation of hydrogen peroxide by iron(III) chelates for abiotic degradation of herbicides and insecticides in water. *J. Agric. Food. Chem.* **41**(2), 308–312.

Symons, M.C.R. and Gutteridge, J.M.C. (1998) *Free Radicals and Iron: Chemistry, Biology, and Medicine.* Oxford University Press, Oxford, UK.

Tang, W.Z. and Huang, C.P. (1996) 2,4-Dichlorophenol oxidation kinetics by Fenton's reagent. *Environ. Technol.* **17**, 1371–1378.

USEPA (1998) *Handbook on Advanced Photochemical Oxidation Processes.* EPA/625/R-98/004, United States Environmental Protection Agency, December 1998.

Utset, B., Garcia, J., Casado, J., Domènech, X. and Peral, J. (2000) Replacement of H_2O_2 by O_2 in Fenton and photo-Fenton Reactions. *Chemosphere* **41**, 1187–1192.

Valentine, R.L. and Wang, H.C.A. (1998) Iron oxide surface catalysed oxidation of quinoline by hydrogen peroxide. *J. Environ. Eng.* **124**(1), 31–38.

Vandevivere, P.C., Bianchi, R. and Verstraete, W. (1998) Treatment and reuse of wastewater from the textile wet-processing industry: review of emerging technologies. *J. Chem. Technol. Biotechnol.* **72**, 289–302.

Venkatadri, R. and Peters, R.W. (1993) Chemical oxidation technologies: ultraviolet light/ hydrogen peroxide, Fenton's reagent, and titanium dioxide-assisted photocatalysis. *Haz. Waste Haz. Mat.* **10**(2), 107–149.

Ventura, A., Jacquet, G., Bermond, A. and Camel, V. (2002) Electrochemical generation of the Fenton's reagent: application to atrazine degradation. *Water Res.* **36**, 3517–3522.

Walling, C. (1975) Fenton's reagent revisited. *Acc. Chem. Res.* **8**, 129–131.

Wang, Q. and Lemley, A.T. (2001) Kinetic model and optimization of 2,4-D degradation by anodic Fenton treatment. *Environ. Sci. Technol.* **35**, 4509–4514.

Wink, D.A., Wink, C.B., Nims, R.W. and Ford, P.C. (1994). Oxidizing intermediates generated in the Fenton reagent: kinetic arguments against the intermediacy of the hydroxyl radical. *Environ. Health Perspect.* **102**(Suppl 3), 11–15.

Winterbourn, C.C. (1995) Toxicity of iron and hydrogen peroxide: the Fenton reaction. *Toxicol. Lett.* **82/83**, 969–974.

6

Semiconductor photocatalysis

Andrew Mills and Soo-Keun Lee

6.1 INTRODUCTION

The last decade has seen the emergence of a major new initiative in the area of water and wastewater treatment, namely semiconductor photocatalysis. A number of major reviews of the academic literature associated with this field have been reported elsewhere (Bhatkhande *et al.*, 2001; Hoffmann *et al.*, 1995; Linsebigler *et al.*, 1995; Mills and Le Hunte, 1997; Ollis and Turchi, 1990; Ollis *et al.*, 1991). There is also the recently published, excellent online book entitled '*Solar Detoxification*' (by semiconductor photocatalysis) (Galvez and Rodriguez, 2002). All these sources and many more have been used to create this introduction to semiconductor photocatalysis for water purification, which begins with the following outline of process fundamentals.

6.2 PROCESS FUNDAMENTALS

In semiconductor photocatalysis, the light-absorbing species is a semiconducting material. The electronic structure of most semiconductor materials comprises a

 Advanced Oxidation Processes for Water and Wastewater Treatment. Edited by Simon Parsons. ISBN: 1 84339 017 5

highest occupied band full of electrons called the valance band (VB), and a lowest unoccupied band called the conductance band (CB). These bands are separated by a region that is largely devoid of energy levels, and the difference in energy between the two bands is called the bandgap energy, E_{bg}.

Ultra-bandgap illumination of such semiconductor materials produces electron–hole pairs, h^+e^-, which can either recombine to liberate heat, Δ, or make their separate ways to the surface of the semiconductor material, where they have the possibility of reacting with surface absorbed species. Recombination, either at the surface, or in the bulk of the semiconductor material, is the usual fate of photogenerated electron–hole pairs. Thus, the efficiencies of most processes involving semiconductor photosensitisers is low; typically <1%. This is particularly true in the case of amorphous semiconductor materials, since electron–hole recombination is promoted by defects. As a result, in most of the current work involving photocatalysis, the semiconductor is comprised of microcrystalline or nanocrystalline particles and used in the form of a thin film or as a powder dispersion.

Photogenerated electrons that are able to make their way to the surface of these semiconductor particles can either react directly, or indirectly through slightly less energetic trap surface states, with absorbed species. Thus, if there is an electron donor, D, adsorbed on the surface of the semiconductor particles, then the photogenerated holes can react with it (directly or indirectly) to generate an oxidized product, D^+. Similarly, if there is an electron acceptor present at the surface, that is A, then the photogenerated CB electrons can react with it (directly or indirectly) to generate a reduced product, A^-. In such a situation, the overall reaction can be summarized as follows:

$$A + D \xrightarrow[h\nu \geqslant E_{bg}]{\text{semiconductor}} A^- + D^+ \tag{6.1}$$

A schematic representation of the energetics associated with the various processes involved in reaction (6.1) are illustrated in Figure 6.1. In reaction (6.1), the

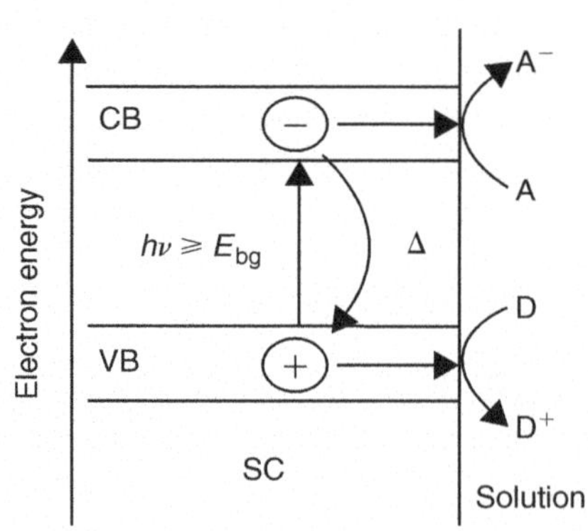

Figure 6.1 Schematic illustration of the energetics of semiconductor photocatalysis.

semiconductor acts as a photosensitizer, but is more often referred to as a photocatalyst; that is, upon absorption of ultra-bandgap light, it is able to promote thermodynamically feasible reactions without undergoing any overall change itself (Serpone and Emeline, 2002).

In the application of semiconductor photocatalysis to the purification of water, the electron acceptor, A, is invariably dissolved oxygen, and the electron donor, D, is the pollutant. Under these circumstances, the overall process is the semiconductor photocatalysed oxidative mineralization of the pollutant by dissolved oxygen, and can be represented by the following equation:

$$\text{Pollutant} + O_2 \xrightarrow[h\nu \geqslant E_{bg}]{\text{semiconductor}} \text{minerals} \tag{6.2}$$

A schematic representation of the various processes associated with reaction (6.2) that occur on a particle of a semiconductor photocatalyst are illustrated in Figure 6.2.

Thus, from Figure 6.2, after photogeneration, recombination can occur in the bulk (process a) or at the surface (process b). Photogenerated electrons that make their way to the surface may reduce absorbed oxygen, ultimately to water (process c); and photogenerated holes may oxidize any adsorbed pollutant, ultimately to its mineral form, that is usually CO_2 and H_2O (process d).

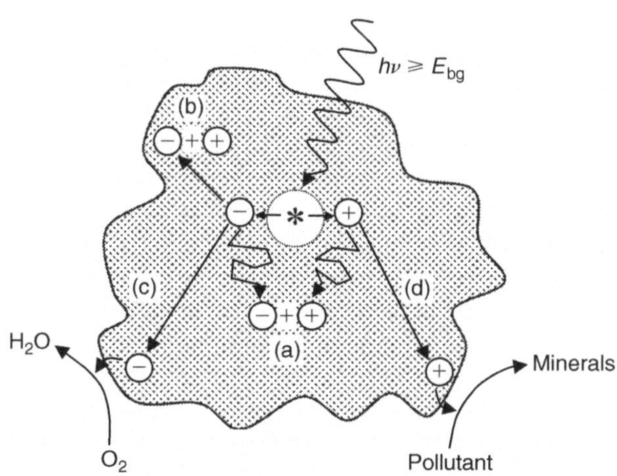

Figure 6.2 Schematic illustration of the major processes that occur on a semiconductor photocatalyst particle upon ultra-bandgap excitation in an aqueous solution containing dissolved oxygen and an oxidizable pollutant. The processes include: (a) electron–hole recombination in the bulk, (b) electron–hole recombination at the surface, (c) direct or indirect (through trap sites) reduction of oxygen, or oxidizing intermediates by the photogenerated electron at the surface of the semiconductor and (d) direct or indirect (through trap sites) oxidation of the pollutant, or an oxidized intermediate by the photogenerated hole at the surface of the semiconductor, leading eventually to the mineralization of the pollutant.

Ideally, a semiconductor photocatalyst for the purification of water should be chemically and biologically inert, photocatalytically active, easy to produce and use, and activated by sunlight. It is hardly surprising that no semiconductor fits this demanding list of requirements perfectly, although the semiconductor titanium dioxide (TiO_2) comes close but does not absorb visible light. Indeed, TiO_2 has a large bandgap energy, $E_{bg} \cong 3.2$–$3.0\,eV$, and, as a consequence, is only able to absorb ultraviolet (UV) light (typically <380 nm), which represents a small fraction, *ca.* 6%, of the solar spectrum. However, the very positive features of TiO_2 as a semiconductor photocatalyst far outweigh the limitations of its spectral profile and thus, it has become *the* semiconducting material for *research* in the field of semiconductor photocatalysis for water purification. Although TiO_2 exists in three crystalline forms, namely anatase, rutile and brookite, the common form used in semiconductor photocatalysis is anatase, as this is the most photocatalytically active and easiest to produce. As a consequence of its wide range of attractive features as a semiconductor photocatalyst, almost without exception, anatase TiO_2 is *the* semiconductor used in *commercial* photocatalytic systems for water purification.

As noted above, in the photomineralization of pollutants sensitized by TiO_2, that is reaction (6.2), the photogenerated electrons reduce oxygen to water and the photogenerated holes mineralize the pollutant. The latter process appears to involve the initial oxidation of surface hydroxyl groups ($>Ti^{IV}OH$) on the TiO_2 to hydroxyl radicals ($>Ti^{IV}OH^{\bullet +}$), which then oxidize the pollutant, and any subsequent intermediate or intermediates. The eventual final product is usually the mineral form of the pollutant, that is typically carbon dioxide, water, and, if a heteroatom is present in the pollutant under test, mineral acid. Photogenerated electrons can be trapped by surface sites, such as $Ti^{IV}OH$, to form Ti^{III} species, such as $Ti^{III}OH$, which can then react with dissolved oxygen to form superoxide, $O_2^{\bullet -}$. The latter species can be further reduced to hydrogen peroxide (H_2O_2) as an intermediate in the overall reduction of oxygen to water. H_2O_2 is also a possible source of hydroxyl radicals, and it appears likely that during the course of reaction (6.2), some of the mineralization of the pollutant is brought about by oxidizing species, such as hydroxyl radicals, that originate from the oxygen reduction process. The primary processes and associated characteristic time domains for semiconductor photocatalysis involving TiO_2 are listed in Table 6.1 (Hoffmann *et al.*, 1995).

From this data, it appears that the rate-determining step in the overall photocatalytic process is the reduction of dissolved oxygen by trapped photogenerated electrons, that is by $Ti^{III}OH$. Certainly, the supply of oxygen to the particles can control the overall kinetics of the process (Foster *et al.*, 1996; Galvez and Rodriguez, 2002; Upadhia and Ollis, 1997). However, the direct or indirect oxidation step involving the pollutant and the photogenerated holes is also a slow step and, under certain circumstances, may be rate determining. Thus, the kinetics of semiconductor photocatalysis for water purification can be complex and will tend to vary from one pollutant to another. Thus, when treated using semiconductor photocatalysis, the bouillabaisse of pollutants found in most waste streams is likely to produce a bouillabaisse of observed kinetics.

Interestingly, the kinetics of destruction of most simple, single pollutant systems by semiconductor photocatalysis exhibit very similar features. Thus, the

Table 6.1 Primary processes and associated characteristic time domains in the TiO_2-sensitized photomineralization of organic pollutants.

Primary processes	Characteristic times
Charge-carrier generation	
$TiO_2 + h\nu \rightarrow h^+ + e^-$	fs (very fast)
Charge-carrier trapping	
$h^+ + {>}Ti^{IV}OH \rightarrow \{{>}Ti^{IV}OH^{\bullet +}\}$	10 ns (fast)
$e^- + {>}Ti^{IV}OH \leftrightarrow \{{>}Ti^{III}OH\}$	100 ps (shallow trap, dynamic equilibrium)
$e^- + {>}Ti^{IV} \rightarrow {>}Ti^{III}$	10 ns (deep trap)
Charge-carrier recombination	
$e^- + \{{>}Ti^{IV}OH^{\bullet +}\} \rightarrow {>}Ti^{IV}OH$	100 ns (slow)
$h^+ + {>}Ti^{III}OH \rightarrow {>}Ti^{IV}OH$	10 ns (fast)
Interfacial charge transfer	
$\{{>}Ti^{IV}OH^{\bullet +}\}$ + organic pollutant $\rightarrow {>}Ti^{IV}OH$ + oxidized pollutant	100 ns (slow)
$\{{>}Ti^{III}OH\} + O_2 \rightarrow {>}Ti^{IV}OH + O_2^{\bullet -}$	ms (very slow)

initial kinetics of photomineralization of general pollutant, P, by oxygen, sensitized by TiO_2, upon steady-state illumination, generally fit a Langmuir–Hinshelwood kinetic scheme, that is

$$R_i = \frac{-d[P]_i}{dt} = \frac{k(P)K(P)[P]_i}{1 + K(P)[P]_i} \tag{6.3}$$

where R_i is the initial rate of the substrate removal, $[P]_i$ is the initial concentration of the pollutant under test and, traditionally, $K(P)$ is taken to be the dark Langmuir adsorption constant of species P on the surface of TiO_2. However, in general, it is usually found that the constant, $K(P)$, is not directly equivalent to the dark Langmuir adsorption constant for P on the semiconductor (values for the latter parameter are usually much smaller). Instead, semiconductor photocatalysis exhibits saturation-type kinetics, which exhibit the same tendencies as Langmuir–Hinshelwood kinetics, but do not usually follow the traditional mechanism associated with the equation (Mills and Le Hunte, 1997; Turchi and Ollis, 1989).

The parameter $k(P)$ is a proportionality constant which provides a measure of the intrinsic reactivity of the photo-activated surface with P. The $k(P)$ is usually found to be proportional to I_a^{θ}, where I_a is the rate of light of absorption and θ is a power term which is equal to 0.5 or unity at high or low light intensity, respectively. The value of 0.5 for θ at high light intensities is attributed to the dominance of electron–hole recombination as the likely fate of the photogenerated electrons and holes. Under these conditions, the steady-state concentrations of electrons and holes are proportional to $I_a^{0.5}$ and thus it is not surprising to find that R_i is proportional to $I_a^{0.5}$. At very low-incident light intensities, the likelihood of electron–hole recombination is negligible and the steady-state concentration of electrons and holes is proportional to I_a and thus R_i is proportional to I_a. Note that as a consequence of the variation of θ with light intensity, the reaction becomes less photon efficient the higher the light intensity. At very high light intensities, the kinetics become mass transfer controlled and independent of the intensity of the ultra-bandgap light; thus, R_i becomes independent of I_a.

The parameter $k(P)$ is also proportional to the fraction of oxygen adsorbed on TiO_2, that is $f(O_2)$ is given as follows:

$$f(O_2) = \frac{K(O_2)[O_2]}{1 + K(O_2)[O_2]} \tag{6.4}$$

where $K(O_2)$ is usually taken as the Langmuir adsorption coefficient for oxygen on the semiconductor, which appears to be non-competitively adsorbed.

A rough but useful guide to efficiency in semiconductor photocatalytic systems is photonic efficiency, δ, which is defined as follows:

$$\delta = \frac{\text{Overall rate of photocatalytic process}}{\text{Intensity of incident light}} \tag{6.5}$$

In semiconductor photocatalysis, the value of δ has often been found to approach unity at low-incident light levels of ultra-bandgap light. However, at the typical light levels generated by black light bulbs or germicidal lamps (*ca.* $1\,\mathrm{mW\,cm^{-2}}$ for an 8 W lamp), the value of δ is usually *ca.* 0.01 for most semiconductor photocatalysis systems. This low efficiency is due to a number of factors including reflection and scattering losses and significant electron–hole recombination at the light intensities typically found in most photoreactors.

Several kinetic models of semiconductor photocatalysis have been developed with equations for the initial rate of semiconductor photocatalysis that conform to Equations (6.3) and (6.4). Two of the most favoured mechanistic schemes involve either the attack of an adsorbed hydroxyl radical on the adsorbed pollutant, or the reaction of an adsorbed hydroxyl radical with a free pollutant molecule as the rate-determining step (Turchi and Ollis, 1990). However, recent work carried out by our group (Mills and Wang, 1999) and others (Emeline *et al.*, 2000) has shown that the parameters $K(P)$ and $K(O_2)$ in Equations (6.3) and (6.4) are light intensity dependent, that is

$$K(P) = K(P)' I^{\beta} \tag{6.6}$$

where β is typically -1. If, as is believed, this additional kinetic feature is found to be generally applicable to semiconductor photocatalysis, then some modification of the current kinetic models used to describe this process is required (Emeline *et al.*, 2000).

In semiconductor photocatalysis, an often employed test pollutant is 4-chlorophenol (4-CP) and a typical test semiconductor photocatalyst is a dispersion of Degussa P25 TiO_2, since the latter has a high specific surface area and high photocatalytic efficiency.

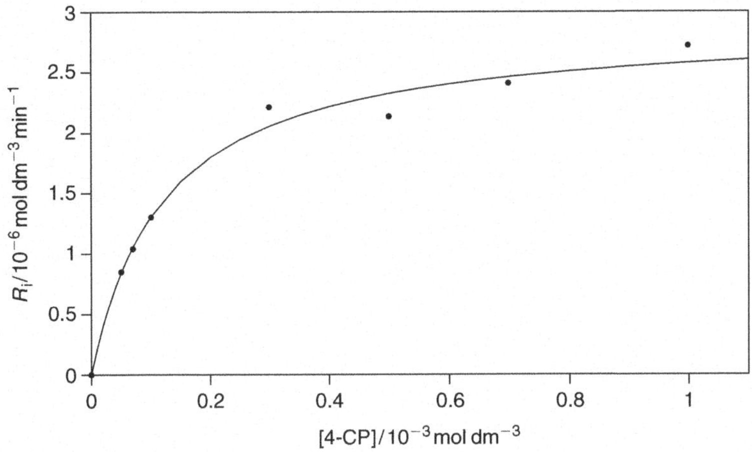

Figure 6.3 Plot of the measured initial rate of disappearance of 4-CP, R_i, as a function of [4-CP]. The results were obtained using a batch reactor containing 4-CP (100 cm^3), $[O_2]$ = 100% saturated, TiO_2 (Degussa P25, 0.5 mg cm^{-3}), pH = 2 and T = 30°C (Mills and Wang, 1999).

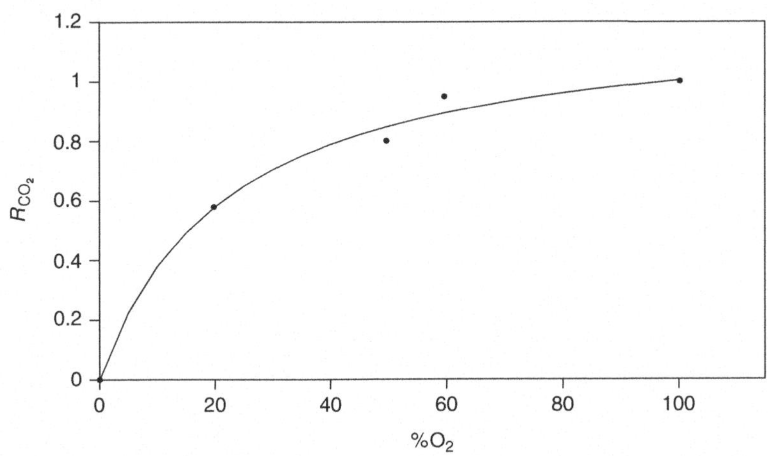

Figure 6.4 Plot of the measured relative rates of carbon dioxide production as a function of $[O_2]$. The results were obtained using a batch reactor containing 4-CP (100 cm^3, 10^{-3} mol dm^{-3}), TiO_2 (Degussa P25, 0.5 mg cm^{-3}), pH = 2 and T = 30°C (Mills *et al.*, 1993).

Figures 6.3 and 6.4 illustrate the observed Langmuir–Hinshelwood-type dependence of the initial rate of destruction of 4-CP by semiconductor photocatalysis, that is R_i, as a function of [4-CP] and $[O_2]$, respectively. Figure 6.5 illustrates the observed variation in R_i as a function of incident light intensity for the same system and Figure 6.6 illustrates the variation in the calculated value of K(4-CP) as a function of incident light intensity. The solid lines in Figures 6.3–6.6 are the

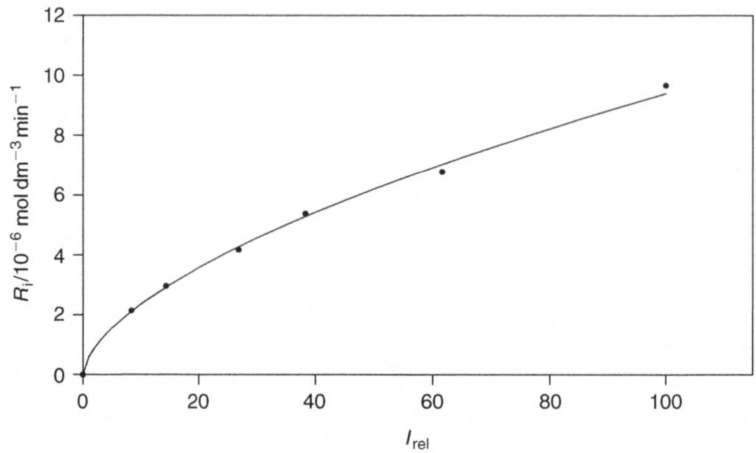

Figure 6.5 Plot of the measured initial rate of disappearance of 4-CP, R_i, as a function of relative incident light intensity. The results were obtained using a batch reactor containing 4-CP (100 cm^3, 5×10^{-4} mol dm^{-3}), [O_2] = 100% saturated, TiO_2 (Degussa P25, 0.5 mg cm^{-3}), pH = 2 and T = 30°C (Mills and Wang, 1999).

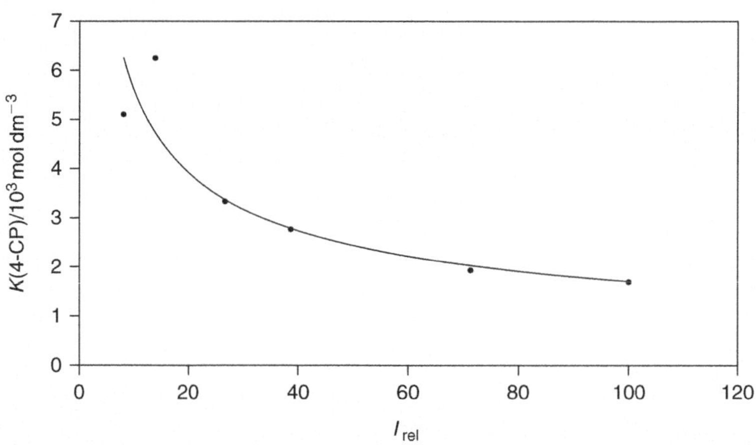

Figure 6.6 Plot of the measured value of K(4-CP) as a function of incident light intensity. The values of K(4-CP) were obtained from plots of R_i versus [4-CP] and subsequent analysis using Equation (6.3) (Mills and Wang, 1999).

lines of best fit to the data using Equations (6.3)–(6.6) and values for K(4-CP), k(4-CP), θ and β of 8.2×10^3 dm^3 mol^{-1}, 2.89×10^{-6} mol dm^{-3} min^{-1}, 0.60 and -1, respectively. The good fit of the data to Equations (6.3) and (6.4) are typical of that found in semiconductor photocatalysis for many pollutants using a TiO_2 semiconductor in dispersed powder or film forms.

TiO_2, when used as a semiconductor photocatalyst in reaction (6.2), usually shows little or no appreciable loss in activity with repeated use. This feature of photochemical robustness and longevity is rather nicely illustrated by the data in

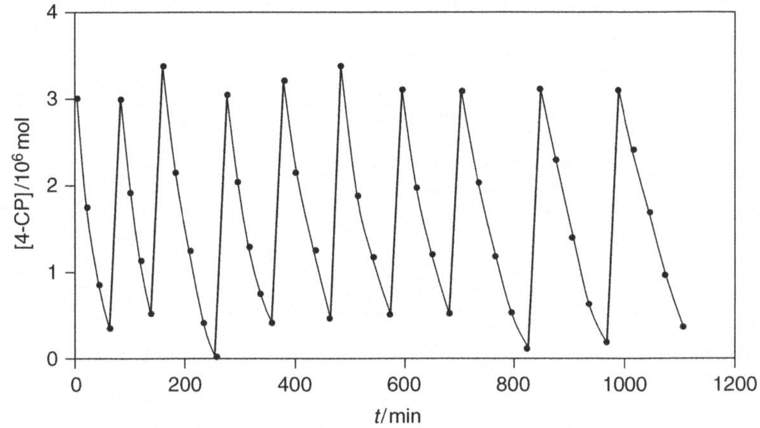

Figure 6.7 [4-CP] versus time profile recorded for 10 successive mineralization cycles using the same TiO_2 photocatalyst dispersion. The dotted vertical lines correspond to the readjustment of the [4-CP] to 1.55×10^{-4} mol dm^{-3}, followed by a 2-min dark period, so that adsorption equilibrium could be obtained (Al-Sayyed *et al.*, 1991).

Figure 6.7, which shows the results of a typical set of successive kinetic runs in which 4-CP was mineralized by oxygen using TiO_2 as the photocatalyst (Al-Sayyed *et al.*, 1991). From the results of this work, it appears that even after 10 successive runs, the TiO_2 semiconductor photocatalyst shows little or no evidence of wear.

However, before getting carried away by such promising results, it should be recognized that the results of experiments conducted in the field might be quite different from those in a laboratory. Most importantly, it is clear that a TiO_2 photocatalyst will not work very well, if at all, if the water under test contains any strongly UV-absorbing pollutants, such as dye stuffs and humic acid. Nor will most photocatalysts work if metal ions, such as ferric and ferrous ions, are present and able to precipitate out as insoluble, and inert oxides and hydroxides, onto the surface of the semiconductor material to form a passivating layer. Although the latter may be removed by using an acid wash, it appears as a clear limitation of the technology. As a consequence, semiconductor photocatalysis, as a method of water treatment, is most likely limited in application to water that has been pre-filtered, to remove any solid material that is likely to deposit on the surface of the semiconductor and any strong UV-absorbing species that may be present, and, preferably deionized, so that it is fairly free of metal ions that may form insoluble oxides or hydroxides on the surface of the semiconductor.

6.3 APPLICATIONS

Research into semiconductor photocatalysis has largely focused on the removal of organic pollutants from water, since it appears that a wide range of organic substrates are prone to complete mineralization by dissolved oxygen, photosensitized by TiO_2.

Table 6.2 Some examples of TiO_2-sensitized photomineralization of organic substrates.

Class	Examples
Alkanes	Methane, *iso*-butane, pentane, heptane, cyclohexane and paraffin
Haloalkanes	Mono-, di-, tri- and tetrachloromethane, tribromoethane and 1,1,1-trifluoro-2,2,2-trichloroethane
Aliphatic alcohols	Methanol, ethanol, *iso*-propyl alcohol, glucose and sucrose
Aliphatic carboxylic acids	Formic, ethanoic, dimethylethanoic, propanoic and oxalic acids
Alkenes	Propenem and cyclohexene
Haloalkenes	Perchloroethene, dichloroethene and 1,1,2-trichloroethene
Aromatics	Benzene and naphthalene
Haloaromatics	Chlorobenzene, 1,2-dichlorobenzene and bromobenzene
Nitrohaloaromatics	3,4-Dichloronitrobenzene and dichloronitrobenzene
Phenols	Phenol, hydroquinone, catechol, 4-methyl catechol, resorcinol, *o*-, *m*- and *p*-cresol
Halophenols	2-, 3-, 4-CP, pentachlorophenol, 4-fluorophenol and 3,4-difluorophenol
Aromatic carboxylic acids	Benzoic, 4-aminobenzoic, phthalic, salicyclic, *m*- and *p*-hydroxybenzoic and chlorohydroxybenzoic acids
Polymers	Polyethylene and polyvinylchloride
Surfactants	SDS, polyethylene glycol, sodium dodecyl benzene sulphonate, trimethyl phosphate and tetrabutyl ammonium phosphate
Herbicides	Methyl viologen, atrazine, simazine, prometron, propetryne and bentazon
Pesticides	DDT, parathion and lindane
Dyes	Methylene blue, rhodamine B, methyl orange, fluorescein and reactive black 5

Table 6.2 provides a list of the wide number of classes of compounds, and examples of each, that have been shown to be completely mineralized by semiconductor photocatalysis using TiO_2 (Bhatkhande *et al.*, 2001; Mills and Le Hunte, 1997).

Notable compounds in Table 6.2 include haloalkanes, such as chloroform and tetrachloromethane, since they are carcinogens and often found in small but detectable amounts in drinking water purified by chlorination. Indeed, it appears likely that semiconductor photocatalysis may find initial commercial application as an 'after-chlorination' step in the production of drinking water. Other compounds listed in Table 6.2 that are highly relevant to the treatment of wastewater include surfactants, hormones, herbicides, pesticides and dyes, all of which can be considered as relatively common pollutants from farms, households and/or industry. One of the very attractive features of semiconductor photocatalysis is its effectiveness against a broad range of pollutants, including those that are not easily removed by other water-treatment processes, including chlorination. Semiconductor photocatalysis is ideal for treating refractory, hazardous, toxic and/or non-biodegradable pollutants.

In addition to the removal of organic pollutants, semiconductor photocatalyst has been used to sensitize the photoconversion of toxic inorganic substrates to harmless or less toxic ones, and a number of examples of such photoreactions, sensitized

Table 6.3 TiO_2-sensitized photosystems for the removal of toxic inorganics.

Overall reaction $A + D \rightarrow A^- + D^+$
$5O_2 + 6NH_3 \rightarrow 2N_2 + N_2O + 9H_2O$
$O_2 + 2NO_2^- \rightarrow 2NO_3^-$
$O_2 + 2SO_3^{2-} \rightarrow 2SO_4^{2-}$
$2O_2 + H_2O + S_2O_3^{2-} \rightarrow 2SO_4^{2-} + 2H^+$
$O_2 + 2CN^- \rightarrow 2OCN^-$
$5O_2 + 4H^+ + 4CN^- \rightarrow 2H_2O + 4CO_2 + 2N_2$
$2BrO_3^- \rightarrow 2Br^- + 3O_2$

by TiO_2, are listed in Table 6.3 (Mills and Le Hunte, 1997). Thus, semiconductor photocatalysis can be used to oxidize nitrite, sulphite and cyanide anions to form relatively harmless products, such as NO_3^-, SO_4^{2-} and CO_2, respectively. TiO_2 is also able to photosensitize the decomposition of bromate, to bromide and oxygen, even when the level of bromate ions is as low as 50 ppb. Bromate is a carcinogen and is found in ppb levels in drinking water that has been purified by chlorinating a source water with a reasonably high background level of bromide ions. The removal of bromate ions at the ppb level represents a major problem to the potable water-production industry, but semiconductor photocatalysis appears to offer a simple and effective solution (Mills *et al.*, 1996).

Semiconductor photocatalysts are also able to mediate the destruction of biological materials, such as bacteria, viruses and moulds. The mode of action appears to be similar to that in the oxidative mineralization of pollutants, that is reaction (6.2), in which the photogenerated holes generate surface hydroxyl radical species, which are then able to damage or destroy the cell walls of biological material, such as bacteria, viruses and moulds. Cells with damaged or destroyed cell walls are usually not viable and die quickly. As in reaction (6.2), the photogenerated electrons reduce oxygen to superoxide and, through the subsequent formation of H_2O_2, this can also act as a source of the antimicrobial hydroxyl radicals.

The photocatalytic destruction of biological material sensitized by semiconductor material, such as TiO_2, is generally referred to as photo-sterilization or photodisinfection. The overall process can be summarized as follows:

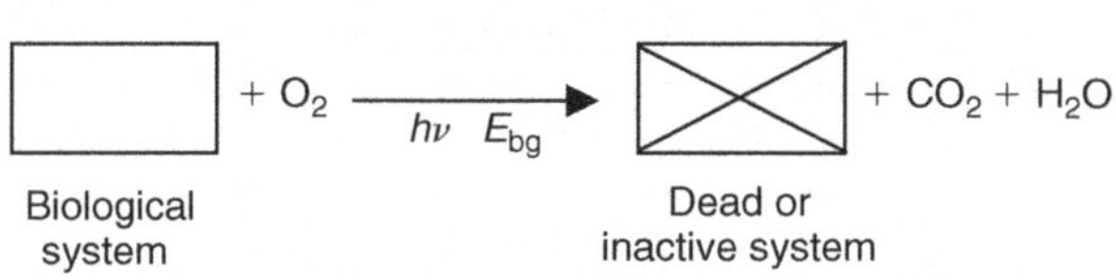

(6.7)

Table 6.4 lists some of the biological systems that have been destroyed using semiconductor photocatalysis. Much of the work in this area has focused on the destruction of simple bacterium, such as *Escherichia coli*, and the application of TiO_2 photodisinfection as a method of destruction of more resistant bacteria, such

Table 6.4 Examples of TiO_2-sensitized photodestruction of bacteria and viruses.

Biological material	Comments	References
Streptococcus mutans, *Streptococcus rattus*, *Streptococcus cricetus* and *Streptococcus sobrinus* AHT	The TiO_2 particles were able to sensitize effectively the photokilling of the bacteria: *Streptococcus mutans*, *Streptococcus rattus*, *Streptococcus cricetus* and *Streptococcus sobrinus* AHT within 3 min of irradiation with UV photons.	Saito *et al.* (1992)
Escherichia coli	A flow system comprising a sleeve of fibreglass mesh, coated with TiO_2, covering the UV lamps. The flow system was also used to treat coloured, algae-containing pond water and it was found that although the total coliform level could be reduced, the heterotrophic plate count could not, unless H_2O_2 was added to enhance the photokilling action of the TiO_2.	Ireland *et al.* (1993)
Escherichia coli	TiO_2 particles immobilized on acetyl cellulose to create a photodisinfecting film for use in a flow system. The flow system was suitable for handling samples containing a low concentration ($<10^3$ cells cm^{-3}) of bacteria, but was less effective at higher bacteria levels. When operated for 1 week, at a bacteria input level of 10^2 cells cm^{-3}, the flow system exhibited no deterioration in performance.	Matsunaga *et al.* (1988)
Escherichia coli	A study of the kinetics of photokilling of the bacteria, sensitized by TiO_2, as a function of $[O_2]$ (cell destruction does not proceed without oxygen), $[TiO_2]$ (half-order) and incident light intensity (first order).	Wei *et al.* (1994)
Escherichia coli	Sunlight used as the light source.	Zhang *et al.* (1994)
Escherichia coli	Diffuse light emitting optical fibres were found to be more effective as a light source than conventional light fibres.	Matsunaga and Okochi (1995)
Pharge MS2 (a single-stranded RNA bacteriophage)	A 2×10^{-6} mol dm^{-3} $FeSO_4$ solution was used to promote (1.7 times) the photodestruction ability of the TiO_2 particles. From the results of this work, it appears that viral components, such as protein capsids and nucleic acids, are susceptible to direct attack by photogenerated $OH^{\bullet}$ radicals. The Fe(II) solution is thought to enhance the steady-state concentration of $OH^{\bullet}$ radicals via a Fenton-type process.	Sjogren and Sierka (1994)
Poliovirus 1 (also coliform bacteria for comparison)	The photodestruction of the poliovirus 1 was complete in 3-min irradiation, whereas it took 150 min to destroy the coliform bacteria. This work was carried out in secondary wastewater effluent.	Watts *et al.* (1995)

as the pathogenic protozoans, that is *Giardia lambia* and *Cryptosporidium*, or persistent algae, has recently been reported (see Chapter 7).

6.4 REACTOR DESIGN

Photochemical reactor engineering is much more complex than chemical reactor engineering. For example, the type of radiation and the arrangement of the light source in the reactor system can dominate reactor design. In addition, the need for at least one of the reactor walls to transmit the chosen radiation requires the use of transparent materials, such as glass or silica, in reactor construction and this can cause limitations on size, and problems with sealing and breakage. The rate equations are also usually much more complex than those of traditional chemical processes and, as a consequence, the overall equations describing the kinetics of the relevant photoprocesses, which are then used to optimize the reactor design, can often only be solved numerically, if at all (Braun, 1988; Cassano and Alfano, 1991; Mukherjee and Ray, 1999; Ollis and Turchi, 1990).

In heterogeneous semiconductor photochemistry, in contrast to the more traditional homogeneous photochemistry, the problem of reactor design is further complicated by the presence of a solid photocatalyst. From Equation (6.2), it is clear that a reactor for semiconductor photocatalysis has to be designed, so that it can handle, in an optimum way, the reaction solution, an oxygen gas stream, a solid photocatalyst and ultra-bandgap light.

With the large number of important parameters and phases associated with semiconductor photocatalysis, the task of designing, or characterizing, a photoreactor for semiconductor photocatalysis can appear quite daunting. However, one starting point is the type of irradiation source, since the latter will influence considerably the subsequent design of the photoreactor itself. In most examples of semiconductor photocatalysis for water purification, UV light is required and this can be obtained from arc lamps in which mercury and/or xenon vapour is activated electrically. Such lamps are usually very inefficient, in terms of UV photons produced per unit of electrical energy input, since a great deal of the electrical energy is converted into either heat or visible light. In addition, arc lamps are both expensive and have very limited lifetimes of operation, typically *ca.* 1000 h. In contrast, UV fluorescent lights are very efficient at converting electrical energy to UV photons, although their power, and therefore emitted light intensity, is usually quite small, up to 150 W. Such lights utilize the 254 nm light emitted when an electrical discharge is passed through a low-pressure vapour of mercury contained in a sealed tube. If no phosphor is coated on the walls of the tube, the product is a germicidal lamp. If a black–blue phosphor is coated onto the walls of the lamp, the product is a black light bulb with an emission output of 365 ± 20 nm. Germicidal and black light bulbs are the most popular artificial light sources used in semiconductor photocatalysis.

The problem with germicidal lamps is that the light emitted is UVC light and as a consequence, dangerous; for example, it can cause skin cancer and blindness.

In addition, the use of a UVC source requires the photoreactor to be built with a quartz window to allow the transmission of the UVC light to the reaction solution. Black light bulbs, on the other hand, emit UVA light and, as a result, pose much less of a risk to biological systems. However, the commonly used photocatalyst, TiO_2, absorbs 365 ± 20 nm light much less (typically by a factor of 10) than it does for germicidal light. As a consequence, TiO_2 photocatalytic systems are more likely to reflect and scatter incident UVA light, as well as absorb it, when compared to UVA light and, as a result, are likely to be less efficient in terms of photons. A major advantage of using black light bulbs over germicidal lamps is that the photoreactor can be constructed from cheap sodium borosilicate glass, which transmits light of wavelengths greater than 300 nm, rather than expensive quartz, which cuts off at 200 nm. The relatively high UV light to electrical energy conversion efficiencies of both germicidal and black light bulbs, typically 10–20%, and their long operational lifetimes, typically 4000–14,000 h, makes these the preferred light sources for use in photoreactors for semiconductor photocatalysis.

It is worth reminding oneself at this point that in all semiconductor photocatalytic systems for water purification, the reaction solution needs to be purged continuously with either air or oxygen for the system to work efficiently, as indicated in Equations (6.2) and (6.3). Failure to do so can lead to oxygen starvation of the system and a catastrophic loss of efficiency for the overall photoreaction (6.2) (Foster *et al.*, 1996; Galvez and Rodriguez, 2002; Upadhia and Ollis, 1997). Thus, the issue of a good and continuous supply of oxygen to the system needs to be addressed when designing any reactor for semiconductor photocatalysis.

Most research work into semiconductor photocatalysis uses batch photoreactors, with a reaction solution, comprising the test pollutant and a dispersion of the semiconductor photocatalyst; and in such systems, oxygen purging is usually carried out in the photoreactor itself. In contrast, in many flow systems, this purging process has to be carried out in a reaction solution reservoir that is separate to the photoreactor. The exception to these general cases of added oxygen is thin-film photoreactors, since the gas exchange between air and the reaction solution is usually sufficiently facile, and the kinetics of photocatalysis sufficiently slow that no additional sparging of the reaction solution with oxygen or air is required. In some closed, usually annular, photocatalytic reactor systems, the level of concern about the oxygen supply may be such that an additional electron scavenger, such as H_2O_2 of sodium persulphate, may be added to improve the overall efficiency of pollutant destruction.

Among the early decisions that need to be made when designing a photoreactor for semiconductor photocatalysis is the form the semiconductor photocatalyst is to take. Most of the research carried out into semiconductor photocatalysis for water purification has been conducted using TiO_2 powder dispersions. The use of such slurries obviously requires a subsequent separation step involving either filtration, centrifugation or coagulation/flocculation. This added step, or steps, increases the complexity of the overall process and decreases its economical viability (Hoffmann *et al.*, 1995). However, slurry reactor photocatalytic systems are usually very efficient in terms of photons (relative to thin-film reactors), and easy to make and maintain.

Table 6.5 Features of dispersed and fixed-film photocatalyst particle reactors.

Feature	Dispersed semiconductor photocatalyst	Fixed-film semiconductor photocatalyst
Ease of photocatalyst preparation	Excellent. Photocatalyst easily prepared by numerous chemical routes, including hydrolysis or air oxidation of simple Ti(IV) precursors.	Can be difficult and usually involves either CVD, sol–gel, sputtering or thermal oxidation methods.
Ease of replacement of the semiconductor photocatalyst	Excellent. Simply add the photocatalyst particles to the polluted water stream.	Often difficult as the usually firmly fixed-film photocatalyst is often attached to the walls of the photoreactor. Photocatalyst on support materials, such as glass beads, are easier to change.
Overall ease of operation as a water-purification system	Not usually very good, since a subsequent particle separation step is required, which usually brings with it the need for regular filter replacement. In addition, ideally, the filtered photocatalyst should be returned to the photoreactor, which poses another technical problem, although a recirculation system with cross-filtration appears to offer a solution. System needs constant sparging with air.	Excellent. No filtering needed. In some 'closed-to-the-ambient air' systems, oxygen sparging is really needed.
Efficiency per m^2	High due to high photocatalyst surface area to reaction solution ratio and no mass transfer effects.	Low, thus, such systems need to be cheap per m^2.
Cost	Low initial outlay, but maintenance costs can be high.	High initial outlay but maintenance costs could be low, provided the water stream does not contain any deactivating contaminants.

Photocatalytic reactors that utilize a fixed bed of semiconductor material are usually less photon-efficient systems for pollutant destruction, due to an intrinsic low surface area to volume ratio; in addition, such systems are difficult to make, can be difficult to maintain (if passivation of the photocatalyst occurs) and are costly to replace. However, the major advantage of such fixed-film photoreactors is that no subsequent separation step is required. These and other key positive and negative features of dispersed and fixed-film photocatalytic reactors are listed in Table 6.5.

As noted earlier, most of the reactors, used to research semiconductor photocatalysis for the purification of water, operate typically in batch, rather than continuous mode. Figure 6.8 illustrates some of the numerous designs that have been used in such photochemical research.

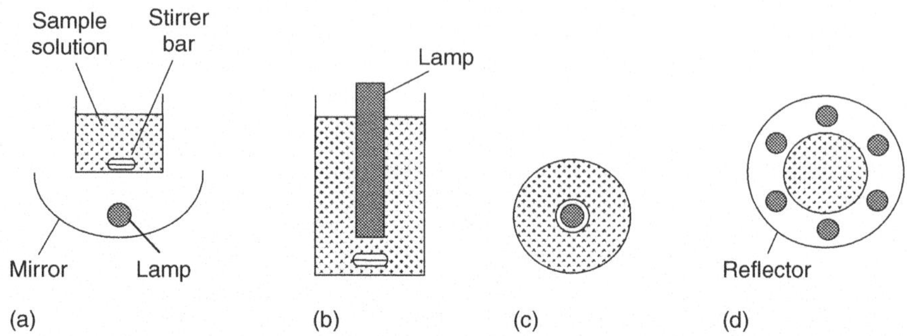

Figure 6.8 Basic common photocatalytic slurry reactor designs including side views: (a) reactor with external illumination; (b) immersion, top views; (c) annular (i.e. tubular with negative illumination geometry) and (d) circular.

In most of the early research work on semiconductor photocatalysis, the semiconductor was used in the form of powder particles dispersed throughout the reaction solution containing the pollutant under test. The simplest and earliest photocatalytic reactors comprised a glass reactor vessel, containing the reaction solution under test, dispersed with semiconductor photocatalyst particles, and either an external light source as illustrated in Figure 6.8a, or a lamp immersed in the reaction solution (see Figure 6.8b). Another photoreactor for semiconductor photocatalysis is an annular system, that is Figure 6.8c, in which the reaction solution passes along the reactor length one or more times. This latter reactor geometry is very popular in the water-disinfection industry. The circular photoreactor, illustrated in Figure 6.8d, is also used in semiconductor photocatalysis research and allows the use of many low-energy, low-intensity UV lamps to achieve a high radiant flux. In this photoreactor, the reaction solution can be flowed, or operated in batch mode and stirred.

As an alternative to using the semiconductor photocatalyst in the form of a dispersion of powder particles, it is possible to fix it to an inert substrate, such as glass, to produce a fixed-bed photoreactor. One of the earliest types of fixed-bed photocatalytic reactors is illustrated in Figure 6.9a (Matthews, 1987a,b). In this system, the semiconductor photocatalyst is coated onto the inner walls of a glass spiral, which has a UV fluorescent lamp along its centre. The contaminated water passes down the glass spiral over the irradiated photocatalyst, where mineralization of any pollutant occurs. Since such a system has a low photocatalyst surface area to reaction solution volume, it is likely to be quite inefficient, and thus, such photoreactors are usually operated in multi-pass or series mode, in order to achieve complete mineralization of the organic pollutant under test. Work by Ollis and his co-workers have shown that this type of spiral fixed-bed photoreactor is prone to exhibit kinetics that are largely mass transfer dependant and, thus, far from optimal (Ollis, 1998). The combination of low surface area to reaction volume ratio and mass transfer kinetics makes this reactor system very unattractive for either research or industrial scale up.

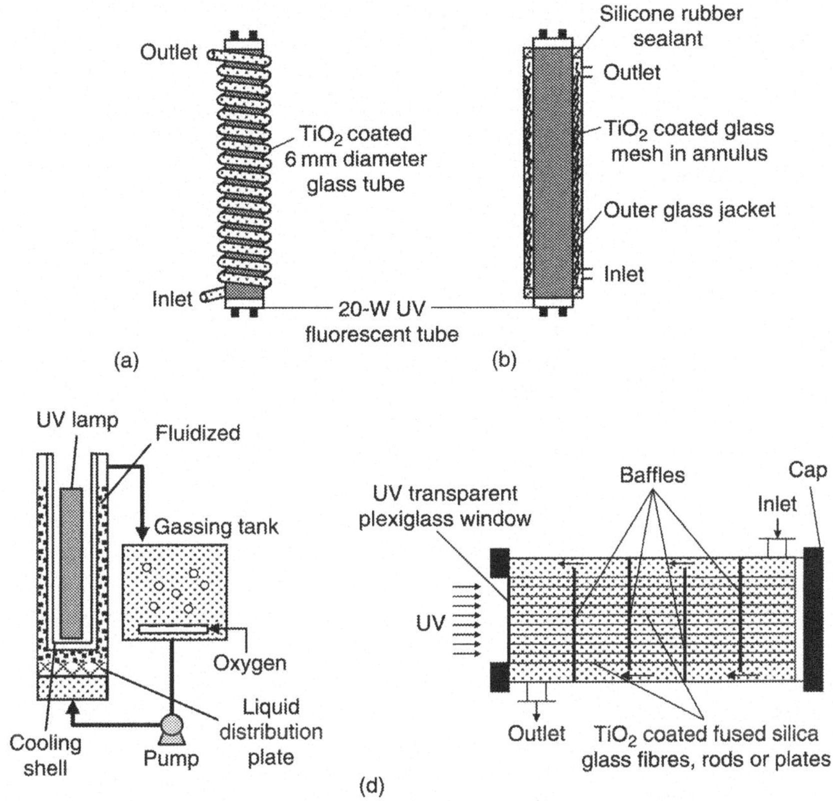

Figure 6.9 Fixed (including supported) photocatalyst film reactors including (a) spiral, (b) annular, (c) fluidized bed and (d) light distribution reactors.

One way to improve the surface area to reaction volume ratio in a fixed photocatalytic-film reactor is to use supported semiconductor photocatalysts, rather than films on the reactor walls (Porter and Yue, 1993). Supported semiconductor photocatalysts comprise a thin layer of the semiconductor material attached, chemically or physically, to a high surface area inert support material, such as alumina pellets, molecular sieve, glass fibre or ceramic membranes. Figure 6.9b illustrates a typical example of a supported fixed-bed system, comprising an annular photoreactor packed with glass fibre impregnated with fine TiO_2 particles (Matthews, 1987b).

Ultimately, the support materials can be made so fine that they behave as a powder dispersion and therefore require a subsequent filtration step. However, a bit before this extreme lies the realm of semiconductor particle sizes that are neither too small nor too large to create problems of either filtration or low photocatalytic efficiency, respectively. Thus, these particles (typically 0.5–5 μm diameter) are sufficiently small that they still offer a reasonable specific surface area, but are sufficiently large to be dense enough to readily settle out under gravity even in the presence of a reasonable flow of reaction solution. The results are photocatalyst particles that can be used in a fluidized bed and which require no subsequent filtration step. An example of such a reactor has been reported recently, in which the support material was 0.3 mm diameter sand particles and the deposited semiconductor was TiO_2 (Haarstrick *et al.*, 1996). The basic features of such a fluidized bed, fixed (onto the support particles) photocatalytic-film reactor, are illustrated in Figure 6.9c.

One of the major problems in semiconductor photocatalysis is the achievement of an even distribution of light in the photoreactor. When the semiconductor photocatalyst is used in the form of a dispersion of powder particles, or coated on an inert support material, such as glass beads, alumina pellets, etc., the distribution of light throughout the reaction vessel is likely to be non-uniform, due to absorption, reflection and scattering by the support as well as the active photocatalyst coating. Obviously, a poor distribution of UV light in a semiconductor photocatalytic reactor will lead to a low overall efficiency of operation. Figure 6.9d illustrates a novel system for distributing the ultra-bandgap light more evenly in a photoreactor (Hoffstadler *et al.*, 1994). In the general form of this system, the photocatalyst material is dispersed throughout the photoreactor via a series of light conductors, such as thin glass plates, rods, fibres or hollow glass tubes (Hoffstadler *et al.*, 1994; Peill and Hoffmann, 1995, 1996; Ray, 1999; Ray and Beenackers, 1998). Such dispersed light systems achieve a high surface area to reaction volume ratio and a more even distribution of the number of photons impinging on the, usually, fixed, catalyst particles in the reactor. One of the problems of such light distribution photoreactors is that the light conductors would have to be made out of expensive quartz, if germicidal lamps rather than black light bulbs were to be used as the light source. An even distribution of light in a photoreactor can also be achieved by using a number of very thin germicidal or black light bulbs, coated with photocatalyst in much the same arrangement of glass rods/fibres as illustrated in Figure 6.9d (Ray, 1999; Ray and Beenackers, 1998). Such light tubes are just entering the market and are likely to find great application in this area provided their cost is low.

One of the most likely manifestation of a reactor for semiconductor photocatalysis for water purification is one in which the semiconductor photocatalyst is fixed onto the walls of the photoreactor and exposed to a thin-falling film of the reaction solution.

A good example of such a photoreactor is illustrated in Figure 6.10a and can be used with the semiconductor either fixed onto the walls of the reaction vessel, or dispersed in the falling film of reaction solution (Puma and Yue, 1998, 1999a,b; Yatmaz *et al.*, 1993).

As noted earlier, with such thin, reaction solution film photoreactors, the rapid exchange of oxygen between air and the thin reaction solution film makes it unnecessary for the latter to be pre-purged with air. In addition, such photocatalytic reactors allow the ready use of germicidal lamps as the irradiation source and so makes them more photon efficient, since the absorption coefficient of TiO_2 at 254 nm is very high. A novel variation of the thin-falling-film photoreactor is illustrated in Figure 6.10b (Toyoda *et al.*, 2000). In this system, the fixed-photocatalytic film is coated on a drum of glass that is half-dipped into the reaction solution under test and

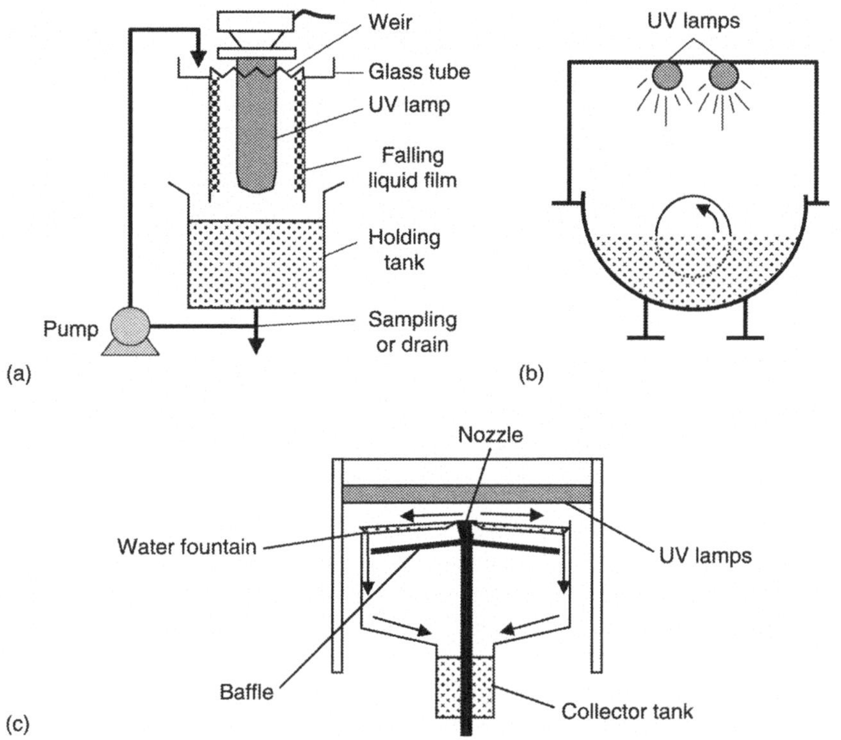

Figure 6.10 Examples of thin falling-film photocatalytic reactors including (a) classic tubular, (b) rotating drum and (c) fountain photoreactors. Note: all can be used with the photocatalyst fixed to the reactor walls (fixed bed) or as a dispersion (moving bed) in the reaction solution.

half in air. The system is irradiated from above. Rotation of the photocatalyst-coated glass drum produces a thin film of reaction solution over the fixed-photocatalytic film coating the drum. The result is a thin-film, fixed-bed photoreactor that allows the ready destruction of any pollutants present by semiconductor photocatalysis using, for example, UVc light. Once again, no pre-purging of the reaction solution with air or oxygen is necessary in such a system.

Recently, Yue and his co-workers have reported a novel fountain photocatalytic reactor for the purification of water (Puma and Yue, 2001a,b). In this system, a fountain of reaction solution, containing the pollutant and a dispersion of semiconductor photocatalyst particles is continuously generated and the fountain head is irradiated with ultra-bandgap light. The basic features of this system are illustrated in Figure 6.10c. Such a system obviously does not require any additional purging with air or oxygen and does have the benefit of a high photocatalyst surface area per unit volume of reaction solution and minimal mass transfer limitations. However, as with all slurry-type reactors, a photocatalysis recovery system is required.

Of course, one way to lessen the cost of a photocatalytic system for water purification is to use that 'free' light source, the sun. Numerous solar reactors for

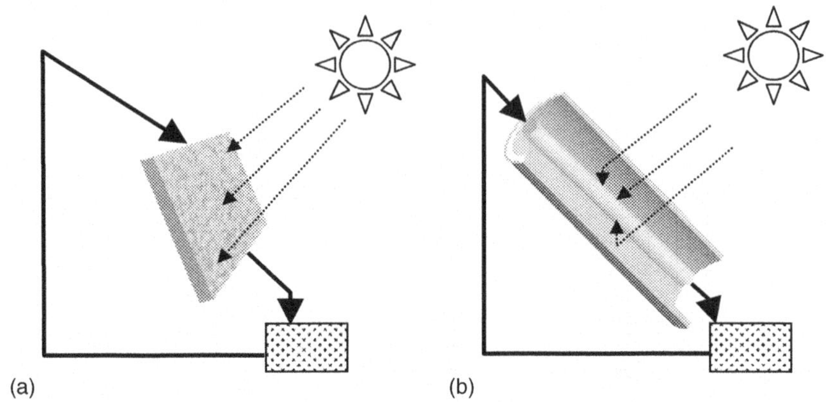

Figure 6.11 Basic solar photocatalytic reactor designs including (a) non-concentrating, thin falling film and (b) concentrating parabolic trough reactors.

water purification via semiconductor photocatalysis have been reported (e.g. Galvez and Rodriguez, 2002; Matthews, 1986; Peill and Hoffmann, 1997; Turchi *et al.*, 1993). Such systems can be non-concentrating and so utilize direct and diffuse sunlight, although typically the UV flux will be limited to that of one sun, that is *ca.* 0.04 mW cm^{-2} (Turchi *et al.*, 1993). Note that this level of UV light is about 25 times less than that from an 8-W UV fluorescent lamp. Figure 6.11a illustrates a typical non-concentrating falling-film solar photoreactor for water purification. Building such a photoreactor is not trivial, since it needs a large area of weather-resistant, chemically inert, cheap, UV-light-transmitting glazing. In addition, careful control of the reaction solution is required, especially if the photocatalyst is used in slurry, rather than fixed bed form (Turchi *et al.*, 1993).

Figure 6.11b illustrates the basic features of a parabolic photoreactor, which allows any incident sunlight to be concentrated on the reaction solution (Matthews, 1986; Turchi *et al.*, 1993). Such parabolic units, when collected together, form a compound parabolic concentrator, that is a CPC. A CPC allows more expensive components, usually supported photocatalysts, to be used in the construction of the photoreactor without adversely affecting the overall cost. When compared to a non-concentrating photoreactor, a CPC has a higher light-captivity efficiency and, therefore, will occupy a smaller area under one sun operation (Turchi *et al.*, 1993). In a CPC, the bank of reflectors are usually inclined, facing south, but do not actively track the sun. Obviously, not all places in the world are sunny enough to justify using a solar-powered system. As a result, most solar-powered water-purification systems will be limited to areas where the solar UV irradiation is >0.015 mW cm^{-2}, that is typically between latitudes 35°N and 35°S.

Both non-concentrating and concentrating solar reactions can be operated using the photocatalyst as a slurry (moving film) or a fixed film. The former often appears to be the preferred embodiment in concentrating solar systems, because of the high surface area to reaction solution volume ratio it affords, despite the need for a subsequent separation step. However, as indicated by Equation (6.3) and noted earlier, the kinetics of semiconductor photocatalysis will be proportional to $I^{0.5}$ at high UV levels. As a result, most concentrating (i.e. high UV-light intensity) solar systems for water purification are not as photon efficient, that is have a lower value of δ, than most non-concentrating solar photoreactors. The higher photon efficiencies of non-concentrating solar photocatalytic reactors, and their added ability to capture diffuse UV sunlight, make these systems the most cost-effective photocatalytic systems for purifying water using sunlight. Any wastewater solar-powered detoxification process will most likely be run in batch mode with the solution recirculated until the levels of pollutants are at an acceptable low level for discharge. All solar-detoxification systems must work independently from the process generating the waste and, as a result, are unlikely to be used for online wastewater treatment.

6.5 COMMERCIAL PROCESSES

Semiconductor photocatalysis is still in its infancy with regard to commercialization and this may appear slightly surprising given the intense level of academic research that has been conducted over the last decade. It appears that, despite its promise as an advanced oxidation process capable of mineralizing a wide range and number of water-borne pollutants, the overall low efficiency of operation of current laboratory- and pilot-scale systems is sufficiently unattractive to prevent adoption by any of the major water-purification companies. This low overall efficiency derives from a number of factors, some of which have been mentioned before, but which are outlined below for completeness.

Thus, most UV lamps have modest (10–20% at best) electrical energy to UV-light energy conversion efficiencies. However, despite this inefficiency, UV-light disinfection of water is already a well-established method in the water-treatment industry. Coating the reactor walls of any commercial annular UV-photodisinfection system with TiO_2 should improve its overall efficiency, given the antibacterial action of TiO_2, and it is slightly surprising that this option has not been the subject of intensive research by the big water UV-disinfection companies.

The overall process of semiconductor photocatalysis is itself intrinsically low (typically $<10\%$), mainly due to significant electron–hole recombination at the typical light intensities used in this work. In practice, the overall photonic efficiencies of most photocatalytic systems are very low (i.e. $<1\%$) due to additional efficiency-lowering processes, such as light scattering and reflection, and mass transfer effects.

The most efficient and cheapest to produce photocatalytic systems use the semiconductor in the form of a powder dispersion, but such systems then require a subsequent filtration step which instantly compromises the system's economical viability. Fixed-bed photocatalytic reactors appear to offer an attractive alternative to dispersed photocatalyst systems, but their initial construction costs are comparatively high. In addition, with such fixed-bed systems, if regular photocatalyst replacement is necessary, due to some passivation process, then maintenance costs are likely to prove prohibitive.

Table 6.6 TiO_2 photocatalyst systems for the purification of water.

Company	Country	Patents	Comments	Web site
ISK	Japan	US5541096	Large TiO_2 manufacturer, Tipaque®, with a patent on water purification using photocatalysis and a novel, porous catalyst.	www.iskweb.co.jp
Hyosung Ebara	Korea	JP 2000237759	Major water-purification company with designs for a semiconductor photocatalyst water-purification systems, but no obvious products.	www.heec.co.kr
Clearwater Industries	USA		Small company with Florida University origins. Flat bed semiconductor photocatalyst reactor can treat $2250\,L\,min^{-1}$.	www.cwirfc.com
Photox Bradford Ltd.	UK		Small company with Bradford University origins. Its semiconductor photocatalyst reactor uses the TiO_2 as a slurry and can treat $168\,L\,min^{-1}$.	www.vcb.co.uk/photox/
Lynntech Inc.	USA	US5779912	A small company selling a fixed-bed semiconductor photocatalyst reactor.	www.lynntech.com
Purifics® Environmental Technologies Inc.	Canada	US6136203, US5589078, US5462674	Manufacturers of the Photo-Cat® water and air automated treatment system and largest supplier of industrial semiconductor photocatalyst treatment systems.	www.purifics.com

Although semiconductor photocatalysis appears to be shunned at present by most major water-purification and -disinfection companies as an economically viable alternative to current methods, such as chlorination and ozonolysis, it has been taken up by companies or government agencies, which have small, specific water-purification needs, such as those found in the electronics industry (high purity water), the dye and car manufacturer industries, and the military and petrochemical industries (effluent or site clean up).

Table 6.6 lists most of the major companies that currently promote semiconductor photocatalysis systems for water purification. The table also includes examples of relevant patents and references to the company web sites. Not surprisingly, a good number of the companies listed in Table 6.6 appears to have 'products' that are still very much in the development stage. Thus, the Japanese company – Ishihara Sangyo Kaisha (ISK) – is a major manufacturer of TiO_2 and has a US patent on water purification, using porous TiO_2 containing inorganic particles, but

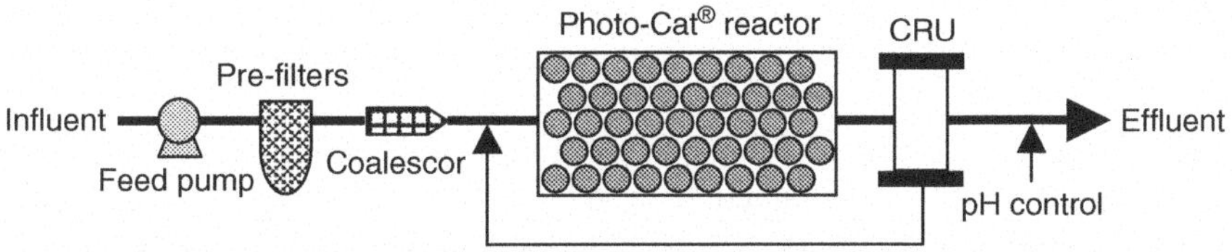

Figure 6.12 Schematic representation of the basic features of a Photo-Cat® reactor sold by Purifics Inc. The unit comprises pre-filters, a coalescor, a photocatalyst unit, a CRU and a pH adjuster feed on the effluent channel. This unit can be fully automated.

does not appear to have a semiconductor photocatalyst water-purification product as yet. Instead, ISK, like many others, has conducted some initial research and may develop a product sometime in the future. Similarly, the environmental engineering company – Hyosung Ebara – based in Korea, has an established track record in the management of wastewater and, on its web site, identifies semiconductor photocatalysis as one of the advanced oxidation processes it has available at its disposal. However, Hyosung Ebara also does not appear to have a major commercial product at the present time.

Given the substantial academic interest in the area of semiconductor photocatalysis, it is not surprising to find that two companies, namely Clearwater Industries and Photox Bradford, have university origins, that is the University of Florida and the University of Bradford, respectively. Clearwater Industries claim to be able to treat over 2250 L min^{-1} of contaminated water using their semiconductor photocatalyst water purification system, R2000. Although there are very few details on the web site concerning the workings of the R2000 solar oxidation system, it appears that it is a fixed photocatalyst-bed reactor and, unlike all other commercial photocatalytic water-purification systems, it utilizes the UV component in sunlight to drive the photocatalytic mineralization process, rather than an artificial UV source.

As with Clearwater Industries, Photox Bradford appears to be a small company that is creating custom-built, small-scale photoreactors for water purification. In contrast to the Clearwater Industries R2000 solar oxidation system, the Photox™ system utilizes a slurry of TiO_2 in which the particles are sufficiently large that a standard, off-the-shelf separator can be used to remove and return to the photoreactor from the purified water. The Photox™ system claims to be able to treat 168 L min^{-1} of contaminated water. Linntech Inc. is yet another small company with a patented small-scale, fixed-bed, PCO reactor for water purification.

All the semiconductor photocatalyst water-purification companies discussed so far are very much overshadowed by Purifics® Environmental Technologies Inc., a Canadian-based water-purification company dedicated to the application of semiconductor photocatalysis for the purification of water. The Purifics® system, called Photo-Cat®, utilizes the TiO_2 in the form of a slurry. The basic components of this system are illustrated in Figure 6.12. The system comprises pre-filters to remove any fine solid material, a coalescor to remove any emulsified oils or greases above 7 μm in size, the Photo-Cat® reactor to photomineralize the pollutants and the catalyst recovery unit (CRU) to continuously separate the photocatalyst particles from the treated water and return the former to the inlet stream of the reactor and the latter to the effluent stream. The TiO_2 is used in unsupported, that is neat powder, dispersed form. The Photo-Cat® reactor uses long-lasting (14,000 h) low-pressure mercury fluorescent lamps, that is germicidal lamps and the TiO_2 slurry plus polluted water is contained in a set of stainless-steel tubes, connected together to form a series of racks. These racks are, in turn, linked together in serial and/or parallel mode depending upon the concentration of contaminants and the required throughput.

From the associated patents, see Table 6.6, it appear that in the Photo-Cat® reactor, each reactor tube is an annulus-type reactor, see Figure 6.8c, with a central germicidal lamp surrounded by a quartz tube and a subsequent outer stainless-steel tube; the TiO_2 slurry and polluted solution are flowed through the annulus gap between the walls of the quartz and stainless-steel tubes. The size of the gap is typically less than 1 mm, to ensure a high degree of turbulence, due to shear stress from the flowing reaction solution, which in turn ensures that mass transfer effects will be minimal. From the associated patents, the key CRU appears to be a cross-flow-type filtration system using a ceramic (alumina) membrane (pore size: 0.2 μm) and a trans-membrane pressure of *ca.* 10 psi. Usually, the pores of such a membrane would be quickly blocked by the photocatalyst particles, but this is prevented by the application to the ceramic filter of a 0.5 s shock wave (150 psi), typically every 5 min from the effluent side. The photocatalyst powder used in the Photo-Cat® is cheap ($1) and can be changed within 30 min. The lamps are relatively inexpensive (lamps costs are estimated at $85 per kW of system size per year) and can be changed within 30 s. Purifics® claim that their Photo-Cat® system is not affected by iron fouling, since by lowering the solution pH to 3 or less, the Photo-Cat® system keeps any iron dissolved in the aqueous phase. Figure 6.13 illustrates the

Figure 6.13 Typical 60-kW Photo-Cat® reactor system, courtesy of Purifics® (Canada) 2002.

appearance of a typical 60-kW Photo-Cat® system. Note the racks of stainless-steel tubes, each containing some of the photocatalyst slurry plus a germicidal lamp, that go to form the Photo-Cat® reactor.

The overall system is essentially a solid-state device that can be operated unattended for long periods of time. The manufacturers claim that the amount of TiO_2 supplied with the system will last the expected lifetime of the unit, although it is not clear what the unit's lifetime is. Purifics® is the largest supplier of industrial photocatalytic treatment systems and has found markets in the USA, Canada and Korea. Most of these systems have been used to treat contaminated groundwater; however, a few have been used to treat industrial wastewater, lagoons and air.

6.6 CASE STUDIES AND ECONOMICS

6.6.1 Case studies

Given the limited number of companies currently promoting semiconductor photocatalysis products, the number of case studies is also limited and usually not well documented. However, the following case studies provide a useful indication of the progress and promise of semiconductor photocatalysis for water purification.

Purifics® are the biggest supplier of photocatalyst systems for water purification, and, not surprisingly, the company reports a number of impressive case histories regarding their Photo-Cat® system. The basic components of the Photo-Cat® system have been described earlier and are illustrated in Figure 6.12. The Photo-Cat® system is driven by germicidal UV light and uses a continuously recirculated TiO_2 slurry in the photoreactor. Two brief case histories for the Photo-Cat® system are as follows.

6.6.1.1 Petrochemical waste site, Glaveston, Texas

Problem Groundwater with contaminant, levels are reduced from 200,000 ppb to 5 ppb. In particular levels of 200 ppm of bis(2-chloroethyl), ether is present which is sufficient to dissolve any plastic components in any treatment system. The groundwater also contained chloride levels of 6000 ppm. A 19-kW Photo-Cat® system was used to reduced the level of bis(2-chloroethyl) ether to 20 ppb. The installed Photo-Cat® unit was designed to operate at 15–30 L min^{-1} in fully automated mode and was operated successfully for over 35,000 h.

6.6.1.2 Contaminated surface water in Ontario, Canada

Problem Surface water with high levels of oil and grease discharges to the Thames river. The Photo-Cat® system was used to treat continuously 30,000 gallons per day from three wells. Contaminants include trichloroethene (TCE), 1,2-dichloroethene (DCE), dichloroethane (DCA), trichloroethane (TCA), vinyl chloride, oil and grease, all present in the low ppm range (wells 1 and 2) and >1000 ppm range (well 3). A low concentration of H_2O_2 was added as an additional scavenger of photogenerated electrons to improve the overall performance of the Photo-Cat® reactor. This system has been operating successfully since 1998. The overall capital cost of this system is reported as \$194,000 and the operating and maintenance costs are typically \$18,000 per year.

6.6.1.3 Gainesville, Florida

Problem A 500 gallon underground aquifier contaminated with benzene, toluene, ethyl benzene and xylene (BTEX) and State legislation requiring that it be reduced to 65 ppb prior to returning the treated water back to the environment. Trials were conducted to measure the effectiveness of treating this water using a conventional carbon absorption method and the Clearwater Industries R2000 solar oxidation facility. The latter uses a modular set of tilted flat bed solar photoreactors, with a slurry TiO_2 photocatalyst. The trials showed that the R2000 was significantly less expensive to install and that it detoxified the polluted water at a fraction of the operating cost, compared with active carbon absorption. Thus, the R2000 system was able to reduce the level of BTEX to well below 65 ppb within 3 h of use on a cloudy day (UV-light intensity $\cong 31\,\mathrm{W\,m^{-2}}$). In their promotional material Clearwater Industries claims that the R2000 system can be used to treat contaminated water at flow rates ranging from 19 to 19,000 $\mathrm{L\,min^{-1}}$.

6.6.2 Economics

Table 6.7 highlights some of the important features of the major destructive methods that are currently used by the big water companies to purify water, and compares them with those of semiconductor photocatalysis (SPC). From this table, it is clear that the latter process has a number of advantages compared with the other, more established destructive water-purification processes, but is neither rapid in action, nor easy to use/maintain, nor cheap to buy.

Table 6.8 provides estimates of the likely operating costs of a number of different sized water-purification systems based on carbon adsorption, UV ozonolysis

Table 6.7 Comparison of destructive water-purification methods.

Method	UV	O_3	UV/O_3	Cl_2	SPC
Features					
Destroys chlorinated hydrocarbons	✗	✓	✓	✓	✓
Total pollutant mineralization usual	✗	✗	✓	✗	✓
Broad compound compatability	✗	✓	✓	✓	✓
Use non-hazardous oxidant	✓	✗	✗	✗	✓
Simple to use	✓	✓	✓	✓	✗
Rapid pollutant destruction	✗	✓	✓	✓	✗
Low operating costs	✓	✗	✗	✓	✓
Low initial equipment costs	✓	✗	✗	✓	✗

and semiconductor photocatalysis (Ollis, 1988). From the data in this table, semiconductor photocatalysis appears very competitive as a method of water purification compared to the other methods listed. However, these estimates are based on an assumed very low maintenance cost, and it is not clear yet that a semiconductor photocatalysis system would exhibit this highly desirable feature over many years of operation when used to purify contaminated water from a variety of different sources, such as groundwater, industrial and household effluent. However, Purifics® reports that their Photo-Cat® reactor system has costs due to electrical consumption, operating and maintenance that are typically a third of those associated with other advanced oxidation processes, such as UV/H_2O_2, UV/O_3 and GACarbon. The capital costs of the Photo-Cat® reactor system are typically on a par with those of UV/H_2O_2 and UV/O_3, but twice those of GACarbon.

The initial pilot plant case studies reported above provide promising evidence that semiconductor photocatalysis systems can be constructed to work very efficiently, with little maintenance, at purifying polluted water for at least 1–2 years. However, given the possibility that a semiconductor photocatalysis system can be passivated by metal oxides or hydroxides, or rendered inactive by highly UV-absorbing pollutants, and that the currently commercially preferred embodiment appears to involve the use of slurries and a subsequent filtration step, it is clear that semiconductor photocatalysis is unlikely to emerge as a frontline, major method of water purification, for raw or just sand-filtered effluent. Rather, semiconductor photocatalysis is likely to find a niche for purifying pre-treated (i.e. at least filtered and possibly deionized) water containing highly toxic, hazardous and non-biodegradable pollutants, such as explosives or some petrochemicals. In addition, it appears likely that solar-powered semiconductor photocatalysis systems will emerge as a viable method for producing purified water in remote locations, provided the initial costs of construction of such systems can be kept low.

Galvez and his co-workers (Galvez and Rodriguez, 2002) have costed a 300 m^2 solar-detoxification plant, treating 6 million litres per year and operating for 3000 h per year, based on a semiconductor photocatalysis system containing 5–10 mM sodium persulphate as an added sacrificial electron donor. The capital costs are estimated at \$140,000 and the operating and maintenance costs are estimated at \$66,000 per year. From these figures, it appears that this system is much more

Table 6.8 Estimated operating costs*.

System size (MGD)	0.029	0.058	0.115	0.23	0.46	0.92	2.44
Carbon adsorption[a]	\$4.9	\$3.4	\$2.7	\$2.0	\$1.7	\$1.4	\$1.2
UV ozone[b]	\$8.0	\$5.3	\$4.0	\$3.1	\$2.7	\$2.4	\$2.0
UV photocatalysis	\$6.2	\$4.0	\$2.7	\$2.0	\$1.6	\$1.5	\$1.3

*\$/litre; data derived from estimates made by Ollis in 1987 (Ollis, 1988) coupled with an assumed 10% rise per year that is a multiple of 4.18.

[a] The cost of an incinerative regenerator is included in the largest size only and no carbon disposal costs are included for the smaller units.

[b] Based on the assumption that the operating costs will be similar to those for a UV-ozone process, without the running costs for the ozone generator.

expensive to run than the artificial light-driven Photo-Cat® system sold by Purifics®. However, of the $66,000 running costs associated with the solar-powered system, $54,000 is due to the cost of added persulphate. Obviously, if the system was run without added electron scavenger, both the capital and operating costs for the solar-powered photocatalytic system would be less than those for the Purifics® system used in treating contaminated surface water in Ontario.

Such comparisons are of very limited use, since they are based on very different sources of information and separate studies. More useful comparisons need to be made, using independent data obtained from side-by-side studies of both solar and artificial light-driven semiconductor photocatalytic systems, and various popular conventional wastewater-treatment processes. Only when such work has been carried out, will the real potential of semiconductor photocatalysis as a commercially viable alternative method of water purification be understood and possibly appreciated. As it stands, the research results look promising, but the transformation from a brilliant concept to a vibrant commercial reality appears to be slow. Over the next few years, it will become apparent if semiconductor photocatalysis has a real future in water purification or is doomed to stay on the peripheries as an example of an advanced cost, advanced oxidation process.

REFERENCES

Al-Sayyed, G., D'Oliveira, J.C. and Pichat, P. (1991) Semiconductor-sensitized photodegradation of 4-chlorophenol in water. *J. Photochem. Photobiol. A: Chem.* **58**, 99–114.

Bhatkhande, D.S., Pangarkar, V.G. and Beenackers, A.A.C.M. (2001) Photocatalytic degradation for environmental applications – a review. *J. Chem. Technol. Biotechnol.* **77**, 102–116.

Braun, A.M. (1988) Technical development and industrial prospects of preparative photochemistry. In: *Photocatalysis and Environment*, Schiavello (ed.). Kluwer Academic Publishers, Amsterdam, pp. 601–622.

Cassano, A.E. and Alfano, O.M. (1991) Photochemical engineering design methods. A rigorous approach for homogeneous systems. In: *Photochemical Conversion and Storage of Solar Energy*, Pelizzetti, E. and Schiavello, M. (eds). Kluwer Academic Publishers, Amsterdam, pp. 623–652.

Emeline, A.V., Ryabchuk, V. and Serpone, N. (2000) Factors affecting the efficiency of a photocatalysed process in aqueous metal-oxide dispersions: prospect of distinguishing between two models. *J. Photochem. Photobiol. A: Chem.* **133**, 89–97.

Foster, N.S., Koval, C.A., Sczechowski, J.G. and Noble, R.D. (1996) Investigation of controlled periodic illumination effects on photo-oxidation process at titanium dioxide films using rotating ring disk photoelectrochemistry. *J. Electroanal. Chem.* **406**, 213–217.

Galvez, J. and Rodriguez, S.M. (2002) Solar detoxification, UNESCO, on-line publication: http://www.unesco.org/science/wsp/publications/solar.htm.

Haarstrick, A., Kut, O.M. and Heinzel, A. (1996) TiO_2-assisted degradation of environmentally relevant organic compounds in waste water using a novel fluidised bed photoreactor. *Environ. Sci. Technol.* **30**, 817–824.

Hoffmann, M.R., Martin, S.T., Choi, W. and Bahnemann, D.W. (1995) Environmental applications of semiconductor photocatalysis. *Chem. Rev.* **95**, 69–96.

Hoffstadler, K., Bauer, R., Novalic, S. and Heisler, G. (1994) New reactor design for photocatalytic wastewater treatment with TiO_2 immobilised on fused-silica glass fibers: photomineralisation of 4-chlorophenol. *Environ. Sci. Technol.* **28**, 670–674.

Ireland, J.C., Klostermann, P., Rice, E.W. and Clark, R.M. (1993) Inactivation of *Escherichia coli* by titanium dioxide photocatalytic oxidation. *Appl. Environ. Microbiol.* **59**, 1668–1670.

Linsebigler, A.L., Lu, G. and Yates, J.T. (1995) Photocatalysis on TiO_2 surfaces: principles, mechanisms and selected results. *Chem. Rev.* **95**, 735–758.

Matsunaga, T. and Okochi, M. (1995) TiO_2-mediated photochemical disinfection of *Escherichia coli* using optical fibers. *Environ. Sci. Technol.* **29**, 501–505.

Matsunaga, T., Tomoda, R., Nakajima, T., Nakamura, N. and Komine, T. (1988) Continuous-sterilization system that uses photosemiconductor powders. *Appl. Environ. Microbiol.* **54**, 1330–1333.

Matthews, R.W. (1986) Photo-oxidation of organic material in aqueous suspensions of titanium dioxide. *Water Res.* **20**, 569–578.

Matthews, R.W. (1987a) Photooxidation of organic impurities in water using thin films of titanium dioxide. *J. Phys. Chem.* **91**, 3328–3333.

Matthews, R.W. (1987b) Solar-electric water-purification using photocatalytic oxidation with TiO_2 as a stationary phase. *Solar Energy* **38**, 405–413.

Mills, A. and Le Hunte, S. (1997) An overview of semiconductor photocatalysis. *J. Photochem. Photobiol. A: Chem.* **108**, 1–35.

Mills, A. and Morris, S. (1993) Photomineralisation of 4-chlorophenol sensitised by titanium dioxide: a study of the initial kinetics of carbon dioxide generation. *J. Photochem. Photobiol. A: Chem.* **71**, 75–83.

Mills, A. and Wang, J. (1999) The kinetics of semiconductor photocatalysis; light intensity effects. *Z. Phys. Chem.* **213**, 49–58.

Mills, A., Belghazi, A. and Rodman, D. (1996) Bromate removal from drinking water by semiconductor photocatalysis. *Water Res.* **30**, 1973–1978.

Mukherjee, P.S. and Ray, A.K. (1999) Major challenges in the design of a large-scale photocatalytic reactor for water treatment. *Chem. Eng. Technol.* **22**, 253–260.

Ollis, D.F. (1988) Process economics for water purification: a comparative assessment. In: *Photocatalysis and Environment*, Schiavello (ed.). Kluwer Academic Publishers, Amsterdam, pp. 663–677.

Ollis, D.F. and Turchi, C. (1990) Heterogeneous photocatalysis for water purification: contaminant mineralisation kinetics and elementary reactor analysis. *Environ. Prog.* **9**, 229–234.

Ollis, D.F., Pelizzetti, E. and Serpone, N. (1991) Destruction of water contaminants. *Environ. Sci. Technol.* **25**, 1523–1529.

Pacheco, J.E. and Holmes, J.T. (1990) Falling-film and glass-tube solar catalytic reactors for treating contaminated water. In: *Emerging Technologies in Hazardous Waste Management*, Tedder, D.W. and Pohland, G.F. (eds). American Chemical Society, Washington, pp. 40–51.

Peill, N.J. and Hoffmann, M.R. (1995) Development and optimisation of a TiO_2-coated fiber-optic cable reactor: photocatalytic degradation of 4-chlorophenol. *Environ. Sci. Technol.* **29**, 2974–2981.

Peill, N.J. and Hoffmann, M.R. (1996) Chemical and physical characterisation of a TiO_2-coated fiber-optic cable reactor. *Environ. Sci. Technol.* **30**, 2806–2812.

Peill, N.J. and Hoffmann, M.R. (1997) Solar-powered photocatalytic fiber-optic cable reactor for waste stream remediation. *J. Solar Eng.* **119**, 229–236.

Porter, J.F. and Yue, P.L. (1993) Simple supported photocatalysts. In: *Photocatalytic Purification and Treatment of Water and Air*, Ollis, D.F. and Al-Ekabi, H. (eds). Elsevier Science, Amsterdam, pp. 759–764.

Puma, G.L. and Yue, P.L. (1998) A laminar falling film slurry photocatalytic reactor. Part II – experimental validation of the model. *Chem. Eng. Sci.* **53**, 3007–3021.

Puma, G.L. and Yue, P.L. (1999a) Enhanced photocatalysis in a pilot laminar falling film slurry reactor. *Ind. Eng. Chem. Res.* **38**, 3246–3254.

Puma, G.L. and Yue, P.L. (1999b) Comparison of the effects of photon-based oxidation processes in a pilot falling film photoreactor. *Environ. Sci. Technol.* **33**, 3210–3216.

Puma, G.L. and Yue, P.L. (2001a) A novel fountain photocatalytic reactor: model development and experimental validation. *Chem. Eng. Sci.* **56**, 2733–2744.

Puma, G.L. and Yue, P.L. (2001b) A novel fountain photocatalytic reactor for water treatment and purification: modelling and design. *Ind. Eng. Chem. Res.* **40**, 5132–5169.

Ray, A.K. (1999) Design, modelling and experimentation of a new large-scale photocatalytic reactor for water treatment. *Chem. Eng. Sci.* **54**, 3113–3125.

Ray, A.K. and Beenackers, A.A.C.M. (1998) Development of a new photocatalytic reactor for water purification. *Catal. Today* **40**, 73–83.

Saito, T., Iwase, T., Horie, J. and Morioka, T. (1992) Mode of photocatalytic bactericidal action of powdered semiconductor TiO_2 on mutans streptococci. *J. Photochem. Photobiol. B: Biol.* **14**, 369–379.

Serpone, N. and Emeline, A.V. (2002) Suggested terms and definitions in photocatalysis and radiocatalysis. *Int. J. Photoenerg.* **4**, 91–131.

Sjogren, J.C. and Sierka, R.A. (1994) Inactivation of phage MS2 by iron-aided titanium dioxide photocatalysis. *Appl. Environ. Microbiol.* **60**, 344–347.

Toyoda, A., Zhang, L., Kanki, T. and Sano, N. (2000) Degradation of phenol in aqueous solution by TiO_2 photocatalyst coated rotating-drum reactor. *J. Chem. Eng. Jpn.* **33**, 118–191.

Turchi, C.S. and Ollis, D.F. (1989) Mixed reactant photocatalysis: intermediate and mutual rate inhibition. *J. Catal.* **119**, 483–496.

Turchi, C.S. and Ollis, D.F. (1990) Photocatalytic degradation of organic-water contaminants – mechanisms involving hydroxyl radical attack. *J. Catal.* **122**, 178–192.

Turchi, C.S. and Ollis, D.F. (1998) Photocatalytic reactor design: an example of mass-transfer limitations with an immobilised catalyst. *J. Phys. Chem.* **92**, 6852–6853.

Upadhia, S. and Ollis, D.F. (1997) Simple photocatalysis model for photoefficiency enhancement via controlled, periodic illumination. *J. Phys. Chem.* **101**, 2625–2631.

Turchi, C., Mehos, M. and Pacheco, J. (1993) Design issues for solar-driven photocatalytic systems. In: *Photocatalytic Purification and Treatment of Water and Air*, Ollis, D.F. and Al-Ekabi, H. (eds). Elsevier, Amsterdam, pp. 789–794.

Watts, R.J., Kong, S., Orr, M.P., Miller, G.C. and Henry, B.E. (1995) Photocatalytic inactivation of coliform bacteria and viruses in secondary wastewater effluent. *Water Res.* **29**, 95–100.

Wei, C., Lin, W.Y., Zainal, Z., Williams, N.E., Zhu, K., Kruzic, A.P., Smith, R.L. and Rajeshwar, K. (1994) Bactericidal activity of TiO_2 photocatalyst in aqueous media: toward a solar-assisted water disinfection system. *Environ. Sci. Technol.* **28**, 934–938.

Yatmaz, H.C., Howarth, C.R. and Wallis, C. (1993) Photocatalysis of organic effluents in a falling film reactor. In: *Photocatalytic Purification and Treatment of Water and Air*, Ollis, D.F. and Al-Ekabi, H. (eds). Elsevier Science, Amsterdam, pp. 795–800.

Zhang, P., Scrudato, R.J. and Germano, G. (1994) Solar-catalytic inactivation if *Escherichia coli* in aqueous solution using TiO_2 as catalyst. *Chemosphere* **28**, 607–611.

7

Photoelectrocatalysis processes

T.A. Egerton and P.A. Christensen

7.1 INTRODUCTION

This chapter describes the photoelectrocatalytic disinfection of water by titanium dioxide (TiO_2). As with other broad-band gap oxide semiconductors, TiO_2 absorbs in the ultraviolet (UV) region of the spectrum and this absorbed energy initiates subsequent catalytic activity. The absorption properties of a number of broad-band semiconductor oxides are listed in Table 7.1 (Henrich and Cox, 1998; Faughnan and Kiss, 1968), which shows that the absorption edge – the wavelength at which absorption becomes significant – depends not only on the chemical composition of the oxide, but also on its crystal form. Thus the rutile form of TiO_2 begins to absorb strongly at 405 nm, while the anatase form does not absorb till 385 nm – a consequence of the slightly larger energy gap between the valence and conduction bands of anatase. Although this chapter will focus on TiO_2, photoelectrocatalysis on other UV absorbing, semiconducting oxides is also possible. Mixed oxides, such as $SrTiO_3$ (Faughnan and Kiss, 1968), have shown no consistent advantage over TiO_2 in conventional photocatalysis and have, therefore, not yet been evaluated in

 Advanced Oxidation Processes for Water and Wastewater Treatment. Edited by Simon Parsons. ISBN: 1 84339 017 5

Table 7.1 The band gaps of some representative oxide semiconductors.

Oxide	Crystalline form	Band gap (eV)	Absorption edge (nm)
TiO_2	Rutile	3.05	405
TiO_2	Anatase	3.25	385
$SrTiO_3$		~3.3	~380
SnO_2	Cassiterite	3.6	~345
ZnO	Wurtzite	3.2–3.4	~400

photoelectrocatalysis. Although zinc oxide is, at first sight, an obvious competitor to TiO_2, it is not used for water treatment, because it slowly dissolves under practical conditions. However, the photochemical behaviour of ZnO has been studied in non-aqueous electrolytes (e.g. Bahadur and Roy, 1993). Tungsten oxide (WO_3) has also received significant attention either in its own right (Hepel and Luo, 2001) or as mixed systems, such as WO_3/TiO_2 (Li *et al.*, 2000; Shiyanovskaya and Hepel, 2000).

7.1.1 Mechanism of photocatalysis by TiO_2

Photocatalysis by TiO_2 has been known for more than 50 years mainly because of its adverse effect in paint, polymers and fibres (Christensen *et al.*, 2000; Egerton and Shah, 1971). The more recent application of TiO_2 to the destruction of organic pollutants in water has been extensively reviewed (see e.g. Christensen and Walker, 1996; Mills and Le Hunte, 1997; Rajeshawar, 1995; Tryk *et al.*, 2000).

The main mechanistic steps which underlie both photocatalytic and photoelectrocatalytic detoxification (i.e. the destruction of organic pollutants) and disinfection (the killing of biological pollutants) are well understood (Chapter 6). First, absorption of UV of energy greater than the semiconductor band gap excites an electron, e^-, from the valence to the conduction band and leaves behind an electron deficit, a positive hole h^+, in the valence band. The excited electrons migrate to the TiO_2 surface, where they react, for example, with adsorbed oxygen to form O^{2-}. The positive holes also migrate to the oxide surface, where they oxidize surface OH^- or water to hydroxyl radicals, $OH^\bullet$ (Figure 7.1).

Although the $OH^\bullet$ radical is extremely short lived, the redox potentials listed in Table 7.2 demonstrate that it is highly oxidizing. Indeed, it is significantly more oxidizing than chlorine, the most commonly employed disinfectant used in water treatment, and second only to fluorine. As a result of this, $OH^\bullet$ radicals are capable of killing pathogens, such as *Cryptosporidium parvum* (Simmons, 1998), that are resistant to chlorination and have resulted in an increasing number of water supply contaminations, the most recent of which was in Glasgow, UK, in August 2002.

There are a number of chemical and/or photochemical systems that generate $OH^\bullet$ radicals known collectively as advanced oxidation processes or AOPs (Christensen and Walker, 1996), and these have been found to be more effective

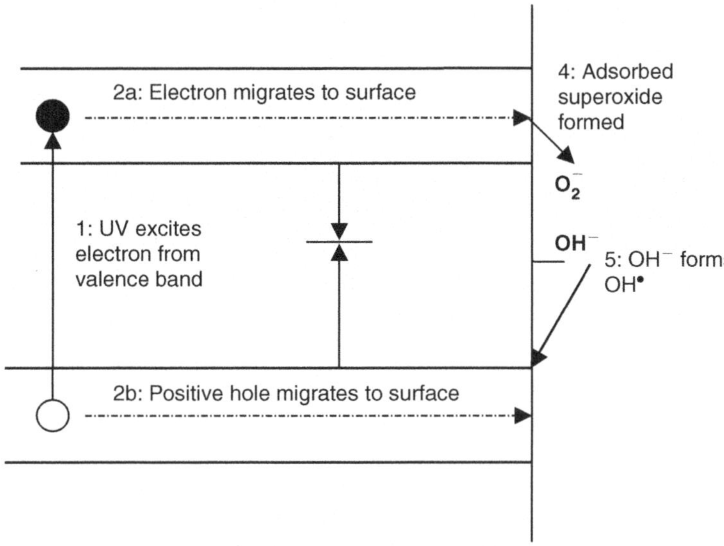

Figure 7.1 UV excitation of TiO_2 generates hydroxyl radicals at the surface of the semiconductor particle.

Table 7.2 The redox potential of a number of chemical systems used for water treatment.

Oxidant	Redox potential (V vs NHE)
F_2	3.03
$^•OH$	2.80
$^•O$	2.42
O_3	2.07
H_2O_2 (acidic)	1.78
Cl_2	1.36

than ozone alone in the oxidation of the precursors of trihalomethanes (THMs) produced *via* the reaction of chlorine with natural organics in water (Ponitus, 1990). Photocatalysis by TiO_2 is an AOP, the two most common of which involve generation of $OH^•$ by UV irradiation of H_2O_2 ($\lambda \leqslant 254$ nm) and/or O_3 ($\lambda \leqslant 310$ nm) (Christensen and Walker, 1996; Ireland, 1993). The UV irradiation of TiO_2 generates $OH^•$ radicals without the use of potentially hazardous short wave radiation or the addition of chemicals, and these are significant advantages. However, there is an associated disadvantage; during photocatalysis by TiO_2, most electrons and holes recombine before they reach the surface of the solid particle (Egerton and King, 1979; Sun and Bolton, 1996). Consequently, much of the absorbed UV does not generate $OH^•$ radicals, but is simply turned into heat and warms the catalyst! Other electrons and holes are trapped either at the surface or within the bulk of the TiO_2.

7.1.2 A simple kinetic model of hydroxyl radical generation

Assuming that the rate of electron transfer to oxygen is faster than the rate of hole trapping to form hydroxyl radicals, Egerton and King (1979) showed that a simple kinetic model may be written for the rate of hydroxyl radical formation.

If I is the intensity of radiation and k_1 is the rate constant for the absorption of the light, we can write

$$\text{Rate of hole formation} = k_1 I \tag{7.1}$$

$$\text{Rate of formation of surface hydroxyls} = k_{s,h}[\mathrm{h}^+] \tag{7.2}$$

$$\text{Rate of charge recombination} = [n]k_n[\mathrm{e}^-][\mathrm{h}^+] = [n]k_n[\mathrm{h}^+]^2 \tag{7.3}$$

where $[\mathrm{e}^-]$ and $[\mathrm{h}^+]$ are the concentration of electrons and holes, respectively, and k_n is the rate of recombination at recombination centres whose concentration is $[n]$. The simplifying assumption has been made that $[\mathrm{e}^-] = [\mathrm{h}^+]$; more exactly that the rates of disappearance of electrons and holes (and, therefore, the formation of surface hydroxyls and surface $O_2^{\bullet -}$) are equal, $k_{s,h}[\mathrm{h}^+] = k_{s,e}[\mathrm{e}^-]$, but this does not alter the conclusions. In practice, there will be many different types of recombination centre, some at the surface of the catalyst and some within the bulk and the overall electron–hole recombination rate may be represented as a summation over all these types:

$$\text{Over recombination rate} = \Sigma[n]k_n[\mathrm{h}^+]^2 = k_2[\mathrm{h}^+]^2 \tag{7.3a}$$

The steady state concentration of holes may then be derived by balancing their rates of formation and removal. This gives

$$\frac{\mathrm{dh}}{\mathrm{d}t} = k_1 I - k_2[\mathrm{h}^+]^2 - k_s[\mathrm{h}^+] = 0 \tag{7.4}$$

When I is large, and this is the practical case, even for solar irradiation in North East England (!), a large number of holes and electrons is generated and hence $k_2[\mathrm{h}^+]^2 >> k_s[\mathrm{h}]$ and the approximate solution is

$$k_1 I = k_2[\mathrm{h}^+]^2 \tag{7.5}$$

and the steady state concentration of holes is $[\mathrm{h}^+] = [k_1 I/k_2]^{0.5}$. Inserting this in Equation (7.2) leads to

$$\text{Rate of hydroxyl radical generation} = k_s\left[\frac{k_1 I}{k_2}\right]^{0.5} \tag{7.6}$$

7.1.3 Consequences of the kinetic model

Although the above treatment is, necessarily, grossly simplified [see e.g. the recent work on the spectral dependence of photocatalytic transformation of phenols by Emeline *et al.* (2000)], it has been shown experimentally (Egerton and King, 1979) that $k_2[h^+]^2 >> k_s[h^+]$ and because of this, the efficiency of hydroxyl radical generation is normally very low. Even in the most favourable cases of photocatalysis, only approximately four hydroxyl radicals are produced for every 100 UV photons that are absorbed by a TiO_2 crystal (Sun and Bolton, 1996). This is because most of the UV-generated electrons and positive holes recombine before they reach the oxide surface (the minority carrier length, L_p, the average distance that the holes can move in a field-free region of TiO_2 before recombination, is $\leqslant 0.1\ \mu m$). A second consequence of charge carrier recombination is that the dependence of photooxidation with light intensity follows an $I^{0.5}$ dependence – a fourfold increase in incident UV increases oxidation rate by only a factor of 2 (Egerton and King, 1979; Emeline *et al.*, 2000). For these reasons, it has been concluded that the most efficient solar detoxification system would not use concentrating collectors; for example, of the type described by Pacheco *et al.* (1993) to collect solar light [see e.g. Christensen and Walker (1996) for a discussion of this].

Clearly, the smaller the particle size of the TiO_2, the greater the chance of charge carriers reaching the surface without prior recombination and one possible way to increase the quantum efficiency is to reduce the size of the TiO_2 particle. This places an upper limit on the particle size of *ca.* 100 nm (Gautron *et al.*, 1980; Hamnett, 1980) and the removal of such fine particles is inherently time consuming and expensive. However, even the use of 100 nm particles may only delay recombination, since the surface itself is well known to act as a recombination centre (Gerischer, 1993). In addition, the TiO_2 must be removed from the treated water, if it is not itself to become a pollutant; see, for example, Matthews (1987) and this removal is more difficult, and more expensive, as the primary particle size becomes smaller. To avoid these downstream filtration problems, the photocatalyst may be immobilized – on carbon (Torimoto *et al.*, 1996, 1997), hollow glass beads (Shin and Kimura, 1999), polymer membranes (Artale *et al.*, 2001), glass fibre (Choi *et al.*, 2001) or metal (Byrne *et al.*, 1998) substrates. The catalyst may be used either in a thin film fixed bed reactor, as an immobilized film on an inclined plane as described by Wyness *et al.* (1994) or in UV-transparent tubes packed with TiO_2 on a particulate support as described by Crittenden *et al.* (1996). The challenges then are to avoid mass transfer problems and to make adhesion to the substrate sufficiently strong to resist scouring. For sequential processes, the influence of mass transfer may be expressed as

$$\frac{1}{V_{obs}} = \frac{I}{V_{MT}} + \frac{1}{V_{kin}} \tag{7.7}$$

where V_{obs} is the observed reaction rate, V_{MT} is the rate of transport of the reactant (organic or pathogen) to the TiO_2 surface and V_{kin} is the actual rate of reaction at the surface. In an ideally dispersed 0.1 wt% slurry of $70\,m^2\,g^{-1}$ TiO_2, since the interparticulate distance is ~200 nm, reactants are always within ~100 nm of a particle. Therefore, V_{MT} is large and the observed rate is close to the actual surface rate. However, when an immobilized TiO_2 layer is used, diffusion distances increase significantly ($\times 10^3$ to 10^4) and, unless specific measures are taken, mass transfer rates to the catalyst surface can be expected to decrease.

Therefore, despite the many advantages of TiO_2, including its wide availability, relatively low cost and non-toxicity (Egerton, 1997), the commercial-scale applications of photocatalysis by TiO_2 have been restricted to a relatively few demonstration units (Bockelman *et al.*, 1995; Christensen and Walker, 1996).

7.2 PHOTOELECTROCATALYSIS: A ROUTE TO REDUCED CHARGE CARRIER RECOMBINATION

In photoelectrocatalysis, the chosen route to reduced charge carrier recombination and consequent increased light utilization is to impose a positive potential bias on a TiO_2 film and exploit the electric field enhancement effect (EFE) (Butterfield *et al.*, 1997a,b; Hamnett, 1987). Providing that the electrolyte concentration is larger than the film dopant density, the application of a positive bias potential generates an electric field to a depth W into the TiO_2 from the surface, the depletion layer, as shown in Figure 7.2. The thickness of the depletion layer increases as the square root of the applied overpotential according to

$$W = \left(\frac{2\varepsilon\varepsilon_o |V - V_f|}{e_o N_d} \right)^{1/2} \tag{7.8}$$

where V is the potential at the conduction band edge, V_f is the flat band potential, N_d is the donor density, ε is the relative permittivity in the direction normal to the surface, e_o is the charge on the electron and ε_o is the relative permittivity of free space. For very low-doped TiO_2, or 'undoped' TiO_2, W will be large, the electric field small and hence any electric field enhancement negligible.

Calculations suggest that, for reasonable dopant concentrations, an applied bias of the order of 1 V should enhance the separation of electrons and holes. Any holes generated within the depletion layer will be efficiently transported to the semiconductor/electrolyte interface, where they can form hydroxyl radicals. The electrons migrate to the back contact, where the potential difference between the TiO_2 photoanode and the cathode then drives them around the external circuit to the counter electrode at which the counter electrode reaction, usually oxygen reduction, is driven separately. Hence, in addition to the improved charge separation, and in contrast to photocatalysis, oxygen reduction is no longer rate limiting (Gersicher, 1993).

Any holes generated at a depth between W and $(W + L_p)$ will have a good chance of diffusing to the edge of the depletion layer and then undergoing efficient transport of the TiO_2 surface. Absorption of radiation at depths $>(W + L_p)$ will result, predominantly, in charge recombination – as in conventional photocatalysis. For anatase

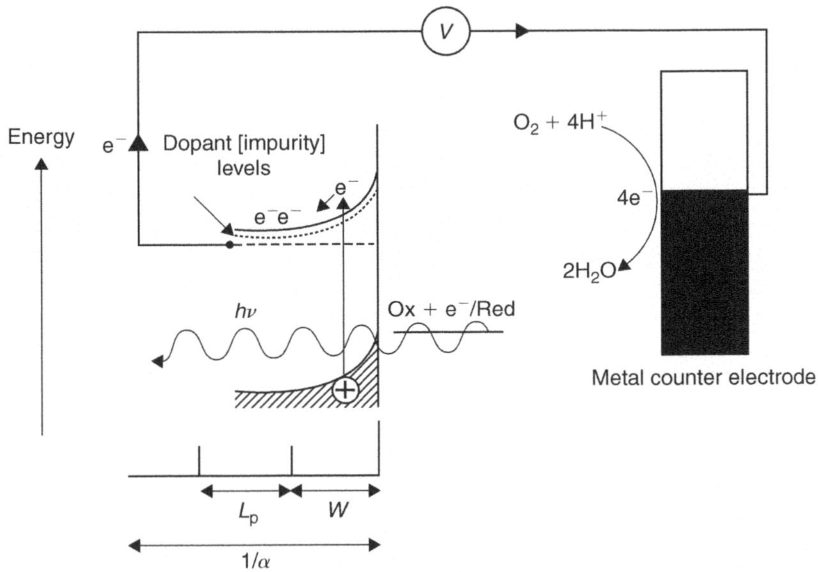

Figure 7.2 The principle of the EFE. See text for details.

thin films, deposited by sputtering, the absorption coefficient at 3.4 eV (~370 nm) is approximately $2 \times 10^4 \text{cm}^{-1}$ (Tang *et al.*, 1994), and 90% of the incident UV would be absorbed in ~200 nm. At 3.6 eV (~345 nm), 90% of the incident UV is absorbed in ~50 nm. The observed photo-current density, J_p, clearly will depend upon L_p, α and W and is described, according to classical semiconductor theory by

$$J_p = e_o I \left\{ 1 - \exp\left[\frac{-\alpha W}{1 + \alpha L_p} \right] \right\} \quad (7.9)$$

7.3 DEMONSTRATION OF THE CONCEPT

7.3.1 Electrode preparation

A wide range of methods may be used to prepare the TiO_2 electrodes used for photoelectrocatalytic water treatment. These include: heating titanium metal to form an oxide layer [a thermal electrode (Butterfield *et al.*, 1997a,b; Torimoto *et al.*, 1996, 1997)], anodizing titanium metal in acid (Hitchman *et al.*, 1996; Li and Li, 2002; Piaza *et al.*, 1997) or alkali (McMillan *et al.*, 1992; Zane *et al.*, 1993), immobilization of pre-formed TiO_2 particles (Byrne *et al.*), sol–gel methods (Yoko *et al.*, 1991), aerosol pyrolysis (Kavan and Gratzel, 1995), sputtering Ti onto a conductive substrate (Rodriguez *et al.*, 2000) and chemical vapour deposition (Doring *et al.*, 1992; Hitchman and Zhao, 1999). The properties of nanocrystalline electrodes have been reviewed by Peter and co-workers (1999). Early, 'proof of concept' experiments have mainly used thermal, sol–gel or deposited-particle electrodes.

7.3.2 Detoxification

Exploratory work, to test the concept described above, investigated the photoelectrocatalytic oxidation of solutions of simple organic molecules which were known to be degraded photocatalytically. Vinodgopal and co-workers (1993, 1994) demonstrated that photochemical degradation of 4-chlorophenol was enhanced by the application of a 0.6 V (vs SCE) electric field across a thin film of particulate TiO_2 (Degussa P25) on a conducting glass plate, and they attributed this to reduced recombination of photogenerated charge carriers. Later, Vinodgopal and Kamat (1995) reported the use of electrochemically assisted photocatalysis for the degradation of aqueous solutions of 4-chlorophenol and Acid Orange 7. Thin nanocrystalline semiconductor films coated on conducting glass surface were used and were reported to be 10 times more effective when coupled with SnO_2 nanocrystallites (at 0.83 V vs SCE).

Kim and Anderson (1994) reported results of the direct comparison of photocatalytic and photoelectrocatalytic degradation of formic acid on porous conducting TiO_2 films. The application of a small positive bias resulted in a small but significant increase in rates of oxidation. In addition, the photoelectrocatalytic degradation was less sensitive to factors, such as the presence of soluble salts. Kim *et al.* (1995) later studied the effect of firing temperature on the photocatalytic and photoelectrocatalytic activity of the TiO_2 thin films and demonstrated that the best efficiency for photocatalytic degradation did not correspond exactly with the best efficiency for photoelectrocatalytic degradation. They also showed (1996) that both the photocatalytic and photoelectrocatalytic approaches gave maximum degradation at pH 3.4, and that no detectable degradation was observed above pH 6. At formic acid concentrations $>7.6 \times 10^{-4}$ M, the rate of photoelectrocatalytic degradation was unaffected by concentration.

Christensen *et al.* (1997) and Christensen and Hutson (1996) subsequently extended work on carboxylic acids to the degradation of 250 μM sodium formate in neutral 0.1 M $NaClO_4$. These reactions were carried out with sol–gel, anodic and thermal films using a 'vortex reactor' consisting of a TiO_2 coated disc (either 700 cm^2 or 1590 cm^2) and nickel mesh counter electrode illuminated from above by a 400 W lamp. The aqueous effluent was injected through peripheral tangential ports and moved across the electrodes to a centrally placed outlet from which it was recycled via an external reservoir. Enhanced degradation of phenol, 250 μM, and 4-chlorophenol, 780 μM, was also demonstrated. These studies were the first of their kind reported on such a scale.

The conclusions of Christensen, Hamnett and co-workers have now been confirmed by a number of other groups. Thus, the photoelectrocatalytic oxidation of oxalic acid by TiO_2 powder immobilized on the surface of Ti–Al–V alloy has been reported by Byrne *et al.* (1998) of methanol at nanoporous TiO_2 film electrodes (Jiang *et al.*, 2001), and of aqueous methyl orange solutions (10 mg dm^{-3}) by anodized electrodes (anatase) under a light intensity of 4 mW cm^{-2} from a 125 W high-pressure mercury lamp in a 165 mL borosilicate glass vessel (Li and Li, 2002). In the last case, the rate of oxidation increased by a factor of ~4 when a 0.6 V bias (vs Cu counter electrode) was applied. Photoelectrocatalytic oxidation of Reactive Brilliant Red X-3B has also been found to be more efficient than the single application of either photocatalytic oxidation or electrochemical oxidation (An *et al.*, 2001). Further work on the photoelectrocatalytic oxidation of aromatics has been reported by Ranchell *et al.* (2002), who investigated the photoelectrochemical sensitized oxidation of methoxybenzyl alcohol and silanes. Photoelectrocatalytic oxidation of nitrophenols has also been studied (Egerton, 2001; Tinlin, 2002). Sun and Chou (1997, 2000) demonstrated photoelectrochemical oxidation of nitrite ions using a TiO_2 (rutile)/Ti photoelectrode and report that the reaction is zero order with respect to both nitrite and hydroxyl ion concentration and varies as $I^{0.8}$.

7.3.3 Disinfection

The photocatalytic disinfection of water by TiO_2 was first reported by Matsunaga *et al.* (1985). Since then, a number of other reports have appeared on the photocatalytic disinfection of water using TiO_2 (see e.g. Belhacova *et al.*, 1999; Ireland *et al.*, 1993; Matsunaga *et al.*, 1988; Okochi *et al.*, 1997; Wei *et al.*, 1994; Zhang *et al.*, 1994), but significantly fewer than those concerning water detoxification. In most of these studies, as in Matsunaga's original work (1985), TiO_2 was added as a slurry to the contaminated water. The optimum loading of catalyst is typically quoted as 0.1 wt%, that is $1\,g\,dm^{-3}$ (Christensen and Walker, 1996; Turchi and Ollis, 1998), for which the Degussa P25 (the most commonly employed commercial catalyst, which is typically ~70% anatase, 30% rutile, surface area $\sim 55\,m^2\,g^{-1}$) corresponds to $5\,m^2$ of catalyst per dm^3. Much of this work uses high cost UV lamps (up to 1 kW) to kill relatively low concentrations of the less robust pathogens, such as *Escherichia coli*, in simple, small-scale cells.

The first reported application of photoelectrocatalytic disinfection of water was by the Newcastle group (Butterfield *et al.*, 1997a,b), and involved the application of a small positive potential bias to adherent sol–gel films of TiO_2 grown on 100 mm diameter titanium plates. Disinfection of both *Clostridium perfringens* spores and *E. coli* was achieved by irradiating the electrode through a nickel mesh counter electrode using two 8 W UVB lamps. The combination of light, electric field and photocatalyst removed over 3-log units of *E. coli* and 2-log units of *Cl. perfringens* spores in 24 min. After 10 min, the survival of the *E. coli* was 10% and of *Cl. perfringens*, 30%. By themselves, neither UV radiation nor the electric field significantly reduced the numbers of indicator bacteria, nor did the combination of light and the fixed TiO_2 film (~90% survival in both cases). More recently (Dunlop *et al.*, 2002) disinfection of *E. coli* has been demonstrated at electrodes prepared by electrophoretic immobilization of Aldrich or Degussa P25 powder, but in this study the application of an electric bias increased the rate of disinfection by only ~40% (P25) to 80% (Aldrich); this is, perhaps, unsurprising as the commercial powders are only very lightly doped, and hence the N_d term in Equation (7.8) is relatively small. Vinodgopal *et al.* (1994) have reasoned that, for such lightly doped particulate TiO_2 films, there is no depletion layer across the film, and electron/hole separation is governed primarily by electrophoretic drift of the electrons to the back contact, as well as trapping of electrons and/or holes at the surface of the particles.

Further work by the Newcastle group (Harper *et al.*, 2001a) on *E. coli* disinfection addressed the problems of electrode choice by comparing the performance of thermal electrodes made by heating titanium metal and 'sol–gel' electrodes. The sol–gel electrodes were made by hydrolysing titanium alkoxide to form a sol and then concentrating this sol to form a gel, which was coated onto a titanium substrate and then heated. Preliminary characterization of these electrodes by cyclic voltammetry was used to optimize their activity (Christensen *et al.*, 1999) and the most active form of each type was compared. The thermal electrodes gave a higher photo-current than the sol–gel electrodes, and, at high *E. coli* levels, were more effective in subsequent disinfection experiments. For the thermal film, the kinetics were approximately first order; but for the sol–gel films, the dependence of rate on concentration was less, approximately half order. One consequence of this is that at low initial concentrations of *E. coli*, the sol–gel electrodes appear more active.

Simple 'proof of concept' experiments to test whether electric field enhancement of photo-oxidation would damage *Cryptosporidium* oocysts were carried out by the Newcastle group using a simple Petri dish reactor (volume ~135 mL) in which were mounted 5 × 5 cm electrodes (Curtis *et al.*, 2002). Experiments were carried out with either sol–gel or thermal electrodes and these electrodes were illuminated from above, through platinum mesh disc counter electrodes with 60%

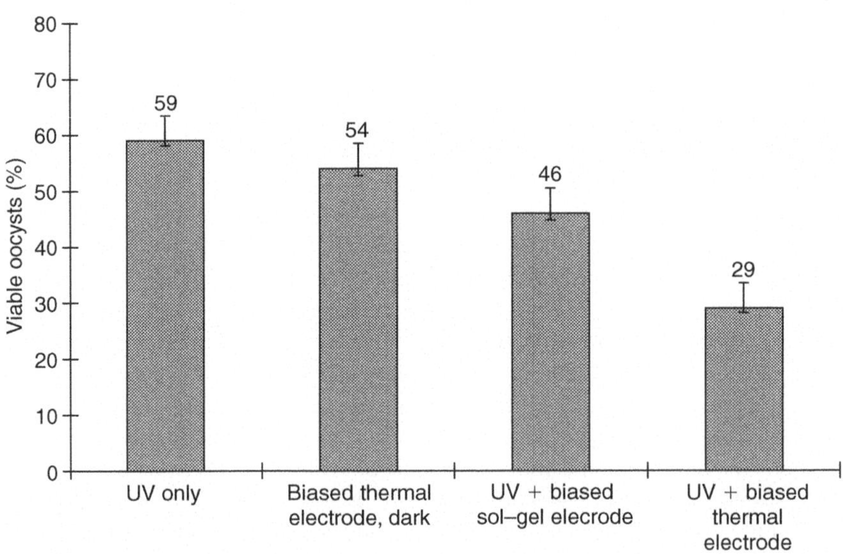

Figure 7.3 Inactivation of *Cryptosporidium* oocysts in an electric field enhanced TiO_2 reactor. Initial seeds for each experimental run had a mean viability of 66% and were statistically indistinguishable.

open area. The UV-irradiation source was deliberately chosen to be low powered; two 8 W disco 'black' lamps with a peak emission at 370 nm were used.

Although statistical analysis showed that, at the 95% confidence level, the electrodes fabricated from sol–gel films just overlap with the light (only) controls, the thermal films were statistically distinguishable from both the sol–gel film and the controls. These results demonstrated that the electric field enhancement is able to damage even the hard-to-treat *Cryptospiridium* oocysts and this is the first time that the photoelectrocatalytic destruction of such robust pathogens has been reported (Figure 7.3). In addition, the results strongly suggest the importance of tailoring the photo-anode characteristics to the targeted pathogens.

7.4 LARGER-SCALE EXPERIMENTS

Although there have been a large number of reports concerned with scale up of photocatalytic reactors for both detoxification and disinfection [see Chapter 6 and Christensen and Walker (1996), Blanc and Malato (1990)], most photoelectrocatalytic work has been done at a relatively small scale.

Two types of larger-scale photoelectrocatalytic reactors have been developed at Newcastle, a vortex reactor (Butterfield *et al.*, 1997a,b) and gas sparged reactors (Harper *et al.*, 2001b) of the type shown in Figure 7.4. This reactor is designed to allow very short UV path lengths – to facilitate operation in strongly absorbing or turbid environments – and yet be highly turbulent. This is achieved by using mesh

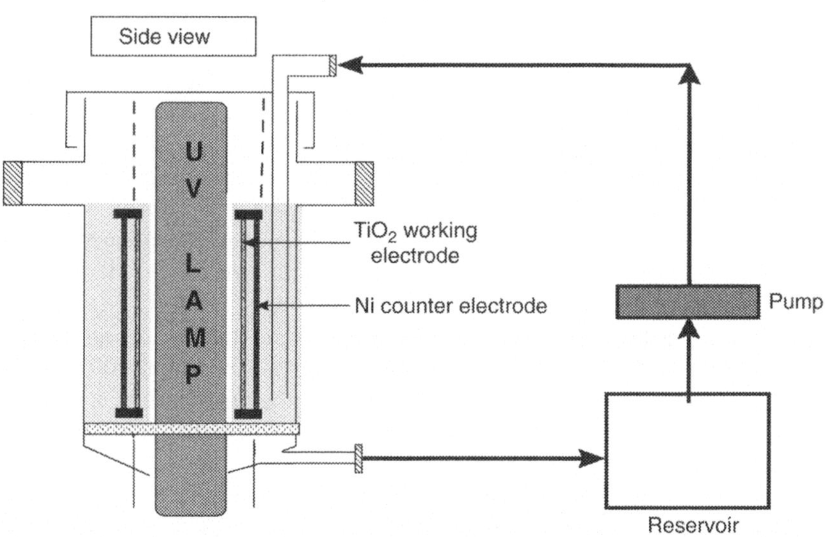

Figure 7.4 A litre-scale gas sparged reactor for photoelectrocatalysis. The reactor may be connected to an external reservoir and used in recirculating batch mode (Harper *et al.*, 2001b).

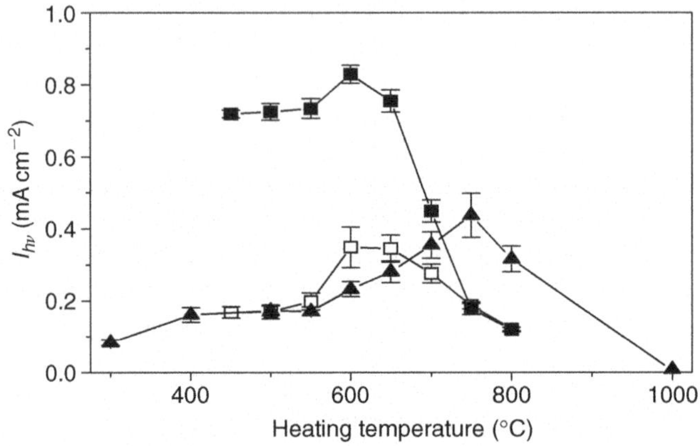

Figure 7.5 Current density at 1.2 V vs Ag/AgCl as a function of the film fabrication temperature of the thermal electrodes (measured in tap water, ▲) and sol–gel electrodes (measured in both tap water □, and in methanol solution ■). Data taken from cyclic voltammogramms using 1 cm^2 electrodes at 100 mV s^{-1}.

electrodes and sparging the reactor with a stream of gas. The reactor can be operated with either one or two electrode cassettes and has been used to study both photoelectrocatalytic detoxification (Tinlin, 2002) and photoelectrocatalytic disinfection (Christensen *et al.*, 2002). The Newcastle group is in the process of commissioning a purpose-built 7 dm^3 photoelectrocatalytic reactor designed for less strongly absorbing liquids.

Although the reactor has been operated with a variety of electrode types, this chapter will, for illustrative purposes restrict itself to results obtained with the two electrode types – thermal electrodes and sol–gel electrodes – already mentioned above. 'Thermal' electrodes were made by oxidation of titanium metal mesh; 'sol–gel' electrodes were made by depositing and then heating a layer of titania gel on titanium mesh (Figure 7.5).

Cyclic voltammetry was used to carry out an initial characterization and optimization of both electrode types using 5×5 cm^2 planar electrodes, see Figure 7.6. The best 'thermal electrodes' (i.e. those with the highest photo-currents) were prepared by heating titanium mesh at ~700°C in air. For sol–gel derived electrodes, optimum performance was obtained by heating at ~600°C. Methanol, an effective hole capture agent, was employed to assess the 'total oxidizing capacity' of the irradiated photo-anodes (Christensen *et al.*, 2002; Tinlin, 2002). Cylindrical electrodes fabricated in the same way were then used in the 0.5 dm^3 gas sparged reactor to disinfect *E. coli* suspensions with an initial concentration of 10^7 colony forming units (cfu) mL^{-1}.

Figure 7.6 shows that the photoelectrocatalytic disinfection is significantly more efficient than the photocatalytic disinfection with the same electrode at open circuit (i.e. without an applied potential). This demonstrates the enhancement in activity associated with the application of an electric field. Even more strikingly, a direct experimental comparison, see Figure 7.7, between photoelectrocatalytic

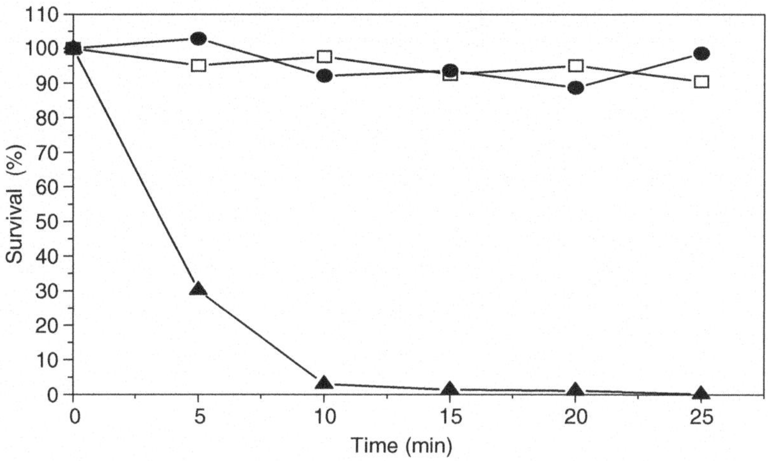

Figure 7.6 *E. coli* destruction by a UV-irradiated thermal electrode at a potential of 1.3 V, ▲; compared with the negligible effect of either the same electrode without UV irradiation ●, or UV irradiation □, either of an open circuit electrode, or UV alone. Initial concentration of *E. coli* is 10^7 cfu mL^{-1}.

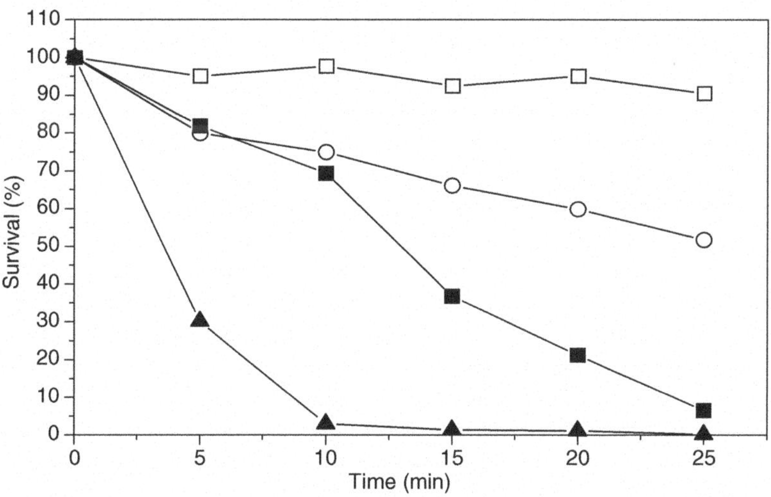

Figure 7.7 A comparison of photoelectrochemical destruction of *E. coli* with photocatalytic destruction by a slurry of P25. UV light (only), ○, optimized sol–gel electrode, UV irradiated at an applied potential of 3.0 V, ■ thermal electrode at 1.3 V; ▲ and UV-irradiated slurry of Degussa P25 at 8 g TiO_2 dm^{-3}, □.

disinfection and photocatalytic disinfection, using a slurry of Degussa P25 in the same reactor with the same lamp, demonstrates that the photoelectrocatalytic system is more efficient than photocatalytic disinfection by an irradiated suspension of TiO_2 (Christensen *et al.*, 2002). In Figures 7.6 and 7.7, 0% survival represents a 5-log deactivation, which is a marked achievement.

In practical systems, the TiO_2 is exposed to a variety of species additional to those that are targeted; the effects of $H_2PO_4^-$ and HCO_3^- ions on disinfection rate have been measured. Phosphate adsorbs strongly on TiO_2 surfaces and could, therefore, be expected to poison both photocatalysis and photoelectrocatalysis. Phosphate poisons both the electrode and particulate-slurry systems and is only partially reversible. Strikingly, the low area electrodes were no more susceptible to phosphate than was the high area slurry system. By contrast, bicarbonate acts as a radical trap and thus prevents photocatalysis by hydroxyl radicals. Comparative studies showed that although bicarbonate does reduce the activity of all three systems, the effects are reversible.

7.5 FUTURE WORK

There are three major areas for future development and research.

The reactors of the type shown in Figure 7.4 were designed to have excellent mass transfer when sparged with air. However, although the excellent mass transfer rates associated with sparging have been demonstrated (Harper, 2001a,b), the rate of oxidation of nitrophenol (a difficult to oxidize molecule, since its total oxidation requires many hydroxyl radicals) does not increase on sparging (Tinlin, 2002). This shows that the measured activity is limited not by mass transfer, but by the electrode efficiency and demonstrates the need for improved electrodes. At Newcastle, preliminary results suggest that the activity of sol–gel electrodes can be significantly improved by iron doping. Parallel work in electrode preparation and characterization of electrode morphology has begun.

The effectiveness of photoelectrocatalytic disinfection has been demonstrated at litre scale. However, to date, ethical considerations, associated with testing, have restricted disinfection studies with recalcitrant bacteria, such as *Cryptospiridium* to small-scale work (Curtis *et al.*, 2002). Since photoelectrocatalytic disinfection is likely to show particular advantages for bacteria of this type, it is essential that results are obtained at larger scale.

The work on electrode poisoning (Christensen *et al.*, 2002) summarized in the previous chapter is the first attempt to evaluate the robustness of photoelectrocatalysis to the presence of potential poisons. For some, but not all electrode systems, the detoxification of nitrophenol solutions was accompanied by electrode deactivation believed to be associated with the formation of polymeric residues at the electrode surfaces (Tinlin, 2002). The next stage in the development of this technology is an economic evaluation and any such techno-commercial evaluation requires a better understanding of the factors which control electrode life.

REFERENCES

An, T.C., Zhu, X.H. and Xiong, Y. (2001) *J. Env. Sci. Health – A Toxic Substances and Environmental Engineering* **36**, 2069–2082.

Artale, M.A., Palmisano, L. and Schiavello, M. (2001) Preparation and characterization of membranes with entrapped TiO_2 and preliminary photocatalytic tests. *Anali di Chimica* **91**, 127–136.

Bahadur, L. and Roy, L. (1999) A binary mixture of dyes for photosensitization of n-ZnO thin film electrodes. *J. Appl. Electrochem.* **29**(1), 109–116.

Belhacova, L., Krysa, J., Geryk, J. and Jirkovsky, J. (1999) *J. Chem. Technol. Biotechnol.* **74**, 149–154.

Blanc, J. and Malato, S. (1990) *Initial Configuration of the PSA Solar Detoxification Loop*, Platforma, Solar de Almeira, Spanien, 1990.

Bockelmann, D., Weichgebe, D., Goslich, R. and Bahnemann, D. (1995) Concentrating v non-concnetrating reactors for solar water detoxification. *Solar Energ. Mat. Solar Cell.* **38**, 441–451.

Butterfield, I.M., Christensen, P.A., Hamnett, A., Shaw, K.E., Walker, G.M., Walker, S.A. and Howarth, C.R. (1997a) Applied studies on immobilized titanium dioxide films as catalysts for the photoelectrochemical detoxification of water. *J. Appl. Electrochem.* **27**, 385–395.

Butterfield, I.M., Christensen, P.A., Curtis, T.P. and GunLazuardi (1997b) Water disinfection using an immobilized titanium dioxide film in a photochemical reactor with electric field enhancement. *Water Res.* **31**, 675–677.

Byrne, J.A. and Eggins, B.R. (1998) Photochemistry of oxalate on particulate TiO_2 electrodes. *J. Electroanal. Chem.* **457**, 61–72.

Byrne, J.A., Eggins, B.R., Browne, N.M.D., McKinney, B. and Rouse, M. (1998) Immobilisation of TiO_2 powder for the treatment of polluted water. *Appl. Cat. B Environ.* **17**, 25–36.

Byrne, J.A., Eggins, B.R., Byers, W. and Brown, N.M.D. (1999) Photoelectrochemical cell for the combined photocatalytic oxidation of organic pollutants and the recovery of metals from waste water. *Appl. Cat. B Environ.* **20**, L85–L89.

Choi, W., Ko, J.Y., Park, H. and Chung, J.S. (2001) Investigation of TiO_2 coated optical fibers for the photocatalytic oxidation of acetone. *Appl. Cat. B Environ.* **31**, 209–220.

Christensen, P.A. and Hutson, G.V. (1996) *Photoelectrochemical Reactor*. European patent 95921936.1, Japanese patent 502904, US patent 5,932,111.

Christensen, P.A. and Walker, G.M. (1996) *Opportunities for the UK in Solar Detoxification.* Report ETSU S/P4/00249/REP, HMSO, London.

Christensen, P.A., Egerton, T.A., Harper, J.C. and Tinlin, J.R. (1999) Photoelectrochemical oxidation of organics on thermal film TiO_2. *Topical Issues in Glasses*, Vol. 3. Photons Glasses and Coating, Society of Glass technology, Sheffield, pp. 75–79.

Christensen, P.A., Dilks, A., Egerton, T.A. and Temperley, J. (2000) Infrared spectroscopic evaluation of the photodegradation of paint. *J. Mater. Sci.* **35**, 5353–5358.

Christensen, P.A., Curtis, T.P., Egerton, T.A., Kosa, S.A.M., Harper and Tinlin, J.R. (2002) Photoelectrocatalytic and photocatalytic disinfection of *E. coli* suspensions by titanium dioxide. *Appl. Cat. B Environmen.* (in press).

Crittenden, J.C., Zhang, Y. and Hand, D.W. (1996) Solar detoxification of fuel contaminated groundwater using fixed bed photocatalysts. *Water Environ. Res.* **68**, 270–278.

Curtis, T.P., Walker, G., Dowling, B.M. and Christensen, P.A. (2002) Fate of cryptosporidium oocysts in an immobilized titanium dioxide reactor with electric field enhancement. *Water Res.* **36**, 2410–2413.

Doring, H., Hashimoto, K. and Fujishima, A. (1992) TiO_2 thin films prepared by pulsed beam chemical vapour deposition from titanium isopropoxide and water. *Ber. Der. Bunsen. – Phys. Chem. Chem. Phys.* **96**, 620–622.

Dunlop, P.S.M., Byrne, J.A., Manga, N. and Eggins, B.R. (2002) *J. Photochem. Photobiol. A-Chem.* **148**, 355–363.

Egerton, T.A. (1997) Titanium compounds. *Kirk Othmer Encyclopedia of Chemical Technology*, 4th edn, Vol. 24. John Wiley, New York, pp. 225–274.

Egerton, T.A. (2001) Casting light on an alternative to chlorination. *The Chem. Eng.* Issue 719, 60.

Egerton, T.A. and King, C.J. (1979) The influence of light intensity on photoactivity in titanium dioxide pigmented systems. *J. Oil Col. Chem. Assoc.* **26**, 386–391.
Egerton, G.S. and Shah, K.M. (1961) *Nature (London)* **202**, 81A.
Emeline, A., Salinaro, A. and Serpone, N. (2000) Spectral dependence and wavelength sensitivity in heterogenous photocatalysis: I Experimental evidence from photocatalysed transformations of phenols. *J. Phys. Chem.* **104**, 11202–11210.
Faughnan, B.W. and Kiss, Z.J. (1968) *Phys. Rev. Lett.* **211**, 331–334.
Gautron, J., Lemasson, P. and Marucco, J.F. (1980) Correlation between the non-stoichiometry of titanium dioxide and its photoelectrochemical behaviour. *Discuss. Faraday Soc.* **70**, 81–91.
Gerischer, H. (1993) Photoelectrochemical catalysis of the oxidation of organic moleulces by oxygen on small semiconductor particles with TiO_2 as an example. *Electrochim. Acta* **38**, 3–9.
Hamnett, A. (1980) *Disc. Far. Soc.* **70**, 124.
Hamnett, A. (1987) *Comp Chem. Kin.* **27**, 61.
Harper, J.C., Christensen, P.A., Egerton, T.A., Curtis, T.P. and Gunlazuardi, J. (2001a) Effect of catalyst type on the kinetics of the photoelectrochemical disinfection of water inoculated with *E. coli. J. Appl. Electrochem.* **31**, 623.
Harper, J.C., Christensen, P.A., Egerton, T.A. and Scott, K. (2001b) Mass transport characterization of a novel gas sparged photoelectrochemical reactor. *J. Appl. Electrochem.* **31**, 267–273.
Henrich, V.A. and Cox, P.A. (1994). *The Surface Science of Metal Oxides.* Table 4.1, Cambridge University Press, Cambridge.
Hepel, M. and Luo, J. (2001) Photoelectrochemical mineralization of textile diazo dye pollutants using nanocrystalline WO_3 electrodes. *Electrochim. Acta* **47**, 729–740.
Hitchman, M.L. and Zhao, J. (1999) The low pressure chemical vapour deposition of rutile at low temperatures. *J. de Physique IV* **9**, 357–364.
Hitchman, M.L., Spackman, R.A. and Agra, C. (1996) Photoelectrochemical study of titanium dioxide films prepared by the anodisation of titanium metal in sulphuric acid. *J. Chem. Soc. Faraday Trans.* **92**, 4049–4052.
Ireland, J.C. (1993) *Photocatalytic Purification and Treatment of Water and Air.* Elsevier, p. 557.
Ireland, J.C., Klosterman, P., Rice, E.W. and Clark, R.M. (1993) Inactivation of *E. coli* by titanium dioxide photocatalytic oxidation. *Appl. Environ. Microbiol.* **59**, 1668–1670.
Jiang, D.I., Zhao, H.J., Jia, Z.B. and John, R. (2001) Photoelectrochemical behaviour of methanol oxidation at nanoporous TiO_2 film electrodes. *J. Photochem. Photobiol. A-Chem.* **144**, 197–204.
Kavan, L. and Gratzel, M. (1995) Highly efficient semiconducting TiO_2 photoelectrodes prepared by aerosol pyrolysis. *Electrochim. Acta* **40**, 643–652.
Kim, D.H. and Anderson, M.A. (1994) Photoelectrocatalytic degradation of formic acid using a porous TiO_2 thin film electrode. *Environ. Sci. Technol.* **28**, 479–483.
Kim, D.H. and Anderson, M.A. (1996) Solution factors affecting the photocatalytic degradation of formic acid using supported TiO_2 thin films. *J. Photochem. Photobiol. A-Chem.* **94**, 221–229.
Kim, D.H., Anderson, M.A. and Zeltner, W.A. (1995) Effects of firing temperature on photocatalytic and photelectrocatalytic properties of TiO_2. *J. Environ. Eng.* **121**, 590–594.
Li, X.Z. and Li, F.B. (2002) *J. Appl. Electrochem.* **32**, 203–210.
Li, F.B., Gu, G.B., Li, X.J. and Wan, H.F. (2000) Preparation, characterization and photocatalytic behaviour of WO_3/TiO_2 nanopowder. *Acta Phys. Chim. Sin.* **16**, 997–1002.
Matsunaga, T., Tomoda, R., Nakajima, T. and Wake, H. (1985) Photoelectrochemical sterilization of microbial cells by semi-conductor powders. *Microbiol. Lett.* **29**, 211–214.
Matsunaga, T., Tomoda, R., Nakajima, T., Nakamura, N. and Komine, T. (1988) Continuous sterilization system that uses semiconductor powders. *Appl. Environ. Microbiol.* **54**, 1330–1333.

Matthews, R.W. (1987) *Sol. Energy* **6**, 405.
McMillan, C.S., Sukamto, J.P.H. and Smyrl, W.H. (1992) *Faraday Disc. Chem. Soc.* **94**, 63.
Mills, A. and Le Hunte, S. (1997) An overview of semi-conductor photocatalysis. *J. Photochem. Photobiol. A* **108**, 1–35.
Mills, A. and Wang, J.S. (1999) The kinetics of semi-conductor photocatalysis: light intensity effects. *Zeit Physik Chem.* **213**, 49–58.
Okochi, M. *et al.* (1997) *App. Microbiol. Tech.* **47**, 118.
Pacheco, J.E. *et al.* (1993) In: *Photocatalytic Purification and Treatment of Water and Air*, Ollis D.F. and El-Ekabi H. (eds). Elsevier Science, p. 547.
Peter, L., Ponomarev, E.A., Franco, G. and Shaw, N.J. (1999) Aspects of the photoelectrochemistry of nanocrystalline systems. *Electrochim. Acta* **45**, 549–560.
Piazza, S., Cala, L., Sunsieri, C. and DiQuarto, F. (1997) Influence of the crystallisation process on the photochemical behaviour of anodic TiO_2 films. *Ber. Bunsen. Phys. Chem.* **101**, 932–942.
Ponitus, F.W. (ed.) (1990) *Water Quality and Treatment – A Handbook of Community Water Supplies.* American Waterworks Association, McGraw-Hill, New York, Ch. 12.
Ranchell, M., Rol, C. and Sebastiani, G.V. (2002) The photoelectrochemical oxidation of benzyl derivatives at Ti/TiO_2 anodes. *J. Chem. Res.* 239–241.
Rajeshawar, K. (1995) Photoelectrochemistry and the environment. *J. Appl. Electrochem.* **25**, 1067–1082.
Rodriguez, G.J., Gomez, M., Lindquist, S.E. and Grandqvist, C.G. (2000) Photoelectrocatalytic oxidation of 4-chlorphenol over sputter deposited Ti oxide films. *Thin Solid Films* **360**, 250–255.
Serpone, N. and Emeline, A.V. (2002) Suggested terms and definitions in photocatalysis and radiocatalysis. *Int. J. Photoenerg.* **4**, 91–913.
Shin, D.Y. and Kimura, K. (1999) Preparation of TiO_2 coated hollow glass microspheres from titania hydrate coated fine volcanic glass. *J. Cer. Soc. Jap.* **107**, 775–779.
Shiyanovskaya, I. and Hepel, M. (1999) Bicomponent WO_3/TiO_2 films as photoelectrodes. *J. Electrochem. Soc.* **146**, 243–249.
Simmons, N.A. (1991) Cryptospiridium and Giardia lamblia. *Biologist* **38**, 147–148.
Sun, L. and Bolton, J.R. (1996) Determination of the quantum yield for the photochemical generation of hydroxyl radicals in TiO_2 suspensions. *J. Phys. Chem.* **100**, 4127–4134.
Sun, C.C. and Chou, T.C. (1998) Kinetics and mechanism of photoelectrochemical oxidation of nitrite ion using the the rutile form of TiO_2/Ti photoelectrode with high electric field enhancement. *Ind. Eng. Chem. Res.* **37**, 4207–4214.
Sun, C.C. and Chou, T.C. (2000) Electrochemically promoted photocatalytic oxidation of nitrite ion by using rutile form of TiO_2/Ti electrode. *J Mol. Cat. A-Chem.* **151**, 133–145.
Tang, H., Prasad, K., Sanjines, R., Schmid, P.E. and Levy, F. (1994) *J. Appl. Phys.* **75**, 2042.
Tinlin, J.R. (2002) PhD Thesis, University of Newcastle, UK.
Torimoto, T., Ito, S., Kuwabata, S. and Yoneyama, H. (1996) Effect of adsorbents used as supports for titanium dioxide on photocatalytic degradation of propyzamide. *Environ. Sci. Technol.* **30**, 1275–1281.
Torimoto, T., Okawa, Y. and Takeda, N. (1997) Effect of activated carbon content in TiO_2 loaded activated carbon on photdegradation of dichlormethane. *J. Photochem. Photbiol. A-Chem.* **103**, 153–157.
Tryk, D.A., Fujishima, A. and Honda, K. (2000) Recent topics in photoelectrochemistry: achievements and future prospects. *Electrochim. Acta* **45**, 2363–2376.
Turchi, C. and Ollis, D.F. (1988) *J. Phys. Chem.* **92**, 6852.
Vinodgopal, K. and Kamat, P.V. (1995) Electrochemically assisted photocatalysis using nano-crystalline semi-conductor thin films. *Sol. Energ. Mat. Solar Cell.* **38**, 401–410.
Vinodgopal, K., Hotchandani, S. and Kamat, P.V. (1993) Electrochemically assisted photocatalysis – TiO_2 particulate film electrodes for photocatalytic degradation of 4-chlorphenol. *J. Phys. Chem.* **97**, 9040–9044.

Vinodgopal, K., Stafford, U., Gray, K.A. and Kamat, P.V. (1994) Electrochemically assisted photocatalysis. 2 The role of oxygen and reaction intermediates in the degradation of 4-chlorophenol on immobilized TiO_2 particulate films. *J. Phys. Chem.* **98**, 6797–6803.

Wei, C., Lin, W.Y., Zulkarnain, Z., Williams, N.E., Zhu, K., Ruzic, A.P., Smith, R.L. and Rajeshwar, K. (1994) *Environ. Sci. Technol.* **28**, 934–938.

Wyness, P., Klausner, J.F. and Goswami, D.Y. (1994) Performance of non-concentrating solar photocatalytic oxidation reactors. 2 Shallow pond configuration. *J. Solar Eng.* **116**, 8–13.

Yoko, T., Yuasa, A., Kamiya, A. and Sakka, S. (1991) *J. Electrochem. Soc.* **138**, 2279.

Zane, D., Decker, F. and Razzini, G. (1993) Characterization of electrodeposited TiO_2 films. *Electrochim. Acta* **38**, 37–42.

Zhang, P., Scrudato, R.J. and Germano, G. (1994) *Chemosphere* **28**, 607–611.

8

Ultrasound processes

T.J. Mason and C. Pétrier

8.1 INTRODUCTION

Research into the use of ultrasound in environmental protection has received a considerable amount of attention with the majority of investigations focusing on the harnessing of cavitational effects for the destruction of biological or chemical pollutants in water and the processing of sewage (Mason and Tiehm, 2001). The field is much broader than this however and a summary of topics is given in Table 8.1. This chapter will concentrate on the various ways in which ultrasound can be used for advanced oxidation processes (AOPs).

8.2 PRINCIPLES OF SONOCHEMISTRY

The use of ultrasound in water treatment is an active subject for research and development. The treatment effects are essentially the result of ultrasonically induced acoustic cavitation (see later) and so the sound ranges employed are those that can

 Advanced Oxidation Processes for Water and Wastewater Treatment. Edited by Simon Parsons. ISBN: 1 84339 017 5

Table 8.1 General uses of ultrasound in environmental protection.

Water decontamination	
Biological	Direct mechanical action, e.g. cell rupture and the break-up of bacterial clumps
	Indirect mechanical action, e.g. increased cell permeability to bactericide
Chemical	Direct oxidation of chemical and pesticide residues
	In combination with other techniques, e.g. ozonation, UV light
Efficient surface cleaning	Removal of surface contamination and biofilms
Washing of soils	Removal of organic and inorganic contamination
Control of air-borne contamination	Agglomeration of smokes and aerosols
	Defoaming of liquids
Sewage treatment	Stabilisation and dewatering of sludge

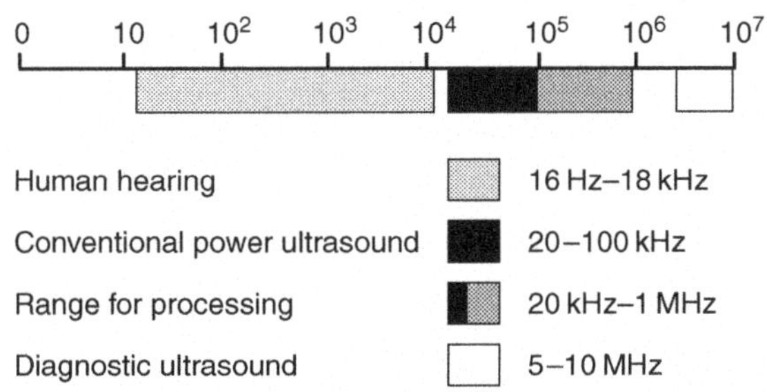

Figure 8.1 Frequency ranges of sound.

generate cavitation, that is generally in the lower frequency range of ultrasound. Large-scale applications are predominantly in the 20–40 kHz (power ultrasound) range but there is an increasing interest in the effect of higher frequencies (Figure 8.1). However this does not normally extend into the use of MHz range (diagnostic ultrasound) where cavitation is difficult to achieve without using high power or some form of focusing (Mason and Lorimer, 2002).

The history of power ultrasound in processing can be traced back to 1927 when a paper was published entitled 'The chemical effects of high-frequency sound waves: a preliminary survey' (Richards and Loomis, 1927) which described the development of power ultrasound for use in a range of applications including emulsification and surface cleaning. By the 1960s the industrial uses of power ultrasound were well accepted (Brown and Goodman, 1965; Frederick, 1965) and this interest has since continued (Abramov, 1998; Mason and Lorimer, 2002).

8.2.1 Acoustic cavitation – the origin of ultrasonically induced effects in processing

Power ultrasound enhances chemical and physical changes in a liquid medium through the generation and subsequent destruction of cavitation bubbles. Like any

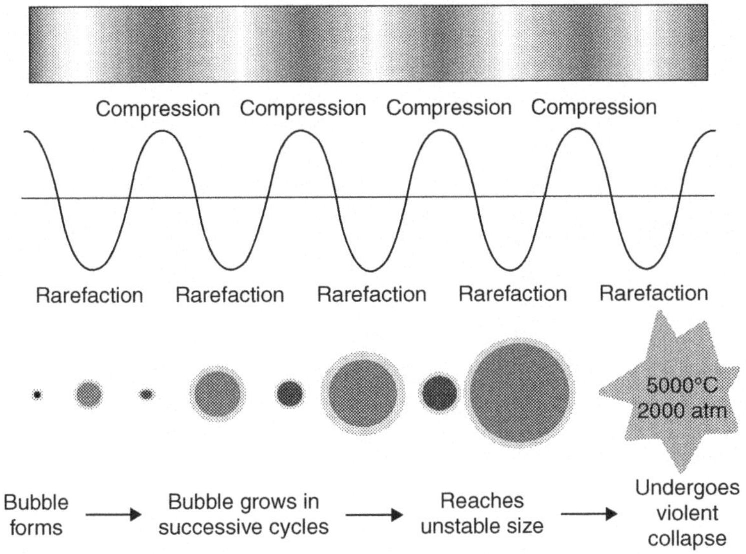

Figure 8.2 The acoustic generation of a cavitation bubble.

sound wave ultrasound is propagated via a series of compression and rarefaction waves induced in the molecules of the medium through which it passes. At sufficiently high power the rarefaction cycle may exceed the attractive forces of the molecules of the liquid and cavitation bubbles will form. Such bubbles grow by a process known as rectified diffusion, that is small amounts of vapour (or gas) from the medium enters the bubble during its expansion phase and is not fully expelled during compression. The bubbles grow over the period of a few cycles to an equilibrium size for the particular frequency applied. It is the fate of these bubbles when they collapse in succeeding compression cycles that generates the energy for chemical and mechanical effects (Figure 8.2). Cavitation bubble collapse is a remarkable phenomenon induced throughout the liquid by the power of sound. In aqueous systems at an ultrasonic frequency of 20 kHz each cavitation bubble collapse acts as a localised 'hotspot' generating temperatures of about 4000 K and pressures in excess of 1000 atmospheres (Henglein, 1987; Neppiras, 1984; Suslick, 1990).

Whatever application is to be studied or developed there are two essential components, a liquid medium and a source of high-energy vibrations. The liquid medium is necessary because sonochemistry is driven by acoustic cavitation and this can only occur in liquids. The source of the vibrational energy is the transducer of which there are three main types: fluid driven (liquid whistle), or electromechanical (magnetostrictive and piezoelectric transducers). The two most common pieces of equipment that are used for the generation of acoustic cavitation in the laboratory are the ultrasonic bath and the more powerful probe system (Figure 8.3) (Mason, 1999). Larger scale equipment will be dealt with later in this chapter.

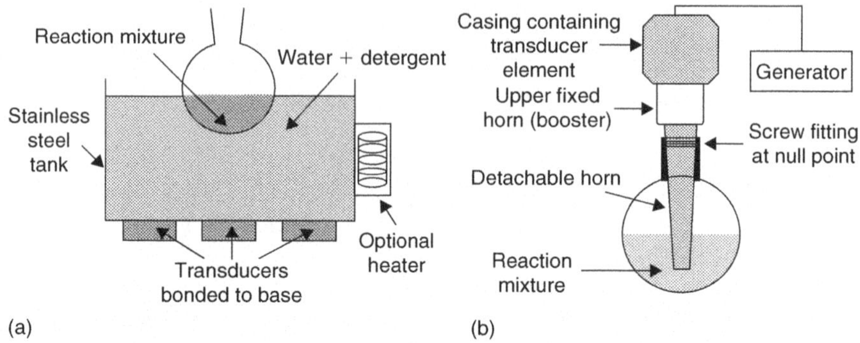

Figure 8.3 Common laboratory ultrasonic equipment: (a) bath and (b) probe.

The cavitation bubble has a variety of effects within the liquid medium depending upon the type of system in which it is generated. These systems can be broadly divided into homogeneous liquid, heterogeneous solid/liquid and heterogeneous liquid/liquid. Each of these classifications is of interest in water treatment.

8.2.1.1 Homogeneous liquid-phase reactions

It is not absolutely correct to describe any system within which cavitation occurs as homogeneous since bubbles must be present. However in this case 'homogeneous' refers to the state of the system before ultrasound is introduced. There are three major zones in which cavitation collapse can influence such systems (Figure 8.4):

- In the bulk liquid immediately surrounding the bubble where the rapid collapse of the bubble generates shear forces, which can produce mechanical effects.
- In the bubble itself where any species introduced during its formation will be subjected to extreme conditions of temperature and pressure on collapse leading to chemical effects.
- There will be a zone at the interface of the bubble and bulk liquid which is also of interest in terms of chemical activity. It is where surfactant species might accumulate and also the region where radicals formed within the bubble begin to react with species in the bulk medium.

However for the purposes of this general description we do not need to consider it separately.

The shear forces in the surrounding bulk liquid are so powerful that they can cause the breaking of chemical bonding in polymeric materials dissolved in the fluid. This was one of the early uses of power ultrasound (Price, 1990). The disturbance caused by cavitation bubble oscillation and collapse also provides extremely efficient mixing that is of use in the more rapid dissolution of chemicals and mixing of gases in water.

If the liquid in which cavitation is induced contains a gas then ultrasonic degassing will take place very readily. In the rarefaction phase any gas in the

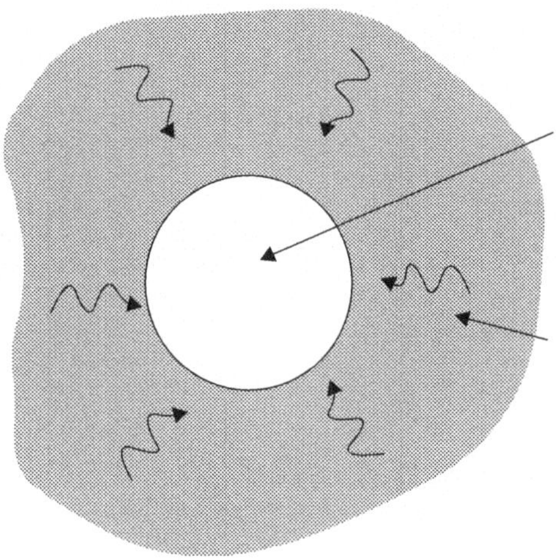

Figure 8.4 Acoustic cavitation in a homogeneous liquid medium.

medium will enter the newly forming cavitation bubble. Such bubbles are not easily collapsed in the compression cycle of the wave due to the fact that they contain gas and they will continue to grow during further rarefaction cycles. As they fill with more gas they eventually float to the surface. Since the rarefaction cycles are taking place extremely rapidly (around 20,000–40,000 times per second using normal power ultrasonic frequencies) the bubbles grow so quickly that degassing appears to occur almost instantaneously. Ultrasonic degassing has found applications in many areas. In its simplest form it is used to degas solvents for use in high performance liquid chromatography (HPLC) but industrially the potential applications are broad from the removal of air from molten glass to the defobbing of beer. In the water industry it has potential for the removal of excess chlorine.

When water is sonicated the extreme conditions generated on collapse of the cavitation bubbles are sufficient to cause rupture of the O—H bond itself. This results in the formation of radical species and the production of oxygen gas and hydrogen peroxide (Scheme 8.1) (Makino *et al.*, 1983).

Any species dissolved in the water is clearly going to be subject to chemical reaction with these ultrasonically produced radicals. This is the reason why ultrasonic irradiation has been extensively investigated for chemical remediation of water.

8.2.1.2 Heterogeneous solid surface–liquid reactions

Unlike cavitation bubble collapse in the bulk liquid, collapse on or near to a surface is unsymmetrical because the surface provides resistance to liquid flow from that side. The result is an inrush of liquid predominantly from the side of the bubble remote from the surface resulting in a powerful liquid jet being formed, targeted at the surface (Figure 8.5). The effect is equivalent to high pressure jetting and this is the reason why ultrasound is used extensively for cleaning and the removal

$$H_2O \rightarrow OH^\bullet + H^\bullet$$

$$OH^\bullet + OH^\bullet \rightarrow H_2O_2$$

$$OH^\bullet + OH^\bullet \rightarrow H_2O + O$$

$$OH^\bullet + OH^\bullet \rightarrow H_2 + O_2$$

$$H^\bullet + O_2 \rightarrow HO_2^\bullet$$

$$HO_2^\bullet + H^\bullet \rightarrow H_2O_2^\bullet$$

$$HO_2^\bullet + HO_2^\bullet \rightarrow H_2O_2 + O_2$$

$$OH^\bullet + H_2O \rightarrow H_2O_2 + O$$

$$H_2O^\bullet + O^\bullet \rightarrow H_2O_2$$

$$H^\bullet + H^\bullet \rightarrow H_2$$

$$H^\bullet + OH^\bullet \rightarrow H_2O$$

Scheme 8.1 Ultrasonically induced decomposition of water and subsequent reactions.

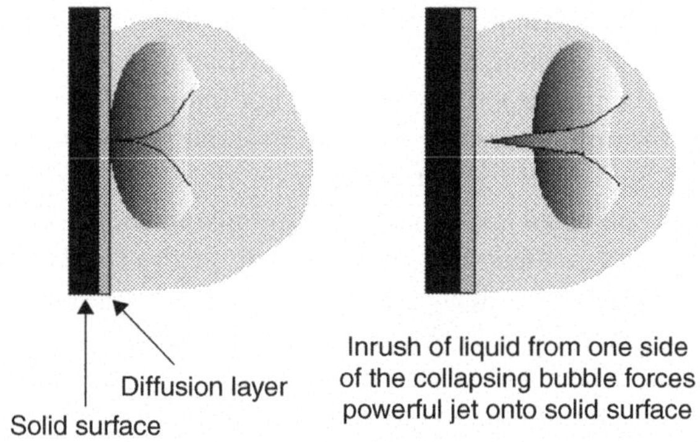

Figure 8.5 Cavitation bubble collapse at or near to a solid surface.

of biofilms. This effect can also increase mass and heat transfer to the surface by disruption of the interfacial boundary layers.

8.2.1.3 *Heterogeneous particle–liquid reactions*

Acoustic cavitation can produce dramatic effects on particulate material or agglomerates in a liquid. In this case the solid surface is not the continuous surface as above but is a small powder fragment suspended in the medium, under these

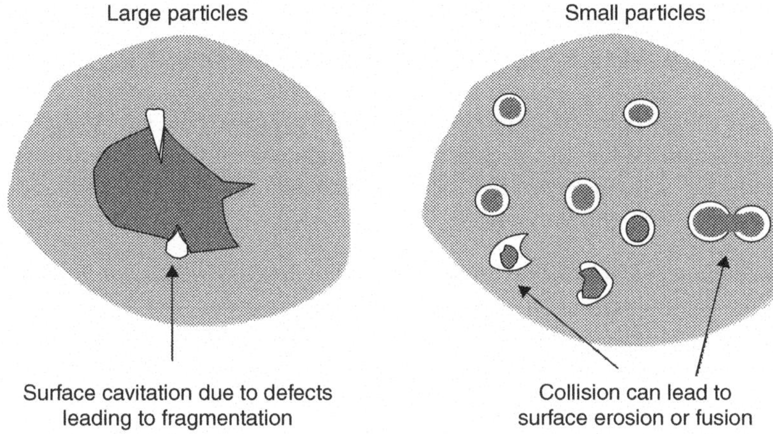

Figure 8.6 Cavitation collapse in the presence of a suspended powder.

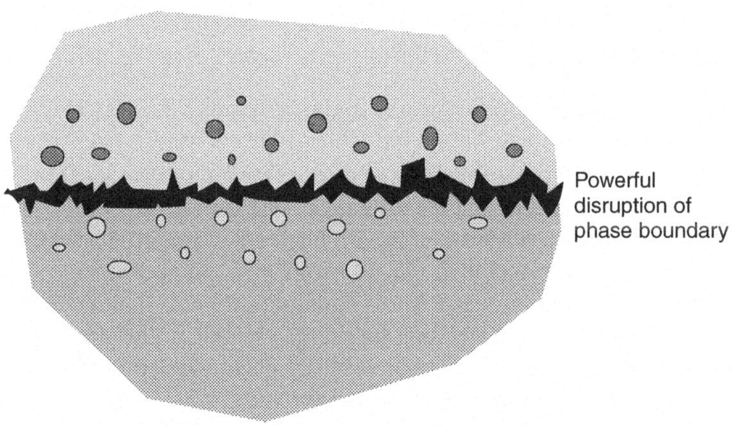

Figure 8.7 Cavitation effects on a heterogeneous liquid–liquid system.

conditions cavitation can achieve both a reduction in particle size and efficient dispersion (Figure 8.6). Surface imperfections or trapped gas can act as the nuclei for cavitation bubble formation on the surface of a particle and subsequent surface collapse can then lead to shock waves which break the particle apart. Cavitation bubble collapse in the liquid phase near to a particle can force it into rapid motion. Under these circumstances the general dispersive effect is accompanied by inter-particle collisions that can also lead to erosion, surface cleaning and wetting of the particles and particle size reduction.

8.2.1.4 Heterogeneous liquid–liquid reactions

In heterogeneous liquid–liquid reactions, cavitational collapse at or near the interface will cause disruption and mixing, resulting in the formation of very fine emulsions (Figure 8.7).

8.3 HISTORICAL INTRODUCTION ON THE OXIDATIVE PROPERTIES OF ULTRASOUND IN WATER

Ever since the earliest observation of ultrasound promoted chemical reactions it has been recognised that the propagation of a sound wave of sufficient intensity in water induced oxidation reactions. Working on aqueous solutions (Schmitt *et al.*, 1929) made several observations:

- Upon ultrasonic irradiation (750 kHz) iodide can be liberated from a potassium iodide solution saturated with air.
- Potassium chloride and potassium bromide solution irradiated in the same manner are able to oxidise a potassium iodide starch reagent.
- Irradiation of titanium sulphate solution giving a positive response, it indicated the presence of hydrogen peroxide that may be involved in the processes.
- Observation of ultrasonic effect required the formation of gas bubbles in the liquid;
- Addition of carbon tetrachloride to the medium accelerates the rate of halide oxidation, the authors suspected the oxygen cleavage in the oxidation process.

Extending these observations, several studies (Flosdorf *et al.*, 1936; Liu and Wu, 1934; Loiseleur, 1944; Schulter and Gohr, 1936) confirmed experimentally that hydrogen peroxide formation and the direct activation of oxygen on irradiation were dependent upon ultrasonic cavitation. The authors also confirmed that carbon tetrachloride or chloroform react with the 'active oxygen' formed.

Further, Prudhomme and Grabar (1949) showed that oxygen is not necessary for the observation of oxidative reaction. Oxidation also occurs in deoxygenated water saturated with argon. Water was however essential in the process. Under these conditions, the ultrasonic treatment of water was found to produce molecular oxygen and hydrogen in addition to hydrogen peroxide.

Frenkel (1940) postulated that the phenomenon of cavitation produces electrical discharges and in agreement with the hypothesis of Weiss (1944) for the case of ionising radiation, he suggested that such phenomenon would then lead to the dissociation of water.

The overall processes can be depicted as follows:

The effect of radiation: $(OH)^- H^+ \rightarrow (OH)H$

$$
\begin{aligned}
2OH^\bullet &\rightarrow H_2O + O \\
2O &\rightarrow O_2 \\
2H &\rightarrow H_2 \\
H + O_2 &\rightarrow HO_2^\bullet \\
2HO_2^\bullet &\rightarrow H_2O_2 + O_2
\end{aligned}
$$

The electrical theory was the main physico-chemical support that was used to explain sonochemical effects up until 1953. The descriptions of the mechanism of the phenomena were considered to be similar to those observed in the case of γ-irradiation. To complete the description, Lindström (1955) postulated the formation of a cluster of species composed of two molecules of water and oxygen ion. In this description, electrical microdischarges lead to the ionisation of molecular oxygen in the cavitation bubble:

$$O_2 \rightarrow O_2^+ + e^-$$

The cluster dissociates upon neutralisation inducing the simultaneous formation of hydrogen peroxide and hydroxyl radical:

$$(HOH{\cdot}O_2{\cdot}HOH)^+ + e^- \rightarrow 2HO^\bullet + H_2O_2$$

Griffing (1950) suggested that water dissociation could occur upon thermal reaction that occurs during the final step of the adiabatic collapse of the cavitation bubble. This explanation 'Hot Spot theory' can account to radicals production upon ultrasonic action and hence to the promotion of oxidation processes. This theory has received experimental supports and currently prevails upon electrical theory (Fitzgerald *et al.*, 1954; Henglein, 1987; Noltingk and Neppiras, 1950; Storey and Szeri, 2000).

In addition to the support obtained from theoretical calculations and chemical experiment the Hot Spot theory has also received strong corroboration from the observations of the phenomenon of sonoluminescence (Walton and Reynolds, 1984). Sonoluminescence, which can be regarded as the transformation of acoustical energy into electromagnetic radiation, accompanies the ultrasonic wave propagation in a liquid and was observed in water for the first time by Frenzel and Schultes (1934). The experimental results of Chambers (1937) showed that sonoluminescence could be observed in other liquids.

In water saturated by a noble gas (Ar, Ne and Kr) the sonoluminescence spectrum exhibits a continuum extending from 240 to 800 nm with three main zones: 270–290 nm (I), 310 nm (II), under 400 nm (III) (Figure 8.8). Seghal *et al.* (1980) and, Didenko and Pugach (1994) attributed this main emission to water and hydroxyl radical excited states: such observation was found consistent with the emission of excited state of water $H_2O^* = H_2O\ (^1A_1)$, excited hydroxyl radical $OH^* = {}^2S^+$ and H, OH combination:

$$H_2O + M \rightarrow H_2O\ [^1B_1] + M$$
$$H_2O\ [^1B_1] \rightarrow H_2O\ [^1A_1] + h\nu\ (1 > 270\text{–}290\ \text{nm})$$
$$H_2O\ [^1B_1] \rightarrow H + OH[^2S^+]$$
$$OH\ [^2S^+] \rightarrow OH\ [^2P] + h\nu\ (1 > 310\ \text{nm})$$

This simplified scheme describing the generation of hydroxyl radical for water saturated with a noble gas is currently the subject of debate. An analysis of the

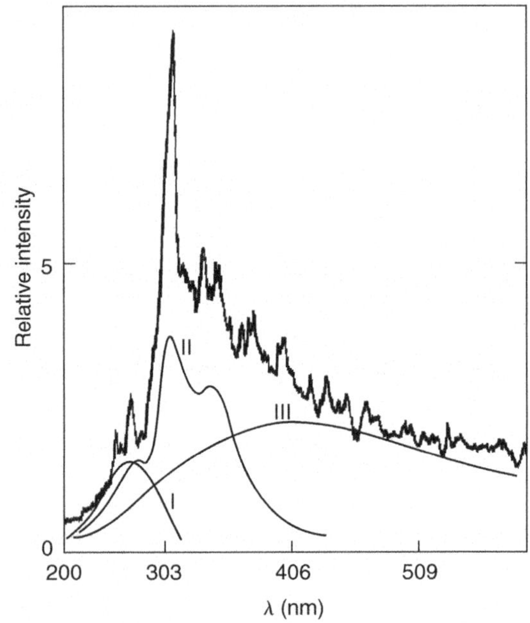

Figure 8.8 Sonoluminescence spectra of water.

high-resolution spectrum suggests the formation of an OH-noble gas excimer to explain the emission in the 310 nm range.

Direct confirmation of the production of radicals in a medium subjected to acoustic cavitation was reported in 1982 by Makino *et al.*, using electron spin resonance. With the use of a non-volatile nitrone spin trap, DMPO (5,5-dimethyl-1-pyrroline N-oxide) they observed the spectrum of the stable spin adducts HO-DMPO and H-DMPO from sonicated water saturated with argon. This was the first time that direct proof was presented for the identities of the radicals produced from the sonochemical splitting of water (Figure 8.9).

Studies of the chemistry involved in hydrogen peroxide formation and potassium iodide oxidation were of importance to elucidate the complementary aspects of the origin and behaviour of hydroxyl radicals. To this end Del Duca *et al.* (1958) investigated the formation of peroxide in water containing slightly enriched $O_2^{18,18}$. Their conclusion was that an oxygen molecule appears to become incorporated as a unit in a single peroxide molecule. It was therefore concluded that the energy release in the bubble is sufficient to dissociate water into H and OH radicals but not sufficient to dissociate molecular oxygen. Hydrogen peroxide formation can then only arise from the reaction between hydrogen atoms and molecular oxygen:

$$H^{\bullet} + O_2 \rightarrow HO_2^{\bullet}$$
$$2HO_2^{\bullet} \rightarrow H_2O_2 + O_2$$

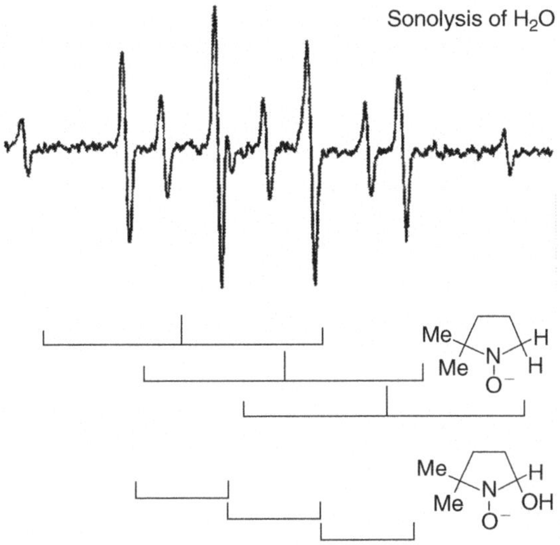

Figure 8.9 Electron spin resonance trapping of radicals from the sonolysis of water.

The origin of hydrogen peroxide was reinvestigated by Anbar and Pecht (1964) who used labelled water H_2O^{18} in presence of hydrogen peroxide $H_2O_2^{16,16}$ in a medium that was saturated either with argon or with labelled molecular oxygen $O_2{}^{18,18}$. Under an argon atmosphere the formation of pure $H_2O_2^{18,18}$ supported the splitting of water. However experiments conducted with water H_2O^{18} saturated with $O_2^{16,16}$, or water H_2O^{16} saturated with $O_2^{18,18}$ lead to the formation of both to $H_2O_2^{18,18}$ and $H_2O_2^{16,18}$. This suggested that in the case of a medium saturated with oxygen that there were three possible sources of H_2O_2: water, molecular oxygen and oxygen atoms. Thus hydrogen peroxide could also originate from the reaction of an oxygen atom with water:

$$O_2 \rightarrow O + O$$
$$O + H_2O \rightarrow H_2O_2$$

Using the same technique, that is using water saturated with various proportions of $O_2^{18,18}$ and argon, Fischer *et al.* (1986) again revisited the question of the source of hydrogen peroxide this time taking into account the stability of the intermediates in the high temperature zone of the bubble. Their discussion emphasised that the formation of the stable final product occurred in the low temperature interfacial zone between bubble and bulk solution mainly through the reaction:

$$OH^\bullet + OH^\bullet \rightarrow HOOH$$

However the reaction between O and H_2O that occurs mainly in the high temperature zone inside the bubble should generate only OH radical:

$$O + H_2O \rightarrow OH^\bullet + OH^\bullet$$

When sonication is carried out in air saturated medium the formation of H_2O_2 is accompanied by the generation of nitrous and nitrate ions. These species originate from the oxidation of molecular nitrogen with $HO^\bullet$ tand $O^\bullet$ tinside the bubble (Hart and Henglein, 1986; Hart *et al.*, 1986). During the first stages of the treatment H_2O_2 and NO_2^- build up simultaneously. As the reaction progresses, the rate of H_2O_2 production decreases and nitrite ions appear with a concentration that increases at the expense of that of nitrous ions. This suggests that H_2O_2 and NO_2^- are the primary products of sonication of air-equilibrated water and that the NO_3^- results from the oxidation of NO_2^- by hydrogen peroxide. The nitrogen containing species are however unlikely to be major participants in sonochemically induced organic oxidations.

8.4 SONOCHEMICAL DECONTAMINATION OF AQUEOUS SYSTEMS

8.4.1 The effect of ultrasound alone

The majority of systems studied have been aqueous solutions of either aromatic compounds or halogenated hydrocarbons. Such materials represent models for the major classes of organic pollutants in waste and ground water. The primary products resulting from the sonochemical treatment of phenol at 541 kHz (27°C with bubbled air) are hydroquinone and catechol (Berlan *et al.*, 1994). These compounds are easy to monitor and are clearly seen to be intermediates that disappear as the reaction progresses (Figure 8.10). Similarly the sonolysis of aqueous 4-chlorophenol leads to products mainly characteristic of oxidation by $^\bullet OH$ radical, for example 4-chlorocatechol but in both cases the final organic products are CO, CO_2 and HCOOH (Figure 8.11) (Pétrier and Casadonte, 2001; Pétrier *et al.*, 1994; Serpone *et al.*, 1994).

After an irradiation time of 300 min (500 kHz, 30 W, 20°C) the hydroxylated intermediates disappear and the chlorine atoms are completely mineralised to chloride ions. The concentrations of the final products (CO and CO_2) rise slowly and after 400 min they represent 21% of the starting chlorophenol and 18% of the starting phenol. These results are in agreement with previous studies in which it was demonstrated that 'the sonochemical reactivity' of an organic compound is related to its vapour pressure and hydrophobicity (Henglein and Kormann, 1985; Von Sonntag *et al.*, 1998). These studies have been extended to the destruction of a number of chloroaromatics (Pétrier and Casadonte, 2001) (Table 8.2).

In general chloroaromatic hydrocarbons decompose faster than the more hydrophilic compounds, such as phenols. The destruction of these two kinds of pollutants occurs at different sites in or around the cavitation bubble and follows different pathways. The difference becomes even more pronounced when more volatile contaminants are used.

This is neatly illustrated in a comparative study of the decomposition of phenol (to carboxylic acids) and carbon tetrachloride (to CO_2 and Cl^-) in water saturated

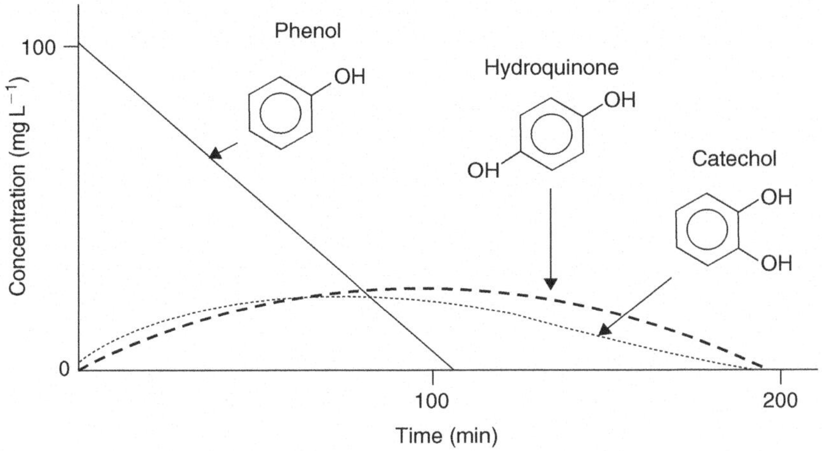

Figure 8.10 Sonochemical decomposition phenol in aerated water.

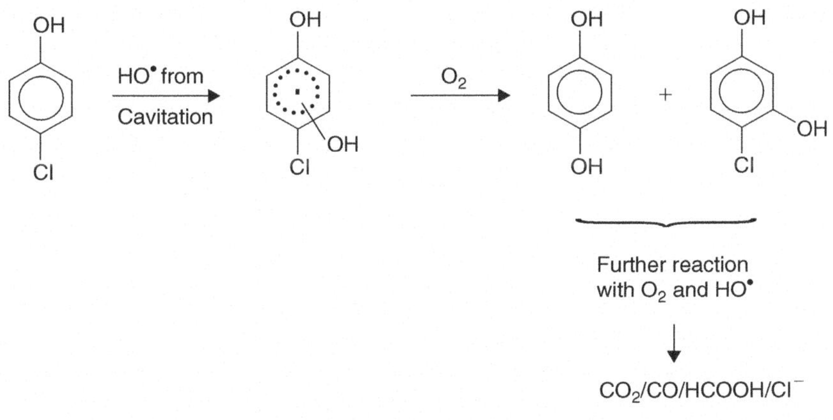

Figure 8.11 Sonochemical decomposition of 4-chlorophenol in aerated water.

Table 8.2 Initial concentration and chloride yield after 150 min irradiation for the degradation of chloroaromatics (volume: 250 cm^3, ultrasonic power: 30 W, electric power: 50 W, frequency 500 kHz).

Chloroaromatic compound	Concentration (mM)	Yield of Cl^- (%)
1,2-Dichlorobenzene	0.4	90
1,3-Dichlorobenzene	0.05	89
1,4-Dichlorobenzene	0.2	95
1,3,5-Trichlorobenzene	0.02	95
1-Chloronaphthalene	0.04	98

Table 8.3 Comparison between H_2O_2 production and the rates of phenol and carbon tetrachloride disappearance at different frequencies ($\mu M\ min^{-1}$).

Frequency (kHz)	20	200	500	800
H_2O_2 formation	0.7	5.0	2.1	1.4
Phenol degradation	0.5	4.9	1.9	1.0
CCl_4 degradation	19	33	37	50

with oxygen at different frequencies (Table 8.3) (Pétrier and Francony, 1997). The results clearly show a difference in the behaviour of the chemical contaminants (original concentration 10^{-3} M) with the phenol degradation mirroring the peroxide formation indicating this reaction proceeding at the bubble interface or outside of the bubble. The volatile CCl_4 however is decomposed within the bubble and increased frequency slightly accelerates the process. An intriguing calculation for these reactions shows that the efficiency for each of these processes over one ultrasonic cycle decreases as frequency increases.

The occurrence of an optimum frequency at 200 kHz was explained through a two-step reaction pathway. In the first step water sonolysis produces radicals within the bubble. In step two the radicals must migrate to the bubble interface or into the bulk aqueous medium to form peroxide or react with the phenolic substrate. The authors suggest that the lower frequencies are the most efficient for the decomposition of molecules inside the bubble but a proportion of the radicals recombine inside the bubble at high temperature to form water thereby reducing the overall yield of H_2O_2 [Equations (8.1) and (8.2)]:

$$2HO^\bullet \rightarrow H_2O + \tfrac{1}{2}O_2 \tag{8.1}$$

$$HO^\bullet + HOO^\bullet \rightarrow H_2O + O_2 \tag{8.2}$$

As the frequency increases the pulsation and collapse of the bubble occurs more rapidly and more radicals escape from the bubble. However as the frequency increases the cavitation intensity decreases and this reduces the yield of radicals and consequently the numbers that reach the interface and bulk solution.

A study of sonochemical treatment of chlorinated hydrocarbons in water demonstrated the homogeneous destruction of CH_2Cl_2, CCl_4, $MeCCl_3$ and $ClCH{=}CCl_2$ in solution at concentrations of 100–1000 ppm by volume (Cheung *et al.*, 1991). The method appears to be potentially quite powerful for the purification of contaminated water. In a separate study when saturated aqueous solutions of CH_2Cl_4 (110 ppm) and $MeCCl_3$ (1300 ppm) were sonicated for 20 min some 75% of the contaminant was degraded (Nagata *et al.*, 1993).

8.4.2 Combined application of ultrasound with ozone

AOPs are a range of water treatments that involve the *in situ* formation of radicals, particularly hydroxyl radicals, in sufficient quantity to affect chemical or biological contaminants. These include ultrasonic and ultraviolet (UV) irradiation but they are sometimes ineffective for the remediation of water that contains a mixture of organic and inorganic compounds. Chemical oxidants can be used to add additional oxidising power to such processes and ozone in conjunction with ultrasound is one such option (Weavers, 2001).

Ozone (O_3) is a very strong oxidant that can oxidise organic compounds to carbon dioxide and water but the reaction rates for many organic pollutants is slow. Nevertheless ozone has been used at water and wastewater treatment plants in a variety of applications (Langlais *et al.*, 1991).

The combination of sonication and ozonation provides three sources of $OH^{\bullet}$ (Weavers and Hoffmann, 1998):

a. from sonochemical decomposition of water,
b. from the normal chemical degradation of ozone,
c. from the thermolytic decomposition of ozone in the acoustic cavitation bubble.

Studies of the combined process involving ultrasound and ozone have shown faster degradation rates for a range of chemical contaminants than either method applied alone. Sonolytic ozonation has also been found effective for the disinfection of water but in these cases sonication also has a number of direct effects on the bacteria and viruses (see above).

An environmental problem that has existed for a long time has been the removal of colour from the effluent streams of textile factories since the presence of residual dyes is rather obvious even at low dilutions. There are several conventional approaches to the solution of this problem including absorption onto activated charcoal, flocculation, chemical oxidation, ozonolysis and irradiation with UV light. Sonication can be added to any of these and the combination of ultrasound with ozonolysis seems particularly efficient. A sonic processor operating at 430 Hz (i.e. within the normal hearing range of frequencies) has been used to study the pilot scale decolourisation of dyes using ozone. The device operates at 20 kW with a 3.2-l flow reactor supplied with ozone from a laboratory scale generator (Russell and Smith, 1999). The study demonstrated that water-soluble textile dyes could be destroyed and decolourised by low frequency sonication combined with ozonation. Basic, acid, direct and even some disperse dyes, which tend to have low solubility, were all rapidly decolourised. For most of the dyes the toxicity of the decolourised dye to micro-organisms was the same as, or less than, the dye solution. This would make them suitable for further treatment on site or for immediate discharge to a publicly operated treatment plant. The amount of ozone required ranged from 1.1 to 1.9 kg of ozone per kg of dye. The rate of dye destruction was about 20 times faster than a simple mechanically stirred gas to liquid mixer. The rate is sufficiently large for short residence times of less than 60 s to be possible in a continuous flow process with industrial ozonisers and for dye concentrations within the range typical of wastewater from textile dyeing plants.

8.4.3 Combined application of ultrasound with UV light

There are a few reports on the combined application of ultrasound and UV light for the destruction of chemical pollutants. A study of the oxidation of humic acid

Table 8.4 Photolysis of aqueous pentachlorophenol (2.4×10^{-4} M) containing 0.2% TiO_2.

Treatment	Cl^- yield after 50 min (%)	Cl^- yield after 120 min
UV alone	40	No change
UV + ultrasound	60	Quantitative

and trihalomethane precursors with ozone revealed that the most effective destruction of the organic carbon compounds was achieved when both UV and ultrasound was used in combination with ozonation (Sierka and Amy, 1985). In other cases, for example the removal of 1,1,1-trichloroethane from aqueous solutions, the combined application of ultrasound and UV proved to be more efficient than the use of either technique individually (Toy *et al.*, 1990).

Photochemical decomposition can also be carried out in the presence of a suspension of photoactive material such as TiO_2 where substrate absorption onto the UV activated surface can initiate chemical reactions, for example the oxidation of sulphides to sulphones and sulphoxides (Sierka and Amy, 1985). This technology has been adapted to the destruction of polychlorobiphenyls (PCBs) in wastewater, of considerable interest in environmental protection. Using pentachlorophenol as a model substrate in the presence of 0.2% TiO_2 UV irradiation is relatively efficient in dechlorination (Table 8.4) (ISRI International, 1992). When ultrasound is used in conjunction with photolysis, dechlorination is dramatically improved. This improvement is the result of three mechanical effects of sonochemistry namely surface cleaning, particle size reduction and increased mass transport to the powder surface.

A similar combination of ultrasound and photocatalysis has also been reported to destroy 2,4,6-trichlorophenol in aqueous solution (Shirgaonkar and Pandit, 1998). An ultrasonic probe (22 kHz) with an UV light source (15 W) was used to examine the effect of changing such operating conditions as ultrasonic intensity, reaction temperature and UV transmission. The experiments involved using 2,4,6-trichlorophenol (100 ppm) and TiO_2 ($0.1\,g\,L^{-1}$) and showed that the degradation rates increased with the temperature of the solution. The cumulative effect was more pronounced at lower ultrasonic intensities with little additional benefit derived at increased ultrasonic powers.

8.4.4 Combined application of ultrasound with electrochemistry

The advantages of using ultrasound in the electrochemistry of organic compounds have been reviewed elsewhere (Walton and Mason, 1998). The combination has been found to be particularly useful in the destruction of chemical pollutants. It has been applied to the destruction of phenols by electrochemical oxidation. Ultrasound (25 kHz, $10^4\,W\,m^{-2}$) when applied to a solution containing phenol ($100\,g\,L^{-1}$) and NaCl ($2\,g\,L^{-1}$) achieves better than 80% oxidation to maleic acid. In the absence

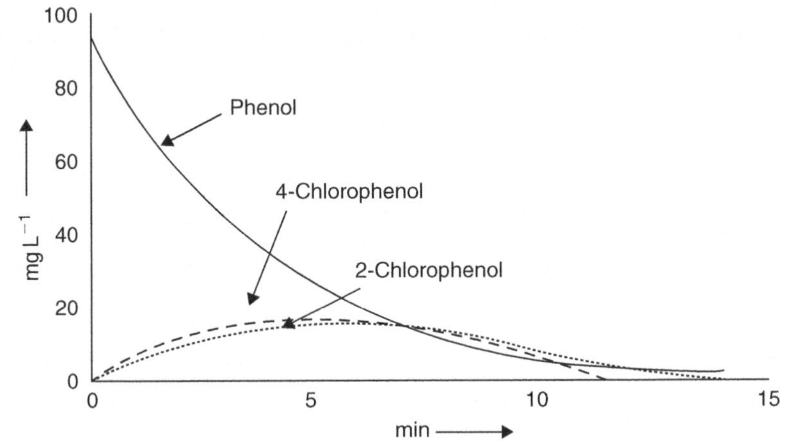

Figure 8.12 Sonoelectrochemical destruction of phenol in saline solution (500 kHz, 1 A, 7 V).

of ultrasound only 50% decomposition was obtained under the same conditions. An almost complete sonoelectrochemical destruction of phenol can be achieved in saline solution (0.5 g L^{-1} NaCl) at pH 6 (Trabelsi, 1995). The cell operated at 1 amp using a nickel mesh cathode (area $115 \times 10^{-4} m^2$) and a platinum mesh anode (area $47 \times 10^{-4} m^2$) and complete destruction of the phenol was achieved within 20 min at 25°C (Figure 8.12). The reaction was shown to proceed via intermediate chlorinated phenols.

Sonoelectrochemical destruction of dyes has also been investigated. Solutions of the acidic dye Sandolan Yellow have been decolourised by a sonoelectrochemical process in aqueous saline solution using platinum electrodes (Lorimer *et al.*, 2000). The process entails the electrolysis of aqueous NaCl solution that involves the liberation of chlorine at the anode and hydroxide ion at the cathode. The overall cell reaction is

$$2OH^- + Cl_2 \rightarrow Cl^- + OCl^- + H_2O$$

The rate of electrochemical decolourisation in the absence of ultrasound was dependent on the electrolyte concentration, reaction temperature and the current density. The rate was enhanced using ultrasound when carried out in a semi-sealed cell, which minimised the effects of ultrasonic degassing. An optimum acoustic power was found when using a source at a frequency of 20 kHz.

8.5 EQUIPMENT DEVELOPMENT

There are essentially two types of large-scale plant: batch and flow types. Sometimes the flow system will form a part of batch processing as a loop attached to the main vat. The results from successful small-scale experiments can be adapted for large-scale work providing that information is available on power input and volume treated. However the majority of reported small-scale studies have not been optimised and thus the calculated powers required for scale-up, based on such studies, will almost certainly be unrealistically high. Several scale-up designs are available which can be broadly divided into batch and flow systems.

8.5.1 Batch systems

Batch systems will generally be based upon the ultrasonic cleaning bath using the whole bath as the reactor. Such systems can be many thousands of litres in capacity and are often purpose built as part of an automated cleaning train. Alternatively it is possible to use a submersible transducer to convert any tank into an ultrasonic reactor (Figure 8.13) – this offers somewhat greater flexibility in use than the purpose-built bath. The submersible is essentially a sealed box containing transducers bonded to the inside of one face. This face then reverberates when the transducers are activated thus providing a large surface area as the source of ultrasound.

Such a batch system would not have wide applications in water treatment but could be ideal for the treatment of the effluent from small factories. In particular this would be useful in the treatment of waste dyes from textile plants. It could also be applied as part of a series of tanks giving sequential treatments in a weir type system.

8.5.2 Flow systems

In situations where higher intensity ultrasound is required for large-scale operations the source of power ultrasound is best utilised as part of a flow system. Even in batch situations a high-energy source can be incorporated as part of a loop outside of the vat. This permits the continuous processing of large volumes and provides major advantages over the batch configuration.

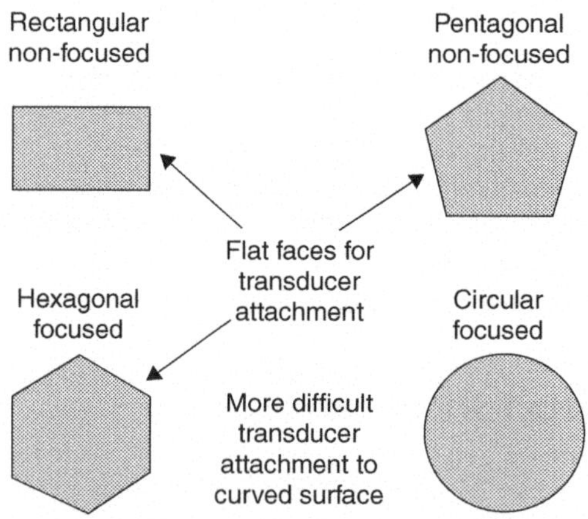

Figure 8.13 Batch system employing a submersible transducer.

- High intensity sonication is provided for a continuously flowing liquid reaction at controllable power by the adjustment of either input power to the transducer or the flow rate through the cell.
- Temperature control is provided through the circulating reaction mixture or a pulse option.

The design options are many and varied but can be summarised in terms of two basic configurations.

8.5.2.1 Resonating tube reactors

This type of system provides the greatest hope for general usage of power ultrasound in a wide range of processing industries. Essentially the liquid to be processed is passed through a pipe with ultrasonically vibrating walls. In this way the sound energy generated from transducers bonded to the outside of the tube is transferred directly into the flowing liquid. Two design engineering problems are associated with this type of sonicator (i) the correct mounting of the transducers on the outer tube and (ii) the length of the tube must be such that the ends are at a null point in the sound wavelength in the unit. This will eliminate vibrational problems associated with the retrofitting of the unit to existing pipework.

Generally the commercial tube reactors are constructed of stainless steel and the choices for pipe cross-section are generally one of four: rectangular, pentagonal, hexagonal and circular (Figure 8.14). Several devices have been developed which incorporate flow systems with vibrating walls many of which have been used in sewage sludge treatment (Mason, 2001). One such example is a system developed by Sonic Process Technologies in the UK (Figure 8.15). It typifies the approach of using an array of transducers bonded to the outside of a steel tube. An alternative approach uses powerful transducers to drive a length of piping in radial mode and

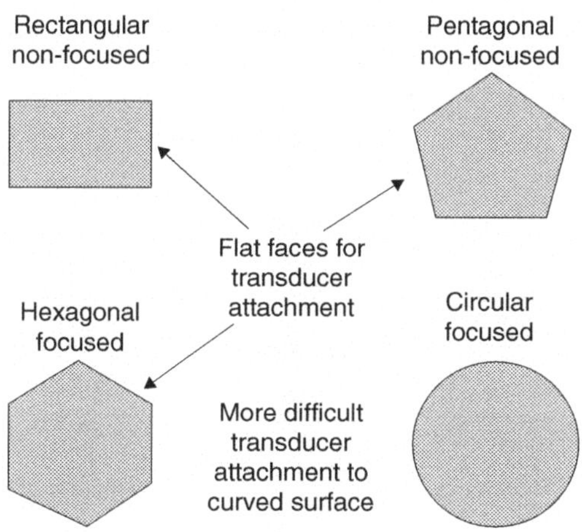

Figure 8.14 Cross-sectional geometries for tubular reactors.

was developed by Sodeva in France. The design is quite simple and comprises of a central disc attached to the transducer (the modular unit) into which stainless steel pipes are screwed (Figure 8.16). The whole system is in resonance with the extreme ends of each pipe being null points for connection into other pipework.

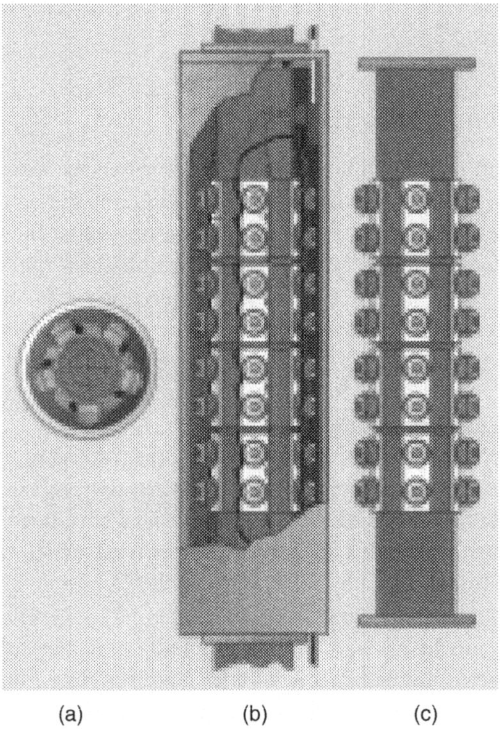

Figure 8.15 Cylindrical tube processor showing, from the left: (a) arrangement of transducers around outside of processing tube; (b) processing tube with outer protective casing; (c) inner processing tube with transducers attached.

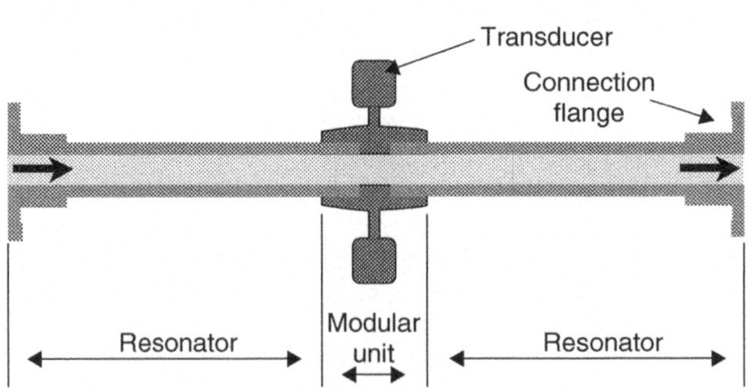

Figure 8.16 Cylindrical tube processor developed by Sodeva.

8.5.2.2 Resonating inserts

In order to introduce ultrasound into a medium flowing through a tube it is only necessary to place a vibrating source (e.g. probe) in contact with that liquid. A neat method of doing this is via the coaxial insertion of a radially emitting bar into the pipe containing the flowing liquid this would require minimal change to existing pipework. This type of system was originally developed as a form of submersible transducer for cleaning the inside of barrels and tubes immersed in a cleaning fluid. One system consists of a hollow, gas filled, tube sealed at one end and driven at the other by a standard piezo transducer (Figure 8.17). This device, developed by Telsonic in Switzerland, looks like a conventional probe system but is significantly different in that the sealed end is at a null point and is not the major source of vibrational energy. Instead the ultrasound is emitted radially at half wavelength distances along its length.

Using a sonicated disc insert within a tube (Figure 8.18) FFR Ultrasonics Ltd have developed a high power system capable of treating flowing liquids. Using 10 discs, each of 3 kW power in series, the resulting 30 kW system can process fluid at a rate of $33\,L\,s^{-1}$.

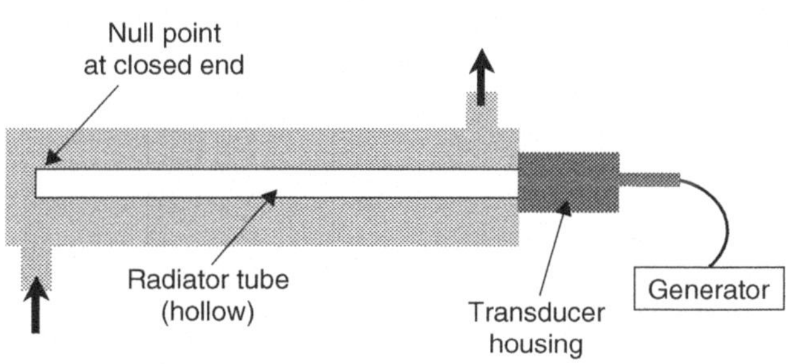

Figure 8.17 The Telsonic tubular insert.

Figure 8.18 Example of a sonicated disc inserted in a tube. (FFR Ultrasonics Ltd.)

8.6 CONCLUSIONS

Laboratory research into the uses of ultrasound for environmental protection is very broad but a number of trends are showing enough promise to be attracting the interest of industry. The study of frequency effects and the combination of ultrasound with other techniques, for example with ozone or UV light would appear to be particularly important for water purification through advanced oxidation. Equipment design is coming on apace both in the laboratory and in the development work necessary for scale-up.

There remain two problems to be solved for high power applications:

- Given that generally the upper limit of acoustic power delivery into a system is decoupling of the emitting face from the liquid load – can this be overcome?
- Can the overall economics of power ultrasound usage be justified in terms of results?

If positive answers to these two questions are found then sonochemistry will be a real technology for environmental protection in the future.

REFERENCES

Abramov, O.V. (1998) *High-Intensity Ultrasound: Theory and Industrial Applications*. Gordon and Breach, London, UK.

Anbar, M. and Pecht, I. (1964) On the sonochemical formation of hydrogen peroxide in water. *J. Phys. Chem.* **68**(2), 352–355.

Berlan, J., Trabelsi, F., Delmas, H., Wilhelm, A.M. and Petrignani, J.F. (1994) *Sonochemistry* **1**, S97.

Brown, B. and Goodman, J.E. (1965) *High Intensity Ultrasonics*. Iliffe Books Ltd, London, UK.

Chambers, L.A. (1937) The emission of visible light from cavitated liquids. *J. Chem. Phys.* **5**, 290–292.

Cheung, H.M., Bhatnagar, A. and Jansen, G. (1991) *Environ. Sci. Technol.* **25**, 1510.

Del Duca, M., Yeager, E., Davies, M.O. and Hovorka, F. (1958) Isotopic techniques in the study of the sonochemical formation of hydrogen peroxide. *J. Acoust. Soc. Am.* **30**(4), 301–307.

Didenko, Y.T. and Pugach, S.P. (1994) Optical spectra of water sonoluminescence. *Ultrason. Sonochem.* **1**(1), S9–S12.

Fischer, C.H., Hart, E.J. and Henglein, A. (1986) Ultrasonic irradiation of water in the presence of $^{18,18}O_2$: isotope exchange and isotopic distribution of H_2O_2. *J. Phys. Chem.* **90**, 1954–1956.

Fitzgerald, M.E., Griffing, V. and Sullivan, J. (1956) Chemical effects of ultrasonics – 'Hot Spot' chemistry. *J. Chem. Phys.* **25**(5), 926–933.

Flosdorf, E.W., Chambers, L.A. and Malisoff, W. (1936) Sonic activation in chemical systems: oxidations at audible frequencies. *J. Am. Chem. Soc.* **58**, 1069–1076.

Frederick, J.R. (1965) *Ultrasonic Engineering*. John Wiley, London, UK.

Frenkel, G. (1940) On electrical phenomena associated with cavitation due to ultrasonic vibrations in liquids. *Acta Physicochim. USSR* **12**(3), 317–322.

Frenzel, H. and Schultes, H. (1934) Luminescenz im ultraschallbeschickten Wasser. Kurse mitteilung. *Z. Phys. Chem.* **27b**, 421–424.

Griffing, V. (1950) Theoretical explanation of the chemical effects of ultrasonics. *J. Chem. Phys.* **18**, 997–998.

Hart, E.J. and Henglein, A. (1986) *J. Phys. Chem.* **90**, 5992.

Hart, E.J., Fischer, C.H., Henglein, A. (1986) *J. Chem. Phys.* **90**, 5989.

Henglein, A. (1987) Sonochemistry – historical developments and modern aspects. *Ultrasonics* **25**, 6–16.

Henglein, A. and Kormann, C. (1985) *Int. J. Radiat. Biol.* **48**, 251.

Langlais, B., Reckhow, D.A. and Brink, D.R. (1991) *Ozone in Water Treatment Application and Engineering.* AWWA Research Foundation and Lewis Publishers, Chelsea.

Lindström, O. (1955) Physico-chemical aspects of chemically active ultrasonic cavitation in aqueous solutions. *J. Acoust. Soc. Am.* **27**(4), 654–672.

Liu, S.-C. and Wu, H. (1934) Mechanism of oxidation promoted by ultrasonic radiation. *J. Am. Chem. Soc.* **56**(5), 1005–1007.

Loiseleur, J. (1944) Sur l'activation de l'oxygène par les ultrasons. *C.R. Acad. Sci. Fr.* **218**, 876–878.

Lorimer, J.P., Mason, T.J., Plattes, M. and Phull, S.S. (2000) *Ultrason. Sonochem.* **7**, 237–242.

Makino, K., Mossoba, M. and Riesz, P. (1982) Chemical effects of ultrasound on aqueous solutions. Evidence for —OH and —H by spin trapping. *J. Am. Chem. Soc.* **104**, 3537–3539.

Makino, K., Mossoba, M.M. and Riesz, P. (1983) Chemical effects of ultrasound on aqueous-solutions – formation of hydroxyl radicals and hydrogen-atoms. *J. Phys. Chem.* **87**, 1369–1377.

Mason, T.J. (1999) *Sonochemistry*. Oxford University Primer Series No. 70, Oxford University Press, UK.

Mason, T.J. (2001) The design of ultrasonic reactors for environmental remediation. In: *Advances in Sonochemistry, Vol. 6, Theme issue – Ultrasound in Environmental Protection*, Mason, T.J. and Tiehm, A. (eds). Elsevier.

Mason, T.J. and Lorimer, J.P. (2002) *Applied Sonochemistry.* Wiley VCH, Weinheim, Germany.

Mason, T.J. and Tiehm, A. (2001) *Advances in Sonochemistry, Vol. 6, Ultrasound in Environmental Protection.* Elsevier, Amsterdam, Netherlands.

Nagata, Y., Kurosaki, Y., Nakagawa, M. and Maeda, Y. (1993) *Chem. Express* **8**, 657.

Neppiras, E.A. (1984) Acoustic cavitation series. 1. Acoustic cavitation – an introduction. *Ultrasonics* **22**, 25–28.

Noltingk, B.E. and Neppiras, E.A. (1950) Cavitation produced by ultrasonics. *Proc. Phys. Soc.* **B63**, 674–685.

Pétrier, C. and Casadonte, D. (2001) The sonochemical degradation of aromatic and chloroaromatic contaminants. In: *Advances in Sonochemistry, Vol. 6, Theme issue – Ultrasound in Environmental Protection,* Mason, T.J. and Tiehm, A. (eds). Elsevier.

Pétrier, C. and Francony, A. (1997) *Ultrason. Sonochem.* **4**, 295.

Pétrier, C., Lamy, M.-F., Francony, A., Benhcene, A. and David, B. (1994) *J. Phys. Chem.* **98**, 10514.

Price, G. (1990) The use of ultrasound for the controlled degradation of polymer solutions. In: *Advances in Sonochemistry*, Vol. 1, Mason T.J. (ed.). JAI Press, London, UK, pp. 231–287.

Prudhomme, R.O. and Grabar, P. (1949) De l'action des ultrasons sur certaines solutions aqueuses. *J. Chim. Phys. Fr.* **46**(7–8), 323–331.

Richards, W.T. and Loomis, A.L. (1927) *J. Am. Chem. Soc.* **49**, 3086.

Russell, J.P. and Smith, M. (1999) Sonic energy in processing: use of a large scale, low frequency sonic reactor. In: *Advances in Sonochemistry*, Vol. 5, Mason, T.J. (ed.). JAI Press, London, UK, pp. 175–208.

Schmitt, F.O., Johnson, C.H. and Olson, H.R. (1929) Oxidations promoted by ultrasonic radiation. *J. Chem. Soc.* **51**, 370–375.

Schulter, H. and Gohr, H. (1936) Über chemische wirkungen des ultraschallwellen. *Zeits. Angew. Chem.* **49**(27), 420–423.

Seghal, C., Sutherland, R.G. and Verrall, R.E. (1980) Optical spectra of sonoluminescence from transient and stable cavitation in water saturated with various gases. *J. Phys. Chem.* **84**(4), 388–395.

Serpone, N., Terzian, R., Hidaka, H. and Pelizzetti, E. (1994) *J. Phys. Chem.* **98**, 2634.

Shirgaonkar, I.Z. and Pandit, A.B. (1998) Sonophotochemical destruction of aqueous solution of 2,4,6-trichlorophenol. *Ultrason. Sonochem.* **5**, 53–61.

Sierka, R.A. and Amy, G.L. (1985) Catalytic effects of ultraviolet light and/or ultrasound on the ozone oxidation of humic acid and trihalomethane precursors. *Ozone Sci. Eng.* **6**, 275–290.

SRI International, 333 *Ravenswood Avenue*, Menlo Park, CA, USA (1992).

Storey, B.D. and Szeri, A.J. (2000) Waper vapour, sonoluminescence and sonochemistry. *Proc. Roy. Soc. Lond. A* **456**, 1685–1709.

Suslick, K.S. (1990) *Sonochem. Sci.* **247**, 1439–1445.

Toy, M.S., Carter, M.K. and Passell, T.O. (1990) *Environ. Technol.* **11**, 837.

Trabelsi, F. (1995) Les ultrason de haute frequence: Etude d'un reacteur, application a la degradation de composes organiques par sonochimie et sonoelectrochimie. PhD Thesis, ENSIGC Toulouse.

Von Sonntag, C., Mark, G., Schuchmann, H.P., Von Sonntag, J. and Tauber, A. (1998) COST meeting: chemical processes and reactions under extreme or non-classical conditions. *COST Action* **D6**, 11–18.

Walton, A.J. and Reynolds, G.T. (1984) Sonoluminescence. *Adv. Phys.* **33**(6), 595–660.

Walton, D.J. and Mason, T.J. (1998) *Organic Sonoelectrochemistry in Synthetic Organic Sonochemistry*, Luche, J.-L. (ed.). Plenum Press, pp. 263–297.

Weavers, L.K. (2001) Sonolytic ozonation for the remediation of hazardous pollutants. In: *Advances in Sonochemistry, Vol. 6, Theme issue – Ultrasound in Environmental Protection*, Mason, T.J. and Tiehm, A. (eds). Elsevier.

Weavers, L.K. and Hoffmann, M.R. (1998) *Environ. Sci. Technol.* **32**, 3941–3947.

Weiss, J. (1944) Radiochemistry of aqueous solutions. *Nature* **153**, 748–750.

9

Radiation processes

W.J. Cooper, P. Gehringer, A.K. Pikaev, C.N. Kurucz and B.J. Mincher

9.1 INTRODUCTION

Innovative, ultimate treatment technologies are currently being developed for polluted water. As a group, these technologies are called advanced oxidation technologies (AOTs) or advanced oxidation processes (AOPs). AOPs seek to mineralise pollutants to simple, often elemental constituents. This is an improvement over many conventional options such as aeration stripping and carbon filtration, which are merely phase transfer processes. Similarly, land disposal merely transfers pollutants that will eventually require further attention. However, mineralisation represents the final and ultimate treatment, which is the goal in many areas of pollutant control. The AOPs use free radicals (principally hydroxyl radical, $^{\bullet}OH$) which attack and decompose pollutants. In some cases oxidation via the $^{\bullet}OH$ is slow and the application of reducing radicals such as

This chapter is dedicated to A.K. Pikaev who passed away after the completion of his contribution to the study.

 Advanced Oxidation Processes for Water and Wastewater Treatment. Edited by Simon Parsons. ISBN: 1 84339 017 5

the solvated electron (e^-_{aq}) or hydrogen atom ($H^\bullet$) is advantageous. An excellent source of both reducing and oxidising radicals for water treatment is ionising radiation.

The simultaneous generation of both reducing and oxidising species allows for a versatile approach to the ultimate treatment of a wide variety of pollutants. Versatility is provided by continuous, rapid treatment potential and a tolerance for feedstocks of varying quality. Radiation sources such as machine-generated electron beams (e-beams), isotopic gamma-ray sources such as ^{60}Co, ^{137}Cs and even spent nuclear fuel have been used. Electron beam sources provide the much higher radiation dose rates necessary for process-scale treatment. Modern electron beam accelerators have excellent reliability, are automated and can be portable.

An electron accelerator is similar to the common television, except that the accelerating potential is significantly higher. This is determined by equipment design. For environmental applications the lowest practical potential is 500 keV. It is likely that potentials of 1–10 MeV will be used. This is because the energy of the electron determines the depth of penetration in water (Figure 9.1). The number of electrons, controlled by the beam current, is determined by the cathode size and configuration. Common cathodes used today result in beam currents of from 50 to 150 mA. The electron beam power is the product of energy (V) times beam current (A). Electron accelerators typically have power ratings of up to 100 kW, and may be as high as 200 kW.

The underlying chemistry of the electron beam process is radiation chemistry. Radiation chemistry is a mature area and the fundamental processes of water radiolysis are reasonably well understood. One of the fundamental techniques used in radiation chemistry is pulse radiolysis. This has been used to evaluate rate constants and mechanisms of the decomposition of organic and inorganic chemicals.

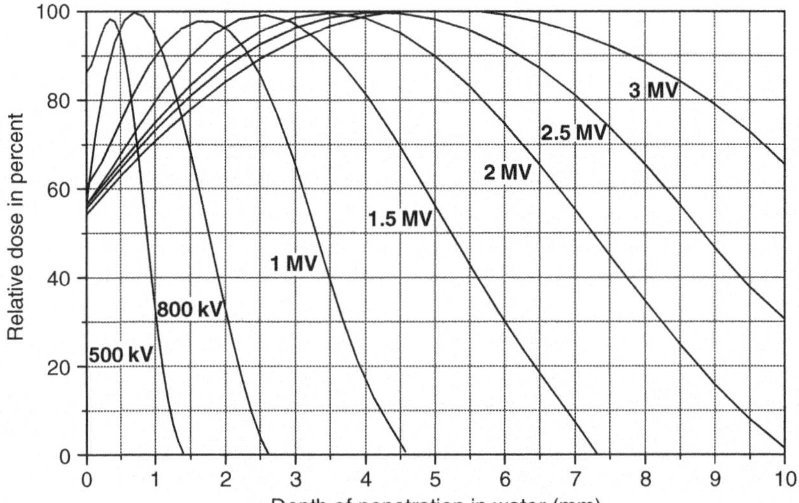

Figure 9.1 The relative energy deposition of electrons, dose, in water at 25°C.

9.1.1 Water radiolysis

Irradiation of water with electron beams or γ-rays results in the formation of electronically excited states, free radicals and ions in the spurs along the track of the incident particle. A continuum of electrons is produced as the initial event dissipates its energy in the irradiated medium. The reactive species produced by irradiation may then diffuse away from the spur, or undergo geminate recombination to recreate their parent species. Those that escape recombination may react with solutes, including pollutants present in the water. This is the essence of waste treatment by radiolysis. At 10^{-7} s after an electron passes through water at neutral pH, the products shown in Equation (9.1) are present (Buxton *et al.*, 1988):

$$H_2O \rightsquigarrow [0.28]^{\bullet}OH + [0.27]\, e^-_{aq} + [0.06]\, H^{\bullet} + [0.07]\, H_2O_2 + [0.27]\, H_3O^+ + [0.05]\, H_2 \tag{9.1}$$

High-energy electrons are capable of initiating several thousand secondary reactions as they dissipate their energy in water. The efficiency of conversion of electron energy to a chemical product is defined as the G value [shown in brackets in Equation (9.1)]. The G value is the μmol of product formed or lost in a system absorbing 1 J of energy. Given that G values are yields of radicals in pure water, it is possible to calculate the concentrations of oxidising and reducing species for any absorbed dose. The SI unit of dose is the gray (Gy), which equals energy deposition of 1 J kg^{-1}. Therefore, as a first approximation, the concentration of $^{\bullet}OH$ produced in pure water when 1 kGy is absorbed is given by

$$[^{\bullet}OH] = 1000\,J\,kg^{-1} \times 0.28\,\mu mol\,J^{-1} = 280\,\mu mol\,kg^{-1} \tag{9.2}$$

This calculated value is a maximum that does not account for reactions that lead to loss of the radical. Table 9.1 provides an estimate of the concentrations of the reactive species in aqueous solution.

The most reactive products in Equation (9.1) are the oxidising hydroxyl radical ($^{\bullet}OH$) and the reducing aqueous electron (e^-_{aq}) and hydrogen atom ($H^{\bullet}$). Oxidising and reducing species are produced in approximately equal amounts, and these three products are responsible for most of the chemistry of interest in waste treatment.

Table 9.1 Estimated concentration of transient reactive species in pure water at several doses using high-energy electron irradiation.

Dose (kGy)	Concentration (mM)			
	e^-_{aq}	$H^{\bullet}$	$^{\bullet}OH$	H_2O_2
1	0.27	0.06	0.28	0.07
5	1.4	0.3	1.4	0.4
10	2.7	0.6	2.8	0.7

The reactions of the aqueous electron (e^-_{aq}) with specific organic and inorganic compounds have been studied extensively (Anbar *et al.*, 1973; Buxton *et al.*, 1988; Hart and Anbar, 1970; Pikaev, 1969; Pikaev and Kabakchi, 1982; Ross, 1973). With a reduction potential of −2.77 V it is a powerful reducing agent that may result in single electron transfer. Of particular interest to the field of waste treatment are the reactions with halogenated compounds. A generalised reaction, dissociative electron attachment, is shown below:

$$e^-_{aq} + RCl \rightarrow R^\bullet + Cl^- \tag{9.3}$$

Reactions involving the e^-_{aq} often result in the dehalogenation of compounds. This falls short of mineralisation, but it may be adequate when the dehalogenation product is considered non-hazardous. The aqueous electrons also contributes to the removal of many non-halogenated compounds from aqueous solution.

Although aqueous electron is produced with a *G* value nearly equal to that of oxidising hydroxyl radicals, they are often less available for reaction due to scavenging by hydronium ion in acidic water, and oxygen in aerated solutions. Scavenging decreases the concentration available for reactions with pollutants.

Reactions of the hydrogen atom ($H^\bullet$) with organic and inorganic compounds have also been summarised (Anbar *et al.*, 1975; Buxton *et al.*, 1988; Pikaev and Kabakchi, 1982). Based on the *G* values shown in Equation (9.1), the hydrogen atom accounts for approximately 10% of the total free radical concentration in irradiated water. It too is a powerful reducing agent with a potential of −2.3 V. The $H^\bullet$ undergoes two general types of reactions with organic compounds, hydrogen addition and hydrogen abstraction. Although less abundant and less reducing than the aqueous electron, the relatively small reaction rate constant of $H^\bullet$ with common radical scavengers found in natural waters makes it important for some pollutants.

The oxidising hydroxyl radical ($^\bullet OH$) is probably the most important of the products of water radiolysis for most applications. Compilations of rate constants for its reactions with many solutes have also been published (Buxton *et al.*, 1988; Dorfman and Adams, 1973; Pikaev and Kabakchi, 1982). The $^\bullet OH$ can undergo several types of reactions including: addition, hydrogen abstraction and electron transfer. Addition reactions occur readily with aromatic and unsaturated aliphatic compounds. The resulting compounds are hydroxylated radicals:

$$^\bullet OH + CH_2{=}CH_2 \rightarrow {}^\bullet CH_2{-}CH_2(OH) \tag{9.4}$$

Hydrogen abstraction occurs with saturated and many unsaturated molecules, for example aldehydes and ketones:

$$^\bullet OH + CH_3{-}CO{-}CH_3 \rightarrow {}^\bullet CH_2COCH_3 + H_2O \tag{9.5}$$

Electron transfer reactions are also common, and occur when aqueous solutions are irradiated with high-energy electrons. For example, reactions involving thioanisole have recently been reported (Tobien *et al.*, 2000):

$$^\bullet OH + CH_3{-}S{-}C_6H_5 \rightarrow [CH_3{-}S{-}C_6H_5]^{+\bullet} + OH^- \tag{9.6}$$

Beside the numerous compilations of bimolecular rate constants for solute reactions with $^{\bullet}OH$, aqueous electron and $H^{\bullet}$ (Anbar *et al.*, 1973, 1975; Buxton *et al.*, 1988; Dorfman and Adams, 1973; Hart and Anbar, 1970; Pikaev, 1969; Pikaev and Kabakchi, 1982; Ross, 1973), there is also a World-Wide-Web site that is updated bi-weekly (http://www.rcdc.nd.edu). Pollutants with large rate constants for reaction with one or more of these species are likely to be amenable to treatment by radiolysis. Table 9.2 summarises the rate constants for several chemicals that have been identified as relatively common pollutants.

While a large rate constant may suggest that treatment is possible, the actual efficiency depends on factors such as the type of ionising radiation (electron beam

Table 9.2 Rate constants ($M^{-1}s^{-1}$) of selected organic chemicals and the free radicals formed in irradiated aqueous solutions (http://www.rcdc.nd.edu).

Compound	e^-_{aq}	$H^{\bullet}$	$^{\bullet}OH$
Benzene	9.0×10^6	9.1×10^8	7.8×10^9
Bromodichloromethane	2.1×10^{10}	NF	NF
Bromoform	2.6×10^{10}	NF	1.1×10^8
Carbon tetrachloride	1.6×10^{10}	3.8×10^7	NF
Chlorobenzene	5.0×10^8	1.4×10^9	5.5×10^9
Chloroform	1.3×10^{10}	1.1×10^7	$\approx 5 \times 10^6$
o-Cresol	NF	NF	1.1×10^{10}
m-Cresol	NF	NF	1.2×10^{10}
p-Cresol	4.2×10^7	1.3×10^9	1.2×10^9
Dibromochloromethane	2.0×10^{10}	NF	NF
o-Dichlorobenzene	4.7×10^9	NF	3.3×10^9
m-Dichlorobenzene	5.2×10^9	F	5.7×10^9
p-Dichlorobenzene	5.0×10^9	NF	4.2×10^9
1,1-Dichloroethane	9.0×10^9	NF	1.3×10^8
1,2-Dichloroethane	6.4×10^9	$\sim 2.3 \times 10^6$	7.9×10^8
trans-1,2-dichloroethylene	7.5×10^9	NF	6.2×10^9
Dichloromethane	5.5×10^9	4.0×10^6	9.0×10^7
1,2-Dichloropropane	NF	NF	3.0×10^8
Ethylbenzene	NF	NF	7.5×10^9
Hexachloroethane	3.9×10^{10}	NF	NF
Nitrobenzene	3.7×10^{10}	1.0×10^9	3.9×10^9
Phenol	2.0×10^7	1.7×10^9	6.6×10^9
Pyridine	1.0×10^9	7.8×10^8	3.1×10^9
Tetrachloroethylene (TCE)	1.3×10^{10}	5.0×10^9	2.0×10^9
Toluene	1.4×10^7	2.6×10^9	3.0×10^9
1,1,1-Trichloroethane	1.4×10^{10}	NF	4.0×10^7
Trichloroethylene	1.9×10^9	7.1×10^8	2.9×10^9
2,4,5-Trichlorophenol	NF	NF	NF
2,4,6-Trichlorophenol	NF	1.1×10^9	1.2×10^{10}
Vinyl chloride	2.5×10^8	NF	1.2×10^{10}
o-Xylene	NF	2.0×10^9	6.7×10^9
m-Xylene	NF	2.6×10^9	7.5×10^9
p-Xylene	NF	2.2×10^9	7.0×10^9

NF: not found.

or γ, energy of radiation, absorbed dose and dose rate), the nature and concentration of pollutant(s), the presence of scavengers and the presence of additives.

For example, there is a decrease in the *G* values of the primary products of water radiolysis for electron beams at low energy (0.3–0.5 MeV) that is not found for γ-radiation. Even at the higher energies normally considered for waste treatment, the high dose rate of electron beams often results in lower yields of radiolysis products (Lin *et al.*, 1995). This results from the enhanced role of radical–radical recombination competition for the radicals at high dose rates. The effect on treatment efficiency is especially noticeable at low pollutant concentrations (Gehringer *et al.*, 1995; Lin *et al.*, 1995). Efficiency is also especially dependent on dose rate when the degradation of pollutants occurs via a chain mechanism. This is typical of a chain radiolysis process.

The role of pollutant structure can be illustrated by $G(Cl^-)$ values from γ-radiolysis of 1×10^{-4}M aqueous solutions of chlorobenzaldehydes. These values are equal to 0.01, 0.006 and 0.02 μmol J^{-1} for 2-chloro-, 3-chloro- and 4-chlorobenzaldehydes, respectively. Often, yields of radiolysis products increase with increasing the pollutant concentration. For instance, $G(Cl^-)$ values for 5×10^{-4} and 1×10^{-3}M aqueous solutions of 1,2-dichloroethane are 0.48 and 0.64 μmol J^{-1}, respectively (Getoff, 1990). However, the increase in *G* value is noticeably less than the increase in the concentration. Due to this, the higher pollutant concentration, the higher the absorbed dose required for its destruction.

Scavengers affect the efficiency of pollutant removal by competition for radicals. The role of oxygen in electron scavenging has already been mentioned. Oxygenated solutions have lower efficiency for the dehalogenation of halogenated hydrocarbons and for the removal of heavy metal ions. The pH of the system may also cause scavenging effects. For instance, in acidic medium e^-_{aq} can be converted to $^\bullet H$ by reaction with H^+. Similarly, $^\bullet OH$ dissociates to $^\bullet O^-$ radical ion in alkaline solution. These products are less reactive than the original species.

Finally, additives are sometimes used to increase radiolysis efficiency. The use of ozone to increase oxidative degradation of organic compounds is discussed in more detail later. Alkaline isopropanol has also been used as an additive to increase the rate of polychlorinated biphenyl (PCB) decomposition in irradiated oil (Mincher *et al.*, 2000).

Irradiation of some systems gives rise not only to the degradation of pollutants but also changes in the physico-chemical properties of the systems. For example, the radiolytic aggregation of fine-disperse particles leads to the formation of precipitates that capture pollutants and act as a removal mechanism. This is often observed upon irradiation of municipal and industrial wastewaters. Commonly, real wastewaters are characterised by the presence of a set of various pollutants at comparatively high concentrations. Large absorbed doses are required for their purification. Combined methods in which irradiation treatment is done along with conventional techniques, such as ozonation, floatation, adsorption, coagulation or biological treatment decrease the dose required and therefore the cost of the process (e.g. Pikaev, 2000a,b,c; Woods and Pikaev, 1994).

In addition to water treatment, radiolytic techniques have also been reported for soil, aqueous sediments and non-aqueous liquids (Arbon *et al.*, 1996;

Curry *et al.*, 1998; Hillarides *et al.*, 1994; Yak *et al.*, 1999). Polyaromatic hydrocarbons, which may be halogenated, have limited water solubility but are found in these matrices. Treatment of sediments may involve irradiation of aqueous slurries or extraction of organic compounds to a non-aqueous phase, such as an oil (Curry *et al.*, 1998) or even supercritical carbon dioxide (Yak *et al.*, 1999). Irradiation of slurries is similar to that described for water, but non-aqueous liquids often result in reducing systems due to $^{\bullet}OH$ scavenging by the solvent (Mincher *et al.*, 1996). Surfactants may also be used to reduce volume and enhance the treatability of soil (Gray and Cleland, 1998).

The balance of this chapter will examine the kinetics and by-products of the irradiation of some important aqueous pollutants. Included are combined radiation methods, in which radiolysis is used in conjunction with other techniques. Modelling of the treatment of these compounds by irradiation will be discussed, and the application of electron beam radiolysis in several case studies will be presented.

9.1.2 Kinetics

The reaction of a solute with $^{\bullet}OH$, e^-_{aq} or $H^{\bullet}$, is a bimolecular process, as shown in Equations (9.3)–(9.6). The overall rate depends on the solute and radical concentrations. Second-order rate constants can be measured using several methods. The most versatile is pulse radiolysis. This technique involves isolating the radical of interest and irradiating solutions of varying solute concentration. The reaction is followed using transient absorption spectroscopy, or time resolved electron paramagnetic resonance (EPR) spectroscopy (Buxton *et al.*, 1988). Examples, using these techniques for chemicals of environmental interest, have recently been published for several compounds (Mezyk *et al.*, 2001; Tobien *et al.*, 2000a,b).

Using Equation (9.2) and the bimolecular rate constants of Table 9.2, it is possible to calculate the relative importance of the three primary products of water radiolysis on the removal of some organic compounds of interest. These calculations are important to consider when predicting the overall efficacy of the process for a particular problem. The extension of these calculations to natural water must also consider scavengers. For example, carbonate ion and dissolved organic matter react with all three reactive species.

For the pure water case, the overall removal of solute (R) can be described by the following kinetic expression:

$$\frac{d[R]_t}{dt} = k_1[R][^{\bullet}OH] + k_2[R][e^-_{aq}] + k_3[R][^{\bullet}H] \tag{9.7}$$

where k_1, k_2 and k_3 are the respective second-order rate constants (Table 9.2). The relative contributions to solute removal by the three species may now be compared. The product of the radical concentration and the bimolecular rate constant (with appropriate unit conversions) is a pseudo-first-order rate constant (k') for reaction with solute R, with units of reciprocal time:

$$G\,(\mu mol\,J^{-1}) \times 1\,J\,L^{-1} \times k\,(L\,mol^{-1}s^{-1}) = k'\;(s^{-1}) \tag{9.8}$$

Table 9.3 Comparison of the reaction rates of e^-_{aq}, $H^\bullet$ and $^\bullet OH$ with chloroform in pure water at an absorbed dose of 1 kGy.

Species	Concentration (M)	k (L mol^{-1} s^{-1})	k' (s^{-1})	%
e^-_{aq}	2.7×10^{-4}	3.0×10^{10}	8.1×10^6	99.98
$H^\bullet$	6.0×10^{-5}	2.4×10^6	1.4×10^2	0.002
$^\bullet OH$	2.8×10^{-4}	5.0×10^6	1.4×10^3	0.020

(For aqueous solutions 1 kg = 1 L and the appropriate value is included to account for 10^6 μmol/mol.) A high first-order rate constant indicates high removal efficiency. The total removal rate is given by:

$$\text{Removal rate}[R] = (k'_1 + k'_2 + k'_3)[R] \tag{9.9}$$

which has units of mol s^{-1}. The rate of removal due to an individual reactive species may be found by dividing the individual pseudo-first-order rate constant by the sum of the three:

$$\%\ \text{Removal}[R] = 100 \times (k'_1/k'_1 + k'_2 + k'_3)[R] \tag{9.10}$$

An example of the relative importance of the three radicals for chloroform removal is shown in Table 9.3.

Reaction with the solvated electron is the dominant pathway for the decomposition of chloroform in irradiated water. A process for chloroform decomposition might be designed to maximise reductive treatment, perhaps by excluding oxygen. An alternative approach would be to increase the carbonate ion to scavenge the $^\bullet OH$, thus reducing the radical/electron recombination, thus increasing the effective concentration of the e^-_{aq}. While the initial decomposition of chloroform is initiated by dissociative electron attachment (loss of chloride ion), the real situation is more complex. As will be shown later, oxygen plays an important role in the decomposition of many organic compounds via peroxyl-radical formation.

In practice, continuous, process irradiations are generally performed at a constant dose rate. Therefore k' may be expressed with respect to absorbed radiation dose (kGy^{-1}) rather than time (s^{-1}). It then represents the concentration of solute removed per unit dose. This 'dose constant', assumes exponential solute removal, which is common in waste treatment applications (Mincher *et al.*, 2000). For example, the concentration of $^\bullet OH$ calculated from Equation (9.2) for absorbed doses between 1 and 10 kGy is 0.28–2.8 mM. Under these conditions the loss of the solutes (at typical solute concentrations in the μM range) is pseudo-first-order, with respect to absorbed dose, and can be described by the following:

$$-\frac{d[R]}{dD} = k_1[R][^\bullet OH] = k_o[R] \tag{9.11}$$

where [$^\bullet$OH] is hydroxyl radical concentration, at dose D, for which the excess concentration remains essentially constant, and $k_o[R]$ is the pseudo-first-order dose constant. The empirical $k_o[R]$ may be obtained from the slope of the plot of ln[R] versus dose (D). In natural water systems the value of $k_o[R]$ is affected by scavengers, and usually must be determined empirically for a given system. It may also vary with the initial concentration of the solute, which suggests that pseudo-first-order conditions are not satisfied. However, deviations in $k_o[R]$ at high initial solute concentrations are often the result of the build up of reaction by-products, and the competition of these by-products for the reactive species.

The dose constant is of value in the application of the electron beam process (Mincher *et al.*, 2000). As an empirical constant (rather than a G value), it may be used to determine dose requirements for treatment of a pollutant to a desired concentration goal. This is calculated using a variation of the first-order rate law:

$$C = C_o e_D^{-k_o[R]} \tag{9.12}$$

Another commonly used figure of merit for exponential systems is the D_{90}, the dose required to reduce the solute concentration to 10% of its initial value. It is found by rearranging Equation (9.12):

$$C/C_o = 0.1 = e_{D_{90}}^{-k_o[R]} \tag{9.13}$$

The above equation is then solved for D to find D_{90}. Alternatively, if the $k_o[R]$ is known, it is possible to determine the dose required for any destruction. For example, the dose required to remove 50, 90 or 99% of the solute is determine using the following relationships:

$$D_{50} = \frac{\ln 2}{k_o} \qquad D_{90} = \frac{\ln 10}{k_o} \qquad D_{99} = \frac{\ln 100}{k_o}$$

A G value for the disappearance of a solute may also be empirically determined. Commonly, G-initial (G_o), at zero absorbed dose is chosen. For treatment purposes, G at dose D is defined by the disappearance of the solute in aqueous solutions, and is determined experimentally using the following equation:

$$G = \frac{[C_o - C]NA}{D} \tag{9.14}$$

where the change in solute concentration is in μmol L^{-1}, D is the dose in Gy and $NA = 6.02 \times 10^{23}$, Avogadro's number. Alternately, the G value at a selected dose is the product of dose constant and the solute concentration at the dose:

$$G_o = k_o C_o \tag{9.15}$$

where the initial G value at zero absorbed dose is the case shown.

9.1.3 Radical scavengers in natural water

Natural waters are extremely variable in composition and in most cases relatively complex. The common chemical species are dissolved organic matter and a number of ions, such as chloride, nitrate, sulphate, carbonates and sodium. The dissolved organic fraction is referred to by various names, such as humic substances,

Table 9.4 Bimolecular rate constants ($M^{-1}s^{-1}$) of chemicals commonly found in natural waters (Buxton *et al.*, 1988).

Compound	$^{\bullet}OH$	e^-_{aq}	$H^{\bullet}$
O_2	NR	1.9×10^{10}	1.2×10^{10}
HCO_3^-	8.5×10^6	$<1.0 \times 10^6$	4.4×10^4
CO_3^{2-}	3.9×10^8	3.9×10^5	NR
Cl^-	3.0×10^9	$<1.0 \times 10^6$	$<1.0 \times 10^5$
NO_2^-	1.1×10^{10}	3.5×10^9	7.1×10^8
NO_3^-	NR	9.7×10^9	1.4×10^6
DOC	2.0×10^8	NR	NR

humic acid, fulvic acid and natural organic matter. These are generally determined analytically as dissolved organic carbon (DOC). Carbonate alkalinity is a measure of the total carbonate concentration common to many natural waters. The carbonate system is complicated by the equilibrium shown below:

$$CO_2 + H_2O \rightarrow H_2CO_3 \tag{9.16}$$

$$H_2CO_3 + H_2O \rightarrow H_3O^+ + HCO^{3-} \quad pK_a = 6.35 \tag{9.17}$$

$$HCO^{3-} + H_2O \rightarrow H_3O^+ + CO_3^{2-} \quad pK_a = 10.3 \tag{9.18}$$

As shown in Equation (9.18), the carbonate/bicarbonate ion ratio is quite different in various waters depending on pH. For example, in neutral water the ratio of carbonate to bicarbonate is 0.00047. Bicarbonate is clearly dominant. However, at pH 10, the ratio is 0.47, meaning the two ions are about equally abundant. The relative effects of these ions on radical scavenging can be found from the appropriate second-order rate constants, shown in Table 9.4. Carbonate ion is the more important hydroxyl radical scavenger. Thus, at constant alkalinity, $^{\bullet}OH$ scavenging is more important at higher pH and the decomposition rate of solutes that depend on reaction with the hydroxyl radical will be reduced.

9.1.4 Modelling

Models using the concepts discussed above have employed a computer code called MAKSIMA-CHEMIST provided by Atomic Energy of Canada, Ltd (Carver *et al.*, 1979; Gear, 1971; Schwarz, 1962). The input to the kinetic model includes a list of all known species, their initial concentrations and the appropriate rate constants. In addition to target solutes, scavengers and reaction by-products are included. Detailed knowledge of the reaction mechanism is necessary to obtain good agreement between empirical and modelled results.

Attempts to model single solute systems have met with reasonable success (Kartasheva *et al.*, 1998c; Kim *et al.*, 1999; Mak *et al.*, 1997; Zele *et al.*, 1998). The destruction of CCl_4 has been modelled using the following equations and rate constants:

$$CCl_4 + e^-_{aq} \rightarrow {}^{\bullet}CCl_3 + Cl^- \quad (1.3 \times 10^{10} M^{-1}s^{-1}) \tag{9.19}$$

$$CCl_4 + H^{\bullet} \rightarrow {}^{\bullet}CCl_3 + Cl^- + H^+ \quad (3.2 \times 10^7 M^{-1}s^{-1}) \tag{9.20}$$

$${}^{\bullet}CCl_3 + O_2 \rightarrow CCl_3OO^{\bullet} \quad (3.3 \times 10^9 M^{-1}s^{-1}) \tag{9.21}$$

$${}^{\bullet}CCl_3 + {}^{\bullet}CCl_3 \rightarrow Cl_3CCCl_3 \quad (3.7 \times 10^8 M^{-1}s^{-1}) \tag{9.22}$$

The model accurately predicts carbon tetrachloride concentration changes over a range of concentrations and solution pH. In experiments conducted at large scale using the Miami Electron Beam Research Facility (EBRF) (Kurucz *et al.*, 1985), methanol was added as a $^{\bullet}OH$ scavenger (Mak *et al.*, 1997; Nickelsen *et al.*, 1994). Methanol scavenges $^{\bullet}OH$ according to Equation (9.23):

$$CH_3OH + {}^{\bullet}OH \rightarrow {}^{\bullet}CH_2OH + H_2O \quad (9.7 \times 10^8 M^{-1}s^{-1}) \tag{9.23}$$

The effects of hydroxyl radical scavenging by methanol were predicted accurately. Computer simulation of the radiolysis of aqueous chlorobenzene solutions was conducted using 61 reactions (Kartasheva *et al.*, 1998c). Among them were the reactions of e^-_{aq}, $H^{\bullet}$ and $^{\bullet}OH$ with chlorobenzene, the formation of peroxyl radicals from reactions of H- and OH-adducts with oxygen, and detachment of HCl from the OH-adduct.

Formic acid is an important component of wastewater from some power plants. Modelling of the electron beam treatment of 1×10^{-2} M aqueous formic acid at high doses (100 kGy) was carried out at varying pH and under different experimental conditions (Kim *et al.*, 1999). Seventy-three possible reactions were used, including reactions of e^-_{aq}, $H^{\bullet}$ and $^{\bullet}OH$ with each other, oxygen, HCOOH and products of its radiolytic conversions. In addition, interaction of free radicals from organic acids with each other, oxygen and hydrogen peroxide were taken into account. From Figure 9.2, it can be seen that the calculated results coincide with the experimental data.

9.1.5 Reaction by-products

A mature treatment technology must demonstrate not only that the pollutant of interest is removed, but that by-products have been identified and quantified. Solute mineralisation cannot be assumed without demonstration of mass balance of the actual products. For example, an analytical scheme that is designed to measure the removal of chlorobenzene from solution must show, not only the decrease in chlorobenzene concentration, but must also monitor the appearance of benzene or phenol, which are likely unacceptable products, as well as inorganic Cl^-. The dechlorination of chlorobenzene to produce the benzene radical proceeds by dissociative electron attachment, as shown in Equation (9.24):

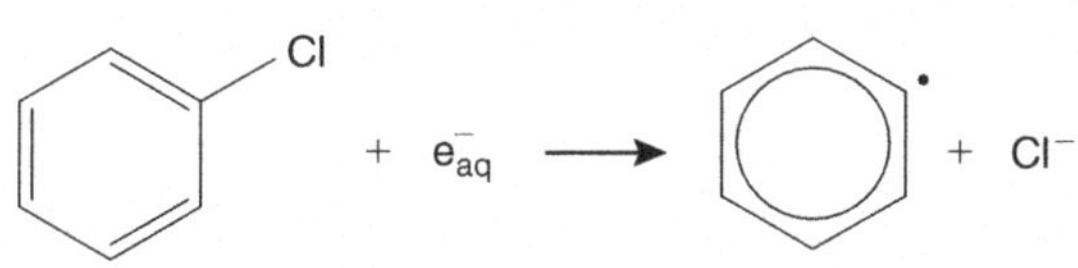

(9.24)

The radical stabilisation occurs via acquisition of a proton from solution. Benzene may also be attacked by the hydroxyl radical to produce the

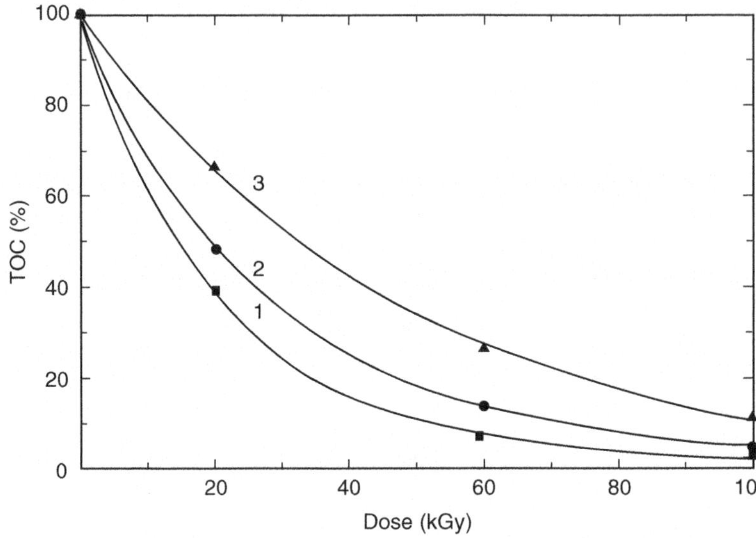

Figure 9.2 Dependence of *TOC* on absorbed dose for the electron-beam treatment of aqueous formic acid saturated with (1) oxygen, (2) air or (3) nitrogen. Experimental points and calculated curves are shown.

hydroxycyclohexadienyl radical (Cooper *et al.*, 1996; Getoff, 1986) [Equation (9.25)], ultimately resulting in either hydroxylation to produce phenol [Equation (9.26)] or decomposition of the phenyl ring via a peroxyl intermediate [Equations (9.27)–(9.29)]. The product of Equation (9.27) is the peroxyl radical, which requires oxygen for its formation.

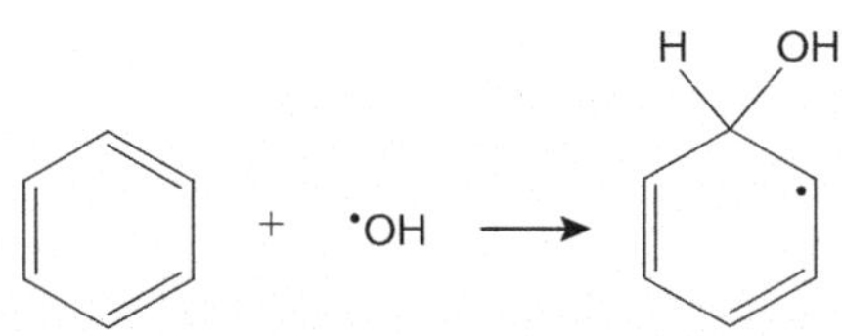
(9.25)

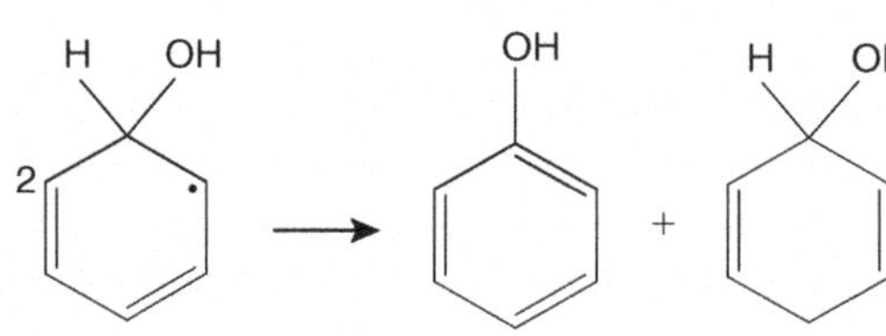
(9.26)

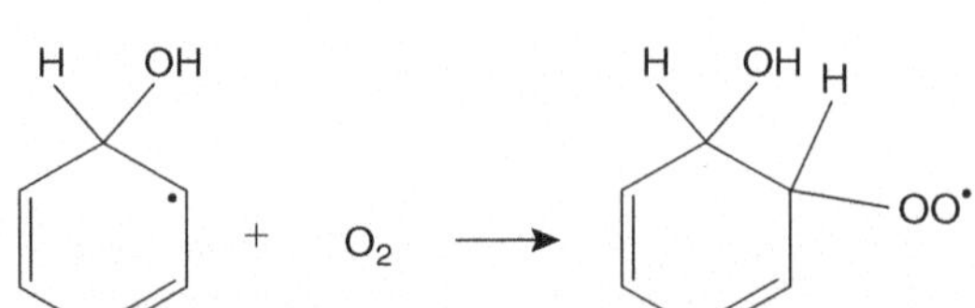
(9.27)

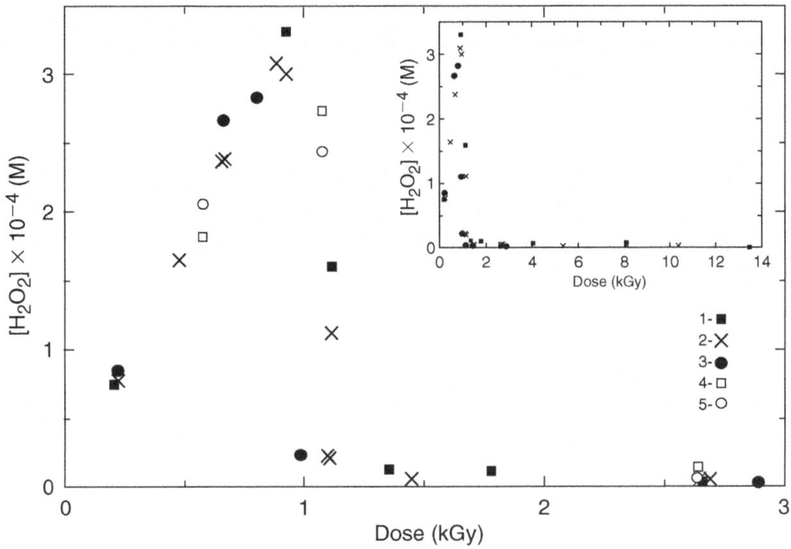

Figure 9.3 Formation of H_2O_2 upon irradiation of aqueous solutions of formic acid: (1) 1×10^{-3} M, γ-radiation; 0.8 Gy s^{-1}; (2) 1×10^{-2} M, γ-radiation, 0.8 Gy s^{-1}; (3) 1×10^{-1} M, γ-radiation, 0.8 Gy s^{-1}; (4) 1×10^{-3} M, electron beam, 10 Gy s^{-1}; (5) 1×10^{-1} M, electron beam, 10 Gy s^{-1}.

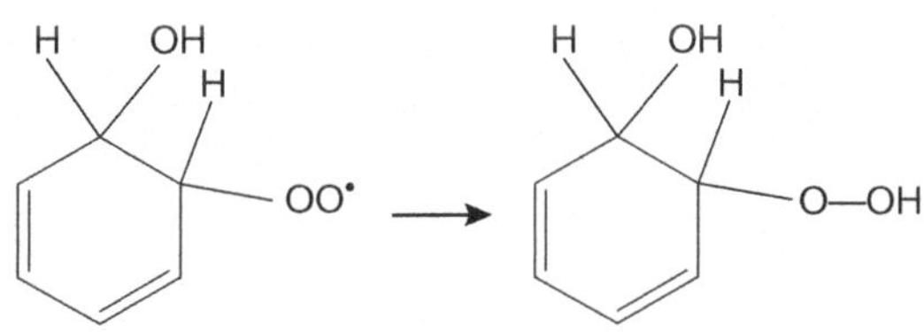

(9.28)

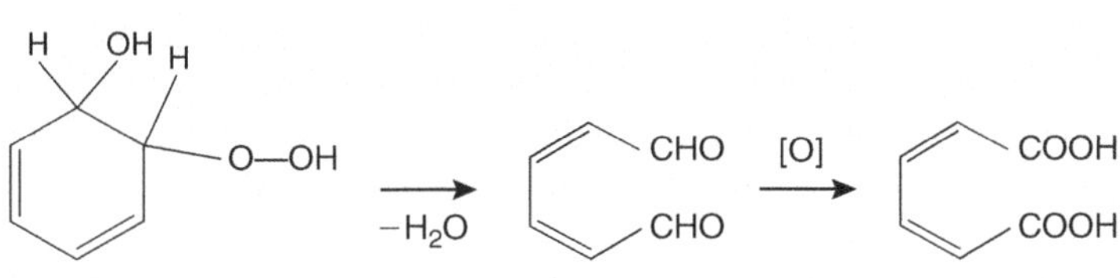

(9.29)

It can be seen that with either reaction pathway, the concentration of chlorobenzene decreases, but only in the presence of sufficient oxygen is the aromatic ring decomposed, leading to mineralisation. In the absence of sufficient oxygen the by-products are phenol, or chlorophenols.

The destruction of formic acid upon electron beam treatment of its aerated aqueous solutions proceeds at comparatively low doses with formation of water, carbon dioxide, molecular hydrogen and hydrogen peroxide (Kartasheva *et al.*, 2000). However, at sufficiently high doses oxygen is consumed and further destruction of formic acid was not observed (Figure 9.3), and the formation of oxalic and other carboxylic acids was also observed. Thus, an understanding of reaction mechanisms is crucial to the prediction and adequate monitoring of by-products. Often, by-products will also be susceptible to radiolytic decomposition. For example, halogenated compounds are dechlorinated by electron attachment in stepwise fashion (Mincher *et al.*, 1994). The generation of an undesirable monochlorobenzene from dichlorobenzene may be regarded as an intermediate since continued irradiation dechlorinates monochlorobenzene as well. Radiation dose requirements for complete treatment must be calculated to include the decomposition of all of the important by-products, that is, for mineralisation of the original pollutant.

9.2 CASE STUDIES

9.2.1 Decolouring of natural water using electron beams

As early as the 1960s, it was reported that a dose of 1 kGy was adequate for decolouring and disinfecting slightly coloured water from reservoirs (Dolin *et al.*, 1973); however, many natural waters are highly coloured and the question that arose is would the process work on such waters. One example is that of the river Segezha (Karelia, north-west Russia), which is used for drinking and service water. Due to neighbouring swamps it is saturated with humic and tannin compounds and contains a large amount of suspended solids. The colour index and turbidity are as high as 100–150° and 7–15 $mg\,L^{-1}$, respectively. This greatly exceeds the maximum permissible levels of 20 and 1, respectively. The existing method for decolouring and disinfection was chlorination. Turbidity was reduced by filtration. This resulted in a decrease in the colour index of only 20–30%. However, an electron beam dose of 1–3 kGy considerably decreased the colour index and the amount of suspended solids (Pikaev *et al.*, 1996). Moreover, irradiation accelerated the sedimentation of suspended solids by 50–100 times. The precipitate that formed upon irradiation also aided in pollutant removal and this process, together with radiolysis, was responsible for water purification. It was concluded that electron beam treatment and subsequent filtration were sufficient for purification of water from the river Segezha (Pikaev *et al.*, 1996).

9.2.2 The Miami Electron Beam Research Facility (EBRF)

One of the largest electron beam facilities that has been used for pollutant treatment is the EBRF in Miami, Florida (Kurucz *et al.*, 1985). Operating since 1988, it has performed process-scale irradiation of many solutes in batch or continuous mode, including mixtures of solutes in natural waters (e.g. Cooper *et al.*, 1993a,b; Kurucz *et al.*, 1985; Lin *et al.*, 1995; Mak *et al.*, 1997; Nickelsen *et al.*, 1991, 1994).

The facility consists of a 1.5-MeV accelerator (Vivirad-High Voltage Corporation) driven by an insulated core transformer (ICT) power supply. The beam current is continuously variable between 0 and 50 mA. The beam may be scanned to provide uniform, continuous irradiation of an area of 122 × 7.6 cm, and is directed onto a stream of water falling over a weir. At the design flow of $460\,L\,min^{-1}$, the thickness of the irradiated falling water stream is 0.38 cm, which results in less than 100% energy deposition of the 1.5-MeV electrons. The absorbed dose is determined by measuring the temperature rise of the water stream (Kurucz *et al.*, 1985). Pre- and post-irradiation grab samples are taken from the facility control room.

Benzene, toluene and xylene radiolysis have been investigated at the EBRF (Nickelsen *et al.*, 1991). These solutes were irradiated at different initial concentrations in batches of differing water types. Decomposition of the solutes was found to observe pseudo-first-order kinetics, facilitating the reporting of results as dose constants. Table 9.5 summarises dose constants for benzene, toluene and xylene(s) in Miami potable water (column 1) and in secondary wastewater (column 2). Wastewater contained higher alkalinity (but lower pH), higher nitrate and higher DOC than potable water.

It can be seen from Table 9.5 that lower dose constants are found for solute treatment in wastewater, especially at low solute concentrations. This was expected due to the high concentrations of radical scavengers in that solvent. Lower initial solute concentrations also result in higher dose constants for a given solute (although the G value would be expected to be lower). At the higher solute concentrations only benzene showed a decreased dose constant in wastewater, while the efficiency of toluene and xylene degradation was unaffected by wastewater scavengers. This may indicate successful competition for radicals by the higher solute concentration.

To determine the influence of the various radicals and radical scavengers on these aromatic solutes, an attempt to model these results was made. Pseudo-first-order rate constants (k') were calculated [Equation (9.8)] and used to determine the percentage of removal of each solute [Equation (9.10)] due to each reactive species. The results are shown for pure water in Table 9.6.

It is obvious from Table 9.6 that the $^{\bullet}OH$ plays the major role in the decomposition of these solutes. Toluene is unique in that it is also reacts relatively rapidly with the hydrogen atom. These data may help to explain the effects shown in

Table 9.5 Pseudo-first-order dose constants for the e beam irradiation of aromatic compounds in water (1) and wastewater (2).

Solute	Concentration (μM)	k_o (kGy^{-1})	
		Water	Wastewater
Benzene	1	0.84	0.26
	17	0.31	0.16
Toluene	1	0.61	0.38
	17	0.22	0.22
m-Xylene	1	0.64	0.43
	12	0.11	0.13
o-Xylene	1.1	0.62	0.40
	12	0.12	0.12

Table 9.6 Percent removal of solutes in pure water due to individual reactive species.

Solute	$^{\bullet}OH$	e^-_{aq}	$H^{\bullet}$
Benzene	97.5	0.1	2.4
Toluene	83.5	0.4	16.1
m-Xylene	92.9	0.1	7.0
o-Xylene	93.7	0.1	6.2

Table 9.5. Benzene decomposition was most severely affected by the changing water quality due to almost complete reliance on $^{\bullet}OH$ attack, and the presence of a six-fold higher DOC concentration competing for $^{\bullet}OH$ in wastewater.

At high concentrations toluene was removed at a rate twice that of the xylenes. This is perhaps due to increased participation by the $^{\bullet}H$ atom. For high concentrations of toluene and the xylenes, $^{\bullet}OH$ scavenging was probably compensated for by $^{\bullet}H$ attack. Information such as this may be important in predicting by-product distributions when treating water of varying quality. An extensive analysis of the by-products of these solutes was conducted using intermediate doses (Nickelsen *et al.*, 1991). Production of phenol and the dihydroxy products hydroquinone, catechol and trace resorcinol was reported. *o*-Cresol was the only hydroxylated product of toluene radiolysis. These were intermediates however, and continued irradiation produced a variety of more highly oxidised products. Among them were low molecular weight aldehydes, which also decomposed upon continued irradiation. Both solutes generated varying amounts of formaldehyde, acetaldehyde, glyoxal and methylglyoxal (toluene only) depending on pH. However, all of the products identified accounted for only about 10% (depending on dose) of the original carbon, suggesting mineralisation to CO_2 and water. More recently, similar studies have been reported for the purification of petroleum-contaminated waters using the electron beam process (Nickelsen *et al.*, 1998; Podzorova *et al.*, 2001; Shin *et al.*, 2002).

9.2.3 Austrian drinking water treatment using ozone and electron beam irradiation

Irradiation of water is the most efficient way to generate $^{\bullet}OH$ in aqueous solution. However, only about half of the energy absorbed, in some waters, is used for $^{\bullet}OH$ radical production. The simultaneously formed reducing species do not contribute to pollutant decomposition in groundwater contaminated at low levels because they are scavenged by natural solutes, such as oxygen and nitrate ion. When water containing high nitrate concentrations ($>5\ mg\ L^{-1}$) is irradiated, reaction with solvated electrons produces nitrite ion. Not only is nitrite ion a $^{\bullet}OH$ scavenger it is also regulated in drinking water. Therefore, radiolysis alone would not be appropriate for purification of nitrate-containing drinking water.

Among the organic pollutants known to occur frequently in groundwater, perchloroethylene (PCE), TCE and atrazine are common. Since these contaminants are usually present in trace amounts ($\sim 0.1\ mg\ L^{-1}$) the simplified model shown in

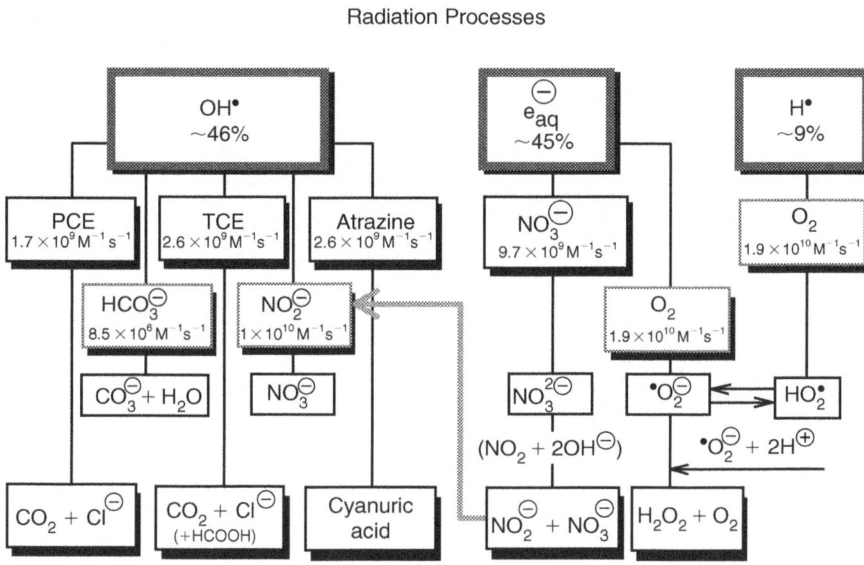

Figure 9.4 A simplified model of the competition for free radicals in natural groundwater. Data shown for some components are bimolecular rate constants. Natural organic matter is not considered.

Figure 9.4 can be used to describe the competition of pollutants and natural solutes for free radicals. The reductive detoxification of the pollutants becomes insignificant despite the high rate constants for electron reactions with TCE and PCE. High oxygen concentrations scavenge most solvated electrons to produce superoxide radical anion ($^{\bullet}O_2^-$). Oxygen also scavenges $^{\bullet}H$, to form hydroperoxy radical ($^{\bullet}HO_2$), which exists in an acid–base equilibrium with $^{\bullet}O_2^-$ (pK_a = 4.8). At groundwater pH, the equilibrium favours superoxide, which is a comparatively inert radical. It does not react with PCE or TCE (Gehringer *et al.*, 1988; Proksch *et al.*, 1987). Its most probable fate is disproportionation to H_2O_2 and O_2. Thus, only $^{\bullet}OH$ remains as an active species for pollutant decomposition in irradiated water. This means that 50% of the radiation energy is wasted.

However, addition of ozone to water prior to irradiation solves both the nitrite ion and energy-efficiency problems discussed in the preceding paragraphs. Figure 9.5 illustrates how ozone converts reducing species into additional $^{\bullet}OH$, and nitrite ion back to nitrate ion (Gehringer *et al.*, 1992). It also eliminates the residual hydrogen peroxide produced by radiolysis (Hoigné and Bader, 1989). The hydroxyl radical is formed as a result of the fast reaction of ozone with the reducing species, and as a result of the reaction with $^{\bullet}O_2^-$ ($k = 1.6 \times 10^9 M^{-1} s^{-1}$). Ozone increases the oxygen concentration in the water, which in turn increases $^{\bullet}O_2^-$, which is then converted to $^{\bullet}OH$. The enhanced oxygen concentration also minimises the production of nitrite from nitrate by scavenging the solvated electrons. Moreover, H_2O_2 formation is abated since the $^{\bullet}O_2^-$ now reacts preferably

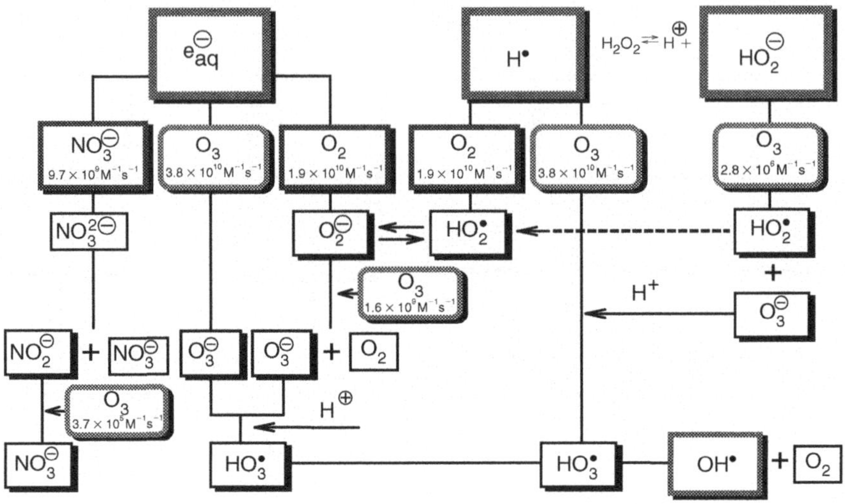

Figure 9.5 Main reaction pathways of the solvated electron (e^-_{aq}), the $H^\bullet$ and hydrogen peroxide in ozone-containing groundwater. Natural organic matter is not considered.

with ozone. The combination ozone/electron beam irradiation represents a unique AOP because hydroxyl radicals originate from two simultaneous sources: (1) directly from the water by radiolysis and (2) from ozone decomposition by the reducing species formed during water radiolysis (Gehringer and Eschweiler, 1996). As a consequence, higher $^\bullet OH$ concentrations result, as compared with the other AOPs. High $^\bullet OH$ concentrations are especially important for the treatment of water when low residual pollutant concentrations are required.

Groundwater has been continuously treated by the combined ozone/electron beam method in a 3-$m^3\,h^{-1}$ pilot plant (Gehringer *et al.*, 1994, 1995). The facility has been described elsewhere (Gehringer *et al.*, 1993). The water contained 251 $mg\,L^{-1}$ bicarbonate, 8 $mg\,L^{-1}$ chloride, 15 $mg\,L^{-1}$ nitrate, $<0.4\,mg\,L^{-1}$ DOC and 61 $\mu g\,L^{-1}$ PCE, as well as genotoxic substances detected by the Trad-MCM (Tradescantia micronucleus, clone #4430) assay (Ma, 1983). It was found that a radiation dose of 200 Gy was sufficient to reduce the PCE to about 1 $\mu g\,L^{-1}$, provided sufficient ozone was present prior to irradiation. The genotoxic components were also completely removed, as shown in Table 9.7.

Figure 9.6 shows a diagram of a planned 108 $m^3 h^{-1}$ treatment plant. Thirteen wt% ozone is made from pure oxygen. The ozone/oxygen stream is compressed and mixed with the polluted water in a static mixing unit. Almost all the ozone, but not oxygen, will dissolve under these conditions. Most of the oxygen is removed by stripping with a gas separator. Trace ozone in the oxygen is destroyed before release to the environment. No stripping of PCE occurs. The ozonated groundwater is then irradiated with 500 keV electrons in a closed chamber. The PCE is mineralised by the electron beam, in the presence of ozone (Gehringer and Eschweiler, 1996).

Table 9.7 Results of Trad-MCN test with water from the well in Bad Fischau-Brunn before and after treatment with ozone/electron beam irradiation.

Sample	Clastogenic concentrations in micronuclei (MCN)/100 tetrads	
	Individual values	Mean value $\pm\ \sigma$
Groundwater before treatment	53.6; 19.2; 9.6; 53.9; 29.7	33.2 ± 20.0
Groundwater after treatment	5.8; 5.4; 11.0; 14.9; 6.3	8.7 ± 4.1
Tap water (negative control)	2.9; 6.9; 2.9; 12.1; 10.1	7.0 ± 4.1
Tap water with 0.25 mM $As_2O_3\,L^{-1}$ (positive control)	31.7; 19.7; 17.0; 16.3; 16.2	20.2 ± 6.5

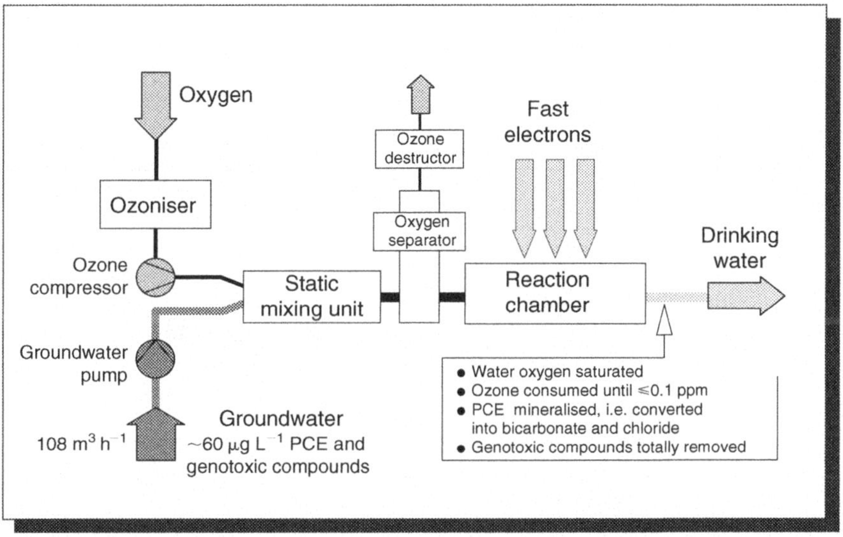

Figure 9.6 Process schematic for an ozone/electron beam treatment plant.

The decrease in PCE concentration versus radiation dose is shown in Figure 9.7 for ozonated irradiated groundwater. Similar data is shown for chlorinated benzenes both with and without ozone addition in Figure 9.8. The improvement in chlorobenzene decomposition upon ozone addition is obvious.

9.2.4 Irradiation of wastewater aerosols

Facilities for the electron beam purification of water are commonly based on electron accelerators with electron energy of approximately 1 MeV or higher. The accelerators are expensive. However, decreasing the density of the irradiated water can allow lower-energy electrons to penetrate the stream, thus allowing the

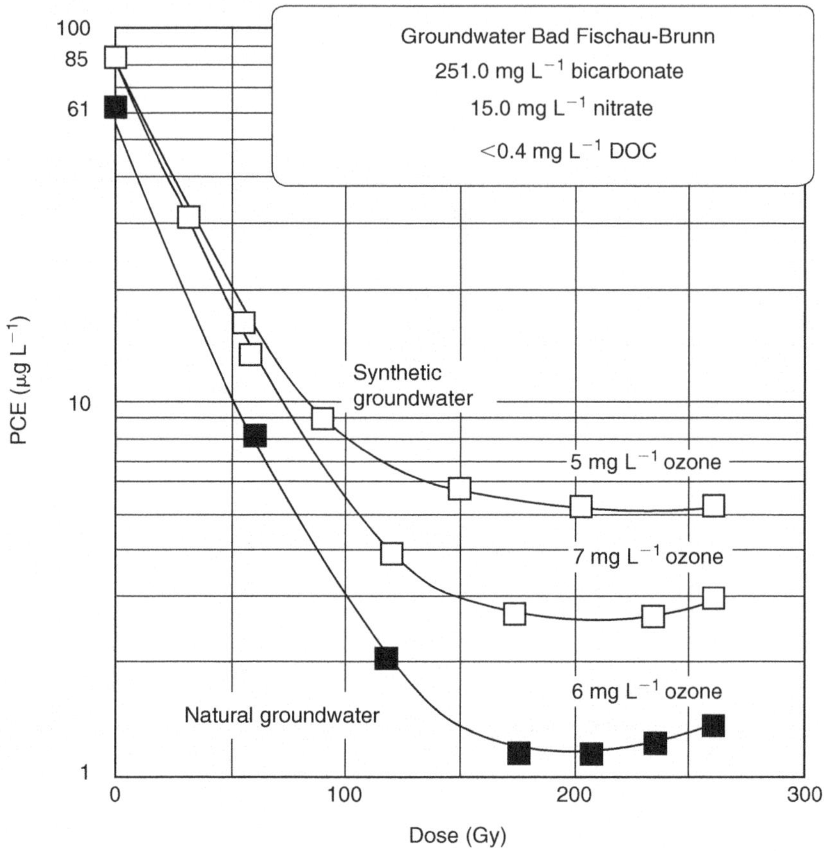

Figure 9.7 Decomposition of perchloroethylene, in Bad Fischau-Brunn groundwater, by ozone/electron beam irradiation treatment.

use of less expensive accelerators. This technique has been used for the purification of municipal wastewater in Raduzhnyi, Russia (Pikaev *et al.*, 1997e; Podzorova *et al.*, 1998). In this method, liquid wastewater is sprayed using injection nozzles and the aerosol is irradiated. The density of the aerosol is 0.02–0.05 g cm^{-3}, and therefore electron penetration is 20–50 times greater than in liquid wastewater. The density of the aerosol is still higher than that of air and the electron energy is primarily absorbed by the water droplets.

The Raduzhnyi plant (Figure 9.9) is equipped with an Astra accelerator. It produces electrons of 0.3 MeV with a maximum beam power of 15 kW. The beam cross section is 700 × 600 mm. The dimensions of the accelerator are 1200 × 800 × 700 mm. The plant processes 500 m^3 day^{-1} with a power consumption of 58 kW at a flow rate of 9.65 m^3 h^{-1}. The plant is also equipped with device for re-circulation of irradiated air containing ozone through the electron beam.

Table 9.8 summarises the results for various water quality parameters measured before and after irradiation (Pikaev *et al.*, 1997e; Podzorova *et al.*, 1998;

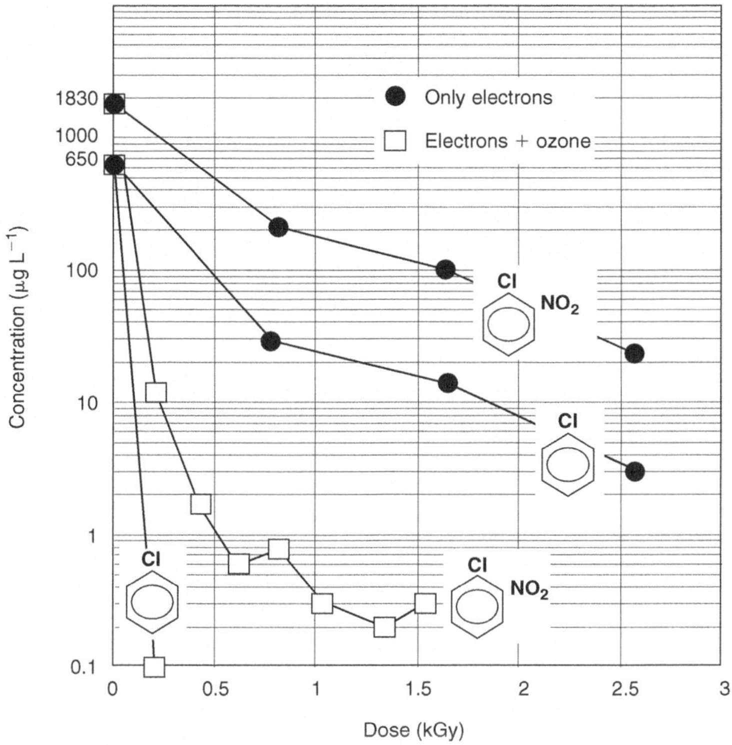

Figure 9.8 Decomposition of *o*-chloronitrobenzene and chlorobenzene in groundwater by electron beam irradiation with and without ozone addition, respectively.

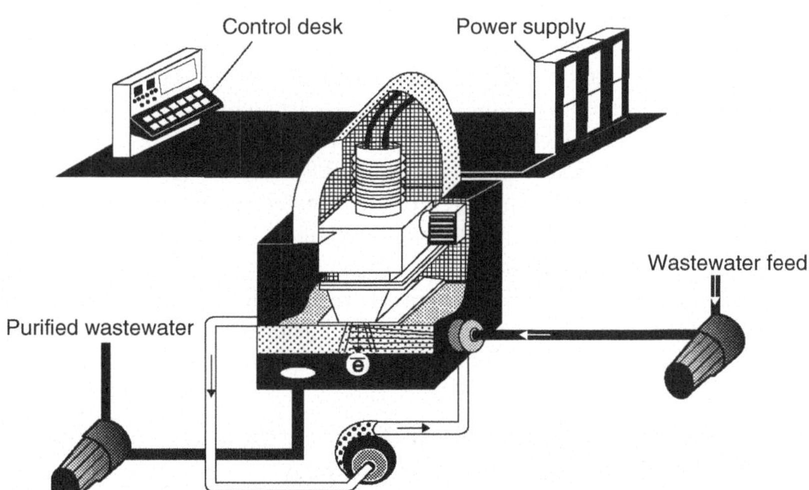

Figure 9.9 Scheme of pilot plant for combined aerosolisation and electron-beam purification of municipal wastewater.

Table 9.8 Parameters of un-irradiated and irradiated aerosolised municipal wastewater.

Parameter	Before irradiation	Dose (kGy)	After irradiation
Colour index (degrees)	160	1.3	20
Transparency (cm)	1.3	1.3	20
Suspended solids ($mg\,L^{-1}$)	156	1.3	5
Odour (force)	3.4	1.4	0
Chemical oxygen demand (COD) ($mg\,L^{-1}$)	40–100	0.9–1.4	4–12
	100–150	0.9–1.4	21–34
	150–450	1.0–1.4	40–1000
	412	2.5–4.6	4–12
BOD_5 ($mg\,L^{-1}$)	18–24	0.8–1.3	7.9–9.2
Synthetic surfactants ($mg\,L^{-1}$)	13	1.7	3.0
	13	2.6	0.2
Sulphates ($mg\,L^{-1}$)	283	0.8–1.3	160
Phosphates ($mg\,L^{-1}$)	8.8	0.8–1.3	8.7
Chlorides ($mg\,L^{-1}$)	48	0.8–1.3	23
Nitrites ($mg\,L^{-1}$)	Not detected	1.0–1.4	Not detected
Nitrates ($mg\,L^{-1}$)	Not detected	1.0–1.4	3.0
Total number of microbes (cell cm^{-3})	10^5	1.3	<1

1999b). It can be seen that dose required for the purification of the water is no higher than 4–5 kGy. This dose is somewhat lower when ozone is used. Irradiation was accompanied by the formation of a precipitate as a consequence of the aggregation of dispersed particles in the water. The precipitate also captures some of the pollutants and leads to a more efficient process. Precipitate formation continued for 1–12 h after irradiation. This post-irradiation effect was responsible for an additional decrease in COD of ~50%. Heavy metals and semi-volatile organic compounds were also removed (Podzorova *et al.*, 1999b).

These results indicate that the pollutants in irradiated wastewater decreased by 2–3 orders of magnitude. The residual concentrations are noticeably less than maximum permissible levels for service water. Irradiation also gives rise to the formation of several new compounds in very low concentrations. Nitrate ion was detected in irradiated wastewater (see Table 9.8). The appearance of nitrate ion was caused by the fact that nitrogen oxides are formed upon irradiation of air, which dissolve in water give nitrite and nitrate ions. Upon continued irradiation, nitrite ions are oxidised to nitrate ions.

Therefore, electron beam treatment of municipal wastewater in the aerosol flow mode to several kGy results in sufficient purification to meet standards. The removal efficiency is increased when treatment is accompanied by ozonation.

9.2.5 Dyeing complex wastewater treatment by combined electron beam/biological method

Wastewater containing dyes is difficult to purify using common methods. Irradiation can solve this problem and the decolouring of dye-containing water

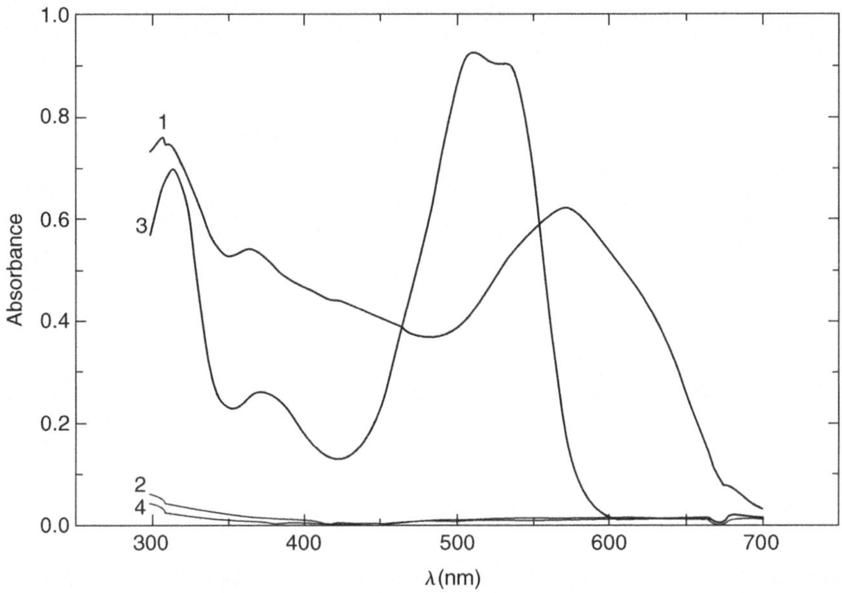

Figure 9.10 Optical absorption spectra of aerated aqueous solutions (50 mg L^{-1}) of dyes Acid Red 172: (1) unirradiated solution, (2) irradiated solution (dose 9 kGy); and Reactive Red 21: (3) unirradiated solution, (4) irradiated solution (dose 7 kGy).

can be accomplished at comparatively low doses (Pikaev, 2000b; Pikaev *et al.*, 1997c). A pilot plant for the combined electron beam/biological purification of wastewater from a dyeing complex at Taegu City (Republic of Korea) has operated since October, 1998 (Han *et al.*, 2001). An example of the treatment of two dyes is shown in Figure 9.10. The plant treats 1000 m^3 day^{-1}, as a bypass from the main wastewater treatment facility. The facility uses a 1 MeV, 40 kW ELV electron accelerator manufactured by Institute of Nuclear Physics, Siberian Division of Russian Academy of Sciences (Novosibirsk). The beam current may be varied within the range of 5–40 mA to change absorbed dose and dose rate as necessary due to changing wastewater flow rate.

This wastewater is a mixture of two flows: raw dyeing complex stream and wastewater from polyester fibre production. The main organic pollutants of the mixed wastewater were terephthalic acid and ethylene glycol. Among the other organic pollutants were hexane, carboxymethyl and hydroxymethyl cellulose, phenols, starch, waxes, detergents and dyes with total content of not over 1% (wt). After pH adjustment, the wastewater is pumped through a 1.5-mm wide injector nozzle and the spray is irradiated by a vertical electron beam, which is scanned across the jet. After treatment, the water is directed to six-step, swing-bed, tower-type biological reactor. Irradiation doses of 1–2 kGy resulted in a decrease in the total biological treatment time to ~8 h, compared to about 17 h for un-irradiated water. A noticeable increase in biodegradability after irradiation treatment was also found for pulp bleaching (Haberl *et al.*, 1991), paper-mill (Shin *et al.*, 2002; Taghipour and Evans, 1996; Wang *et al.*, 1994) and some other effluents.

9.2.6 Nekal decontamination using combined electron beam/biological method

A second process using a combined electron beam/biological system is that used for Nekal decontamination of wastewater. Operating since 1984, the plant was originally equipped with a 0.7-MeV, 50-kW electron accelerator of ELV-3 type. In 1988, this plant was equipped with an additional 1 MeV, 50 kW ELV-4 accelerator (Pikaev, 1998; Pikaev *et al.*, 1997a; Salimov *et al.*, 1998).

Nekal (isobutylnaphthalene sulphonates) is used as an emulsifier in a synthetic rubber factory in Voronezh, Russia (Salimov *et al.*, 1998). It is a toxic, non-biodegradable substance and the wastewater was formerly disposed by underground injection. This resulted in contamination of a groundwater reservoir with the total area of 75 km^2, and caused serious problems with drinking water in the district (Buslaeva *et al.*, 1991). Fortunately, Nekal need not be mineralised to perform adequate treatment, which decreases the radiation dose requirements (Buslaeva *et al.*, 1991). Detachment of an alkyl or sulphonate group renders the product molecule biodegradable. Pulse radiolysis was used to show that hydroxyl radicals play an important role in this process (Gogolev *et al.*, 1991). The *G* value for Nekal degradation was found to be about 0.2 $\mu mol\,J^{-1}$, requiring 2–3 kGy at a concentration of 100–150 $mg\,L^{-1}$. Wastewater containing Nekal is irradiated as a low-density foam produced by intense air bubbling, following which, it is subjected to final biological treatment. The average output of the plant is 12,000 $m^3\,day^{-1}$, and the maximum output is 30,000 $m^3\,day^{-1}$. Saturation of the wastewater with irradiated air containing ozone leads to some increase in the output. The use of this method results a degree of water purification as high as 99.5%. At present, the purification of groundwater from the polluted reservoir is almost completed.

9.2.7 Molasses distillery waste treatment by electron beam

A combination of electron beam and coagulation processes was developed for purification of molasses distillery waste from the fermentation of plant feedstock at Arkhangel'sk distillery in Russia (Pikaev *et al.*, 2001; Ponomarev *et al.*, 1999). This wastewater, pH 4, is dark brown in colour, with a colour index as high as 10,000–15,000°. Humic compounds and lignin derivatives constitute the main part of pollutants. This wastewater is diluted with municipal wastewater prior to treatment, resulting in the initial parameters shown in Table 9.9.

The data presented in Table 9.9 correspond to a treatment level acceptable for discharge to the local river. On this basis, a 7000 $m^3\,day^{-1}$ facility equipped with two accelerators is under development. Similar processes have been proposed for the treatment of paper-mill wastewater and landfill leachates (Pikaev *et al.*, 2001).

Table 9.9 Parameters of the Arkhangel'sk mixed wastewater before and after electron beam/coagulation treatment to 5 kGy.

Parameter	Initial wastewater	Treated wastewater
Colour index (degrees)	900–2500	20–45
Suspended solids ($mg\,L^{-1}$)	320–580	2–5
COD(Cr) ($mg\,L^{-1}$)	800–3500	30–60
BOD_{20} ($mg\,L^{-1}$)	520–2420	8–20
pH	5.6–6.8	6.0–6.8

9.2.8 Radiolytic removal of heavy metals from water

The electron beam treatment of water containing heavy metals is based on radiolytic reduction of ions to the metal state, or to ions in lower oxidation states that can be removed by filtration or centrifugation. Efficient reduction requires that oxygen be absent and that $^{\bullet}OH$ radicals be scavenged. Additives that scavenge hydroxyl radicals include formate ions (Buxton *et al.*, 1988; Kartasheva *et al.*, 1998b; Pikaev and Podzorova, 1999; Ponomarev *et al.*, 1997), ethanol, isopropanol or acetic acid (Chaychian *et al.*, 1998a,b; Schmelling *et al.*, 1998), and reducing agents such as $Na_2S_2O_3$, Na_2SO_3, or Na_2S (Salimov *et al.*, 1998). This technique has been used for the removal of Cd(II) (Kartasheva *et al.*, 1997, 1998a), Hg(II) (Lyu *et al.*, 1983a,b,c), Pb(II), and Cr(VI) from solution. Cd and Pb were reduced to Cd(0) and Pb(0), while Cr(VI) was reduced to less toxic Cr(III). The doses required for the removal of Cd and Pb to maximum permissible concentrations in discharge water were 0.7–4 kGy, depending on the metal concentration.

During electron beam treatment, reduction occurs via reactions of metal ions with solvated electrons and $H^{\bullet}$. In the case of Pb(II), the following reactions can be written as follows (Zhestkova *et al.*, 1997):

$$\mathrm{Pb(II)} + e^-_{aq} \rightarrow \mathrm{Pb(I)} \tag{9.30}$$

$$\mathrm{Pb(I)} + \mathrm{Pb(I)} \rightarrow \mathrm{Pb(0)} + \mathrm{Pb(II)} \tag{9.31}$$

$$n\mathrm{Pb(0)} \rightarrow \mathrm{Pb(0)} \tag{9.32}$$

$H^{\bullet}$ cannot reduce Pb(II) but participates in formation of Pb(0) via Equations (9.33) and (9.34):

$$\mathrm{Pb(I)} + H^{\bullet} \rightarrow \mathrm{Pb(I)H^{\bullet}} \tag{9.33}$$

$$\mathrm{Pb(I)H^{\bullet}} + \mathrm{Pb(I)H^{\bullet}} \rightarrow \mathrm{Pb(0)} + \mathrm{Pb(II)} + H_2 \tag{9.34}$$

Simultaneously with the e^-_{aq} and $H^{\bullet}$, $^{\bullet}OH$ radicals and hydrogen peroxide are formed from water radiolysis. The following reaction mechanism accounts for the treatment:

$$\mathrm{Pb(I)} + {}^{\bullet}OH \rightarrow \mathrm{Pb(II)} + OH^- \tag{9.35}$$

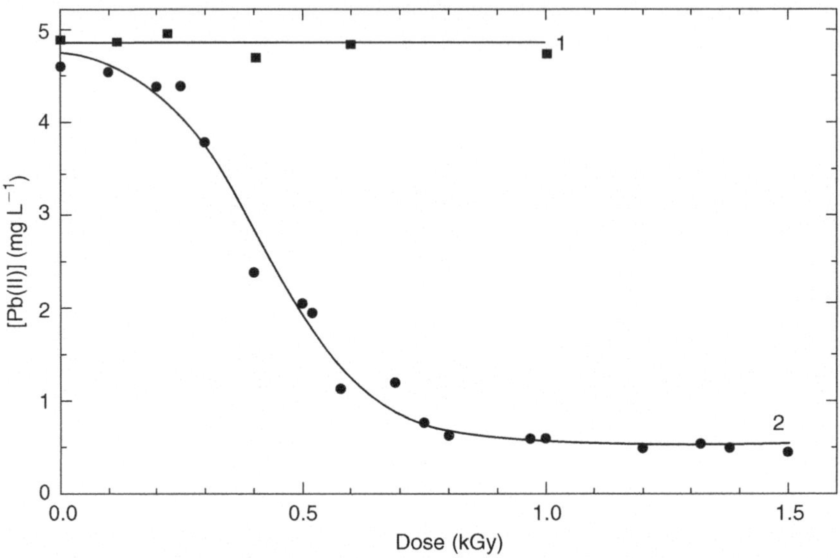

Figure 9.11 Dependence of Pb(II) concentration in aerated aqueous solution in the absence of formate (1) and in the presence of 1×10^{-2} mol L^{-1} formate (2) on dose of electron radiation.

$$Pb(I) + H_2O_2 \rightarrow Pb(II) + {}^{\bullet}OH + OH^- \quad (9.36)$$

$$Pb(II) + {}^{\bullet}OH \rightarrow Pb(III) + OH^- \quad (9.37)$$

$$Pb(III) + Pb(I) \rightarrow 2Pb(II) \quad (9.38)$$

The net result is the regeneration of Pb(II). Since yield of the sum of the yield of $^{\bullet}OH$ and double the amount of hydrogen peroxide is higher than sum of yields of e^-_{aq} and $H^{\bullet}$, the total effect is the absence of Pb(II) reduction even in de-aerated solutions. Experimental data have confirmed this observation (see Figure 9.11).

To exclude the oxidative effect of $^{\bullet}OH$ radicals, formate ion is used as an additive. Figure 9.11 dramatically shows the importance of formate scavenging of hydroxyl radical. In reaction with the $^{\bullet}OH$ (and with $H^{\bullet}$) it gives $COO^{-\bullet}$ radical ion, which can also participate in reductive reactions:

$$HCOO^- + {}^{\bullet}OH(H) \rightarrow COO^{-\bullet} + H_2O(H_2) \quad (9.39)$$

$$Pb(II) + COO^{-\bullet} \rightarrow Pb(I) + CO_2 \quad (9.40)$$

$$Pb(I) + COO^{-\bullet} \rightarrow Pb(0) + CO_2 \quad (9.41)$$

If water is aerated, oxygen reacts with e^-_{aq}, $H^{\bullet}$ and $COO^{-\bullet}$ partially or completely

$$e^-_{aq} + O_2 \rightarrow O_2^{-\bullet} \quad (9.42)$$

$$H^{\bullet} + O_2 \rightarrow HO_2^{\bullet} \quad (9.43)$$

$$HO_2^\bullet \Leftrightarrow H^+ + O_2^{-\bullet} \tag{9.44}$$

$$COO^{-\bullet} + O_2 \rightarrow CO_2 + O_2^{-\bullet} \tag{9.45}$$

suppressing reactions (9.30), (9.33), (9.40) and (9.41). Moreover, the reactions of $HO_2^\bullet$ and $O_2^{-\bullet}$ with Pb(II) are comparatively slow; these radicals can combine forming hydrogen peroxide:

$$HO_2^\bullet + O_2^{-\bullet} + H^+ \rightarrow H_2O_2 + O_2 \tag{9.46}$$

and/or react with reduced metal ions:

$$HO_2^\bullet + Pb(I) + H^+ \rightarrow Pb(II) + H_2O_2 \tag{9.47}$$

Hydrogen peroxide formed oxidises the additional amount of Pb(I) [reaction (9.36)]. Moreover, oxygen can directly react with Pb(I) (Zhestkova *et al.*, 1997):

$$Pb(I) + O_2 \rightarrow Pb(II) + O_2^{-\bullet} \tag{9.48}$$

Therefore, the reduction takes place only after the consumption of oxygen. Due to this, upon electron beam treatment there is an induction period when reduction does not proceed (see Figure 9.11). In order to increase the efficiency of the process, air should be removed from the solution (e.g. by bubbling inert gas) before and/or during irradiation. Carbon dioxide produced in reactions (9.40), (9.41) and (9.45) is not toxic. However, the use of other $^\bullet OH$ scavengers can give rise to the formation of toxic compounds. For example, ethyl and isopropyl alcohols, which are also utilised as scavengers, can form acetaldehyde and acetone, respectively.

The mechanism for the removal of Cd(II) is similar to that for Pb(II). If Cd(II) and Pb(II) ions are present together, the reduction of Cd(II) is observed only after reduction of the main part of Pb(II) (Zhestkova *et al.*, 1997). This can be explained by the fact that Cd(I) quickly reduces Pb(II) ($k_{27u} = 7.5 \times 10^7$ L mol s^{-1}) (Buxton and Sellers, 1978):

$$Pb(II) + Cd(I) \rightarrow Pb(I) + Cd(II) \tag{9.49}$$

Thus, the method allows for the consecutive removal of heavy metals from water (Figure 9.12). The addition of formate to water containing Cr(VI) increases the efficiency of reduction to much less toxic Cr(III) ions (Pikaev *et al.*, 1997b; Yurik and Pikaev, 1999). The degree of reduction depends on the pH of the solution.

9.3 ECONOMICS

A discussion of the economics of radiolysis must be preceded by a choice between the two possible radiation sources available for water treatment. The first is the use of isotope gamma-ray sources; the second is the electron accelerator. Alternatively, bremsstrahlung may be produced by directing a beam of electrons at a suitable target. Each source has technical and economic advantages and disadvantages.

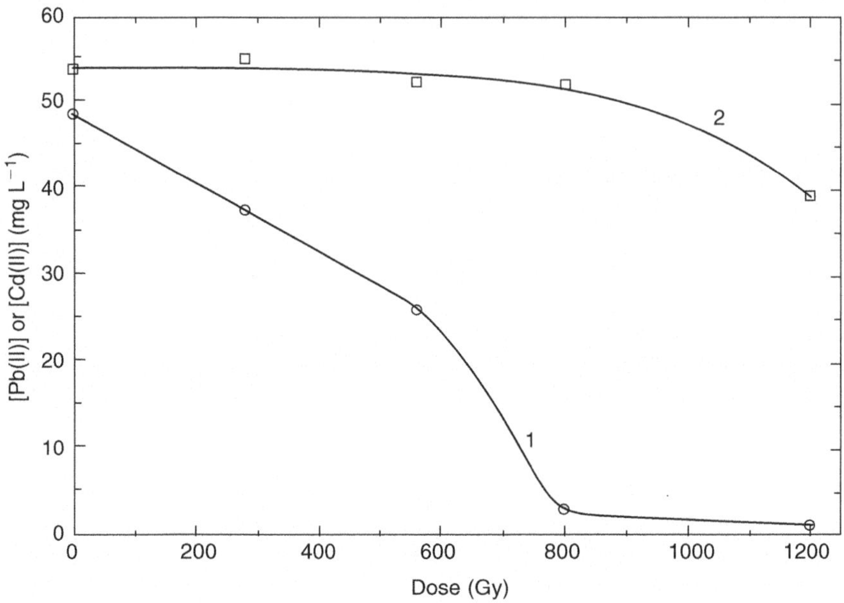

Figure 9.12 Concentrations of (1) Pb(II) and (2) Cd(II) simultaneously present in aerated aqueous solution containing 1×10^{-2} mol L^{-1} formate as functions of dose of electron radiation.

Isotopes are convenient and uncomplicated sources of radiation and by far, most experimental work has been done using isotope gamma-ray sources, especially ^{60}Co. They are ideally suited to small scale, batch irradiations. However, on a process-scale isotopes suffer the disadvantage of low dose rates, which limit process flow. A large cobalt source, for example, delivers about 10 kGy h^{-1}, while rates as high as 1 kGy s^{-1} are common with accelerators. A further problem of isotopes is that the common ^{60}Co source has a 5.27 year half-life, thus losing 12.5% of its activity annually. Economic analyses have shown that when cobalt replacement costs are accounted for, electron beams are more cost effective sources (Cleland *et al.*, 1984). Further, isotope costs are largely driven by regulatory requirements, which are hard to predict or rely on.

Scale-up from isotope experiments to electron beam processing must consider complications arising from differences in oxygen content between batch and process irradiation (process water streams are likely to be irradiated in air); the much higher dose rates of electron beam sources; and the limited penetrating ability of electrons compared to photons. All of these considerations may affect dose constants (and *G* values) and by-product distributions.

It is a misconception that the generation of high-energy electrons results in high-energy (power) costs. In fact, the electron beam accelerator (using ICTs) is an energy efficient means of creating the $^{\bullet}$OH radical (Bolton *et al.*, 1998). Efficiency may be defined as the kilowatt-hours (kWh) of electricity required to

Table 9.10 Typical EE/O values for contaminant destruction by the electron beam process.

Contaminant	EE/O (kWh/1000 US gallons/order)
Benzene	0.5–3
Toluene	0.4–3
Xylene(s)	1–3
Phenol	0.4–7
Methylene blue (dye)	0.2–2
Trichloroethylene	0.5–2
TCE	2–6
Chloroform	2–12
Carbon tetrachloride	0.5–5
Vinyl chloride	0.5–2

reduce the concentration of a contaminant in 1000 gallons of water by one order of magnitude or 90%. This is termed the electrical energy per order (EE/O in kWh/1000 US gallons/order).

For example, if it takes 10 kWh of electricity to reduce the concentration of a contaminant from 100 mg L^{-1} to 10 mg L^{-1} in 1000 gallons of wastewater, then the EE/O is 10 kWh/1000 gallons/order. It would then take an additional 10 kWh to reduce the compound one additional order, from 10 to 1 mg L^{-1}; which would be an overall removal efficiency of 99%.

Since the exponential relationship between the change in pollutant concentration and electron beam radiation dose described by the dose constant is so common, that slope can also usefully be described by the EE/O for economic analyses. Like k_0, EE/O is determined from a feasibility study. It too is specific to the pollutant, the initial concentration and the nature of the water being tested. Water quality may have a great effect on process efficiency, since the presence of various scavenger compounds may remove the desired radicals from solution.

Calculating the slope of the exponential decrease of solute concentration in terms of EE/O allows for a comparison with the energy costs of competing technologies. Care should be taken when comparing various processes that all of the energy costs associated with each are included. For example, if H_2O_2 is added during a treatment, there is an electrical cost associated with the production of the peroxide and it too needs to be taken into consideration, either as an addition to the EE/O calculation or in the operational costs.

Typical EE/O values for common pollutants range from 0.5 to 12 kWh/1000 gallons/order. Those for several compounds of interest are presented in Table 9.10, using the EBRF at a flow rate of 100 gpm. Once an EE/O value has been determined, either through feasibility studies or estimated from a table of values, the electron beam dose (in kW) required for any specific application may be calculated:

$$\text{Dose} = (\text{EE/O}) \times (\log C_o/C) \tag{9.50}$$

where C_0 is the initial contaminant concentration and C is the treatment objective concentration. For waste streams with complex mixtures of contaminants, the energy required for treatment is not additive but is determined for the contaminant with the highest EE/O in the water to be treated.

9.3.1 Cost comparison of ozone/UV and ozone/electron beam

In the drinking water works of Niederrohrdorf near Zürich, Switzerland an ozone/UV-combination system for groundwater remediation contaminated with about 100 $\mu g\,L^{-1}$ TCE was installed by WEDECO, Germany. The groundwater contains about 402 $mg\,L^{-1}$ bicarbonate, 30 $mg\,L^{-1}$ nitrate and about 0.5 $mg\,L^{-1}$ DOC. The plant flow is 180 m^3h^{-1}. For TCE reduction to below 10 $\mu g\,L^{-1}$, 1.3 kg ozone per hour (corresponding to an initial ozone concentration of about 6–7 $mg\,L^{-1}$) and 7.8 kW UVC radiation are needed (Leitzke, 1992). These data have been used to study the TCE decomposition in a simulated Niederrohrdorf water by ozone/electron beam irradiation. Figure 9.13 shows the decomposition of

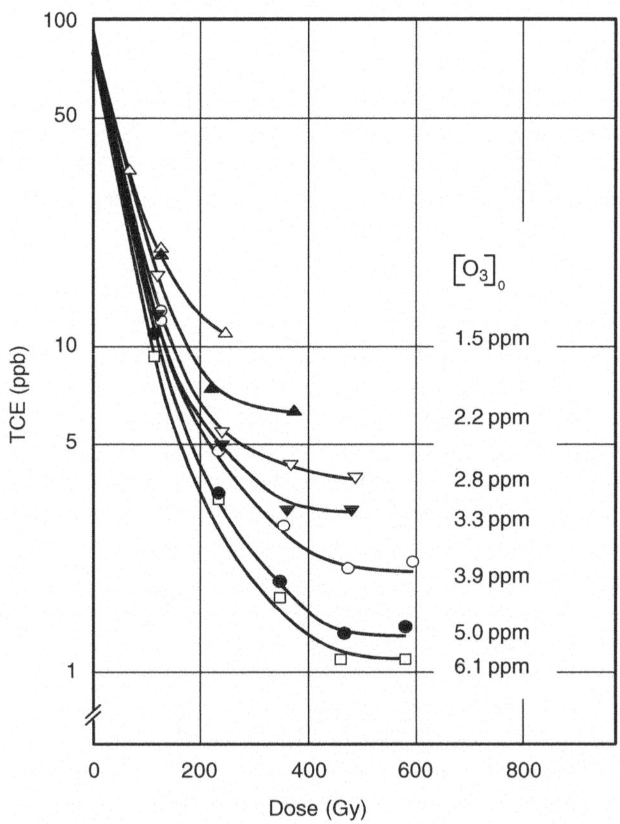

Figure 9.13 Decomposition of TCE in groundwater (402 $mg\,L^{-1}$ bicarbonate, 30 $mg\,L^{-1}$ nitrate, 0.5 $mg\,L^{-1}$ DOC) by ozone/electron beam irradiation treatment.

about 100 μg L^{-1} TCE in simulated Niederrohrdorf water by ozone/electron beam irradiation treatment as a function of the radiation dose and the initial ozone concentration.

The ozone/electron beam treatment met the same TCE reduction goal using about 2 mg L^{-1} initial ozone concentration. Using 5–6 mg L^{-1} initial ozone concentration in the ozone/electron beam combination would result in a residual TCE concentration near 1 μg L^{-1}, almost one order of magnitude better than the ozone/UV-combination. This is due to the higher •OH free radical concentration achieved by the ozone/electron beam combination. Both processes use ozone and radiation in the same way. Therefore, the cost comparison of the two processes is reduced to a cost comparison of the radiation sources and the amount of ozone production. Using a throughput of 5 million gallons a day, the relevant parameters are shown in Table 9.11.

The data show that less energy is consumed by the electron beam process than the ozone/UV. Moreover, the electron beam process needs less energy and oxygen for ozone production due to the low initial ozone concentration required. As a consequence, the total energy demand differs almost by a factor of 2 in favour of the electron beam. The capital cost of the electron beam is substantially higher. The electron beam accelerator is a factor of 3 more expensive than the lamps needed in the UV-process. However, when the operational costs are also considered, the electron beam process is found to be less expensive (Figure 9.14).

9.3.2 Treatment cost estimates for selected applications

Case studies detailing many diverse applications of electron beam treatment both as a stand-alone process and in conjunction with other conventional processes were presented in Section 9.2. Table 9.12 summarises the electron beam systems used in selected applications along with required dose, flow rate and per cubic meter cost estimates.

The high flow and low required dose for the drinking water case results in the lowest cost per cubic meter. Since the 25 kW accelerator assumed in this case has a capacity in excess of 2600 m^3 day^{-1} at 0.2 kGy, it is assumed that it will be run continuously at a reduced beam current as required to deliver 0.2 kGy. The relatively

Table 9.11 Comparison O_3/UV versus O_3/electrons (plant flow 810 $m^3 h^{-1}$ = 5 million gallons/day).

Process parameter	O_3/UV	O_3/electrons
Radiation source	UV-medium pressure	800 kV-accelerator
Energy consumption	35 kW UVC	67.5 kW
	234 kWh	135 kWh
O_3-production	5.85 kg h^{-1}	2.25 kg h^{-1}
O_2-consumption	50 $m^3 h^{-1}$	20 $m^3 h^{-1}$
Energy consumption	50 kWh	20 kWh
Energy (total)	284 kWh	155 kWh

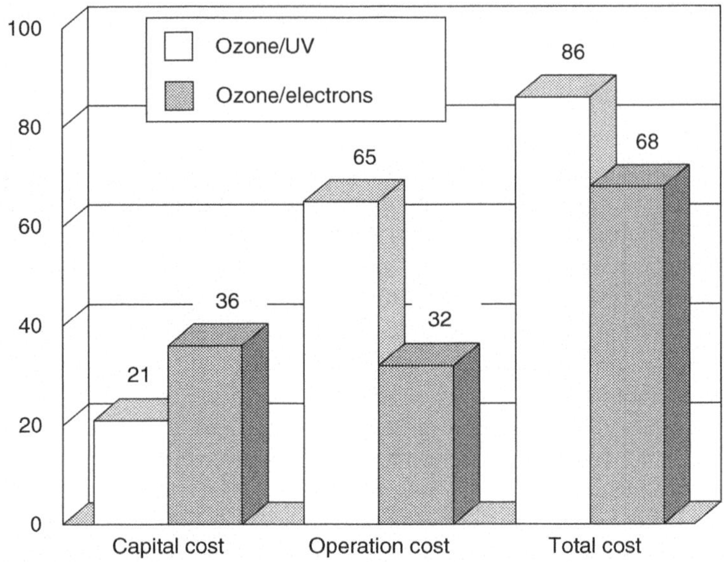

Figure 9.14 Cost comparison Ozone/UV versus Ozone/electrons (US\$ h^{-1}).

Table 9.12 Estimated current US\$ electron beam treatment system costs (not including cost of supplemental treatment processes).

	Austrian drinking water	Municipal wastewater	Dyeing complex wastewater	Nekal (emulsifier) wastewater
Reference	Section 2.3	Section 2.4	Section 2.5	Section 2.6
Flow ($m^3 day^{-1}$)	2600	500	1000	2000[d]
Target dose (kGy)	0.2	2.5	2	3
Accelerator	0.5 MeV, 25 kW	0.3 MeV, 15 kW	1 MeV, 40 kW	0.7 MeV, 50 kW 1.0 MeV, 50 kW
Supplemental treatment	Ozone	Aerosolisation	Biological	Biological
Capital cost[a]	\$600,000	\$325,000	\$1,100,000	\$1,700,000
Amortisation[b]	\$6.97 h^{-1}	\$3.78 h^{-1}	\$12.79 h^{-1}	\$19.77 h^{-1}
Operating cost[c]	\$2.19 h^{-1}	\$1.31 h^{-1}	\$3.50 h^{-1}	\$8.75 h^{-1}
Cost m^{-3}	\$0.084	\$0.244	≤\$0.391	≤\$0.342

[a]Includes estimated cost of accelerator and shielding.
[b]Estimated life of 20 years and interest at 8%.
[c]Includes electricity only at \$0.07 kWh^{-1}.
[d]Liquid influent (prior to foaming).

low cost of municipal wastewater treatment is the result of using a low voltage/ low cost curtain type accelerator in conjunction with irradiating the waste in an aerosol form. This results in a beam utilisation exceeding 96%. Industrial wastewater applications, with a relatively high flow, resulted in the highest treatment costs. In the dyeing complex and Nekal cases, the implied beam utilisation's were 60% and 70%, respectively. The target doses presented in Table 9.12 are the upper limits of the dose ranges given in these cases. Since a lower dose would allow increased flow and thus a lower treatment cost, the indicated costs per cubic meter are the maximums to be expected. Note in all applications the electrical efficiency of the accelerators is assumed to be 80%.

As indicated in Table 9.12, only the estimated cost of the accelerator, required shielding and water presentation device are included in capital costs. Only the cost of electricity for the accelerator is included in operating cost. Many other costs must be included in the determination of total treatment costs. These include the costs of system installation, building, maintenance, operators, etc. Examples of more complete cost analyses are available elsewhere (Kurucz *et al.*, 1985; USEPA, 1997). Since many of these costs are similar for other treatment processes, the estimates in Table 9.12 are useful in comparing treatment alternatives.

REFERENCES

Anbar, M., Bambenek, M. and Ross, A.B. (1973) Selected specific rates of reactions of transients from water in aqueous solution. I. Hydrated electron. *Nat. Stand. Ref. Data Ser., Nat. Bur. Stand. (US)* **43**, 67.

Anbar, M., Farhataziz and Ross, A.B. (1975) Selected specific rates of reaction of transients from water in aqueous solution. II. Hydrogen atom. *Nat. Stand. Ref. Data Ser., Nat. Bur. Stand. (US)* **51**, 56.

Arbon, R.E., Mincher, B.J. and Knighton, W.B. (1996) Gamma-ray destruction of PCBs in isooctane and transformer oil. *Environ. Sci. Technol.* **30**, 1866–1871.

Bolton, J.R., Valladares, J.E., Cooper, W.J., Waite, T.D., Kurucz, C.N., Nickelsen, M.G. and Kajdi, D.C. (1998) Figures-of-merit for advanced oxidation processes – a comparison of homogenous UV/H_2O_2, heterogeneous TiO_2 and electron beam processes. *J. Adv. Oxid. Technol.* **3**, 174–181.

Buslaeva, S.P., Vanyushkin, B.M., Gogolev, A.V., Kabakchi, S.A., Panin, Yu.A., Putilov, A.V. and Upadyshev, L.B. (1991) Purification of aqueous solutions from non-biodegradable surfactants by accelerated electrons. *Khim. Prom.* **6**, 47–50.

Buxton, G.V. and Sellers, R.M. (1978) *Compilation of Rate Constants for the Reactions of Metal Ions in Usual Oxidation State.* NSRDS-NBS 82, Washington, DC.

Buxton, G.F., Greenstock, C.L., Helman, W.P. and Ross, A.B. (1988) Critical review of rate constants for reactions of hydrated electrons, hydrogen atoms and hydroxyl radicals ($^{\bullet}OH/^{\bullet}O^-$) in aqueous solution. *J. Phys. Chem. Ref. Data* **17**, 513–886.

Carver, M.B., Hanlet, D.V. and Chapin, D.R. (1979) *MAKSIMA-CHEMIST, A program for mass action kinetic simulated manipulation and integration using stiff techniques*. Chalk River Nuclear Laboratories Report, Atomic Energy Canada Ltd. 6413, 1–28.

Chaychian, M., Al-Sheikhly, M., Silverman, J. and McLaughlin, W.J. (1998) Radiation-induced reduction and removal of heavy-metal ions from water. In: *Environmental Applications of Ionizing Radiation*, Cooper, W.J., Curry, R.D. and O'Shea, K.E. (eds). Wiley, New York, pp. 353–367.

Chaychian, M., Al-Sheikhly, M., Silverman, J. and McLaughlin, W.J. (1998) The mechanisms of removal of heavy metal ions from water by ionizing radiation. *Radiat. Phys. Chem.* **53**, 145–150.

Cleland, M.R., Fernald, R.A. and Maloof, S.R. (1984) Electron beam process design for the treatment of wastes and economic feasibility of the process. *Radiat. Phys. Chem.* **24**, 179–190.

Cooper, W.J., Cadavid, E.M., Nickelsen, M.G., Lin, K., Kurucz, C.N. and Waite, T.D. (1993a) Removing THMs from drinking water using high-energy electron-beam irradiation. *J. Am. Wat. Works Assoc.* **85**, 106–112.

Cooper, W.J., Meacham, D.E., Nickelsen, M.G., Lin, K., Ford, D.B., Kurucz, C.N. and Waite, T.D. (1993b) The removal of tri- (TCE) and tetrachloroethylene (PCE) from aqueous solution using high energy electrons. *J. Air Waste. Manag. Assoc.* **43**, 1358–1366.

Cooper, W.J., Dougal, R.A., Nickelsen, M.G., Waite, T.D., Kurucz, C.N., Lin, K. and Bibler, J.P. (1996) Benzene destruction in aqueous waste – I. Bench-scale gamma irradiation experiments. *Radiat. Phys. Chem.* **48**, 81–87.

Curry, R.D., Clevenger, T., Oana, S.-C., Miller, W.H., Farmer, J., Mincher, B.J. and Kapila, S. (1998) Decontamination of soil contaminated with Aroclor 1260 using a solvent extraction process and γ-ray radiolysis. *J. Adv. Oxid. Technol.* **3**, 55–62.

Dolin, P.I., Shubin, V.N. and Brusentseva, S.A. (1973) *Radiation Treatment of Water*. Nauka, Moscow, 152 pp.

Dorfman, L.M. and Adams, G.E. (1973) Reactivity of the hydroxyl radical in aqueous solution. *Nat. Stand. Ref. Data Ser., Nat. Bur. Stand. (US)* **46**, 59.

Gear, C.W. (1971) Automatic integration of ordinary differential equations. *Comm. ACM* **14**, 176.

Gehringer, P. and Eschweiler, H. (1996) The use of radiation-induced advanced oxidation for water reclamation. *Water Sci. Technol.* **34**, 343–349.

Gehringer, P., Eschweiler, H. and Fiedler, H. (1994) Advanced oxidation process based on ozone/electron beam irradiation for treating groundwater. In: *Proceedings of the 1994 Annual Conference American Water Works Association*, Water Quality, New York, pp. 633–639.

Gehringer, P., Eschweiler, H. and Fiedler, H. (1995) Ozone-electron beam treatment for groundwater remediation. *Radiat. Phys. Chem.* **46**, 1075–1078.

Gehringer, P., Proksch, E., Szinovatz, W. and Eschweiler, H. (1988) Radiation-induced decomposition of aqueous trichloroethylene solutions. *Appl. Radiat. Isot.* **39**, 1227–1231.

Gehringer, P., Proksch, E., Eschweiler, H. and Szinovatz, W. (1992) Clean-up of polluted groundwater by radiation-induced oxidation. In: *Proceedings of the Symposium on Application of Isotopes and Radiation in Conservation of Environment*, Karlsruhe, Germany, pp. 203–205.

Gehringer, P., Eschweiler, H., Szinovatz, W., Fiedler, H., Steiner, R. and Sonneck, G. (1993) Radiation-induced OH radical generation and its use for groundwater remediation. *Radiat. Phys. Chem.* **42**, 711–714.

Getoff, N. (1986) Radiation induced decomposition of biological resistant pollutants in water. *Appl. Radiat. Isot.* **37**, 1103–1109.

Getoff, N. (1990) Decomposition of biological resistant pollutants in water by irradiation. *Radiat. Phys. Chem.* **35**, 432–439.

Gogolev, A.V., Kabakchi, S.A. and Pikaev, A.K. (1991) Pulse radiolysis of aqueous solutions of Nekal. *Khim. Vys. Energ.* **25**, 531–535.

Gray, K.A. and Cleland, M.R. (1998) Environmental radiolysis for soil and sediment treatment: a review of chemistry, design and economic issues. *J. Adv. Oxid. Tech.* **3**, 22–36.

Haberl, R., Urban, W., Gehringer, P. and Szinovatz, W. (1991) Treatment of pulp-bleaching effluents by activated sludge, precipitation, ozonation and irradiation. *Water Sci. Technol.* **24**, 229–239.

Han, B., Ko, J., Kim, J., Kim, Y., Chung, W., Makarov, I.E., Ponomarev, A.V. and Pikaev, A.K. (2001). Combined electron-beam and biological treatment of dyeing complex wastewater. Pilot plant experiments. *Radiat. Phys. Chem.* (in press).

Hart, E.J. and Anbar, M. (1970) *The Hydrated Electron*. Wiley-Interscience, New York, 267 pp.

He, Y., Chen, H., Sheng, X. and Wu, J. (1993) *Radiat. Phys. Chem.* **42**, 715.

Hillarides, R.J., Gray, K.A., Guzzetta, J., Cortellucci, N. and Sommer, C. (1994) Radiolytic degradation of 2,3,7,8-TCDD in artificially contaminated soils. *Environ. Sci. Technol.* **28**, 2249–2258.

Hoigné, J. and Bader, H. (1989) Ozone/UV process for water treatment; formation of hydrogen peroxyde (H_2O_2)-residual. Summaries Wasser Berlin '89, P-3-1.

Kartasheva, L.I., Chulkov, V.N., Didenko, O.A., Pikaev, A.K., Kim, D.K., Han, B. and Kim, Y. (1997) Purification of water from cadmium in the presence of an OH radical scavenger. *Khim. Vys. Energ.* **31**(5), 349–352.

Kartasheva, L.I., Chulkov, V.N., Didenko, O.A., Pikaev, A.K., Kim, D.K., Kim, Y. and Han, B. (1998a) Combined electron-beam and adsorption method of removal of cadmium from water. *Khim. Vys. Energ.* **32**(2), 152–153.

Kartasheva, L.I., Chulkov, V.N., Didenko, O.A., Kim, D.K., Kim, M.E. and Han, B. (1998b) Combined electron-beam and adsorption method of decomposition of hydrogen peroxide in water. *Khim. Vys. Energ.* **32**, 175–178.

Kartasheva, L.I., Chulkov, V.N., Didenko, O.A., Makarov, I.E. and Pikaev, A.K. (1998c) On the mechanism of radiolysis of chlorobenzene aqueous solutions. *Khim. Vys. Energ.* **32**, 250–254.

Kartasheva, L.I., Chulkov, V.N., Didenko, O.A. and Pikaev, A.K. (2000) On hydrogen peroxide yields upon radiolysis of aerated aqueous solutions of formic acid. *Khim. Vys. Energ.* **34**, 467–469.

Kim, J.C., Kim, D.H., Kim, D.K., Kim, Y., Makarov, I.E., Pikaev, A.K., Ponomarev, A.V., Seo, Y.T. and Han, B. (1999) Deep degradation of formic acid in aqueous solutions under electron-beam treatment. *Khim. Vys. Energ.* **33**, 413–417.

Kurucz, C.N., Waite, T.D. and Cooper, W.J. (1985) The Miami electron beam research facility: a large scale wastewater treatment application. *Radiat. Phys. Chem.* **45**, 299–308.

Leitzke, O. (1992) Eliminierung von Schadstoffen im Wasser (CSB, AOX, CKW and PAK) durch die Kombination von Ozon und UV-Licht. *Brauwelt* **49**, 2516–2523.

Lin, K., Cooper, W.J., Nickelsen, M.G., Kurucz, C.N. and Waite, T.D. (1995) Decomposition of aqueous solutions of phenol using high energy electron beam irradiation – a large scale study. *Appl. Rad. Isot.* **46**, 1307–1316.

Lyu, E., Makarov, I.E., Kuryatnikov, Yu.I., Mazanko, R.V., Dzhagatspanyan, R.V. and Pikaev, A.K. (1983a) Radiation-flotation purification of wastewater from mercury. *Radiat. Phys. Chem.* **22**(3–5), 503–509.

Lyu, E., Makarov, I.E. and Pikaev, A.K. (1983b) Pulse radiolysis of alkaline solutions of Hg(II) compounds. *Khim. Vys. Energ.* **17**(1), 50–55.

Lyu, E., Makarov, I.E. and Pikaev, A.K. (1983c) Pulse radiolysis of aqueous solutions of metallic mercury. In: *Proceedings of V Tihany Symposium on Radiation Chemistry*, Dobo, J., Hedvig, P., Schiller, R. (eds), Vol. 1. Akademia Kiado, Budapest, pp. 215–220.

Ma, T.A. (1983) Tradescantia micronuclei test (Trade-MCN) for environmental clastogens. In: *In Vitro Toxicity Testing of Environmental Agents, Part A*, Kolber, A.R., Wong, T.K., Grant, L.R., Dewoskin, R.S. and Hughes, T.G. (eds). Plenum, New York, pp. 191–214.

Mak, F.T., Zele, S., Cooper, W.J., Kurucz, C.N., Waite, T.D. and Nickelsen, M.G. (1997) Kinetic modeling of carbon tetrachloride, chloroform, and methylene chloride removal from aqueous solution using the electron beam process. *Water Res.* **31**, 219–228.

Mezyk, S.P., Cooper, W.J., Bartels, D.M., O'Shea, K.E. and Wu, T. (2001) Radiation chemistry of alternative fuel oxygenates: substituted ethers. *J. Phys. Chem. Part A* **105**, 3521–3526.

Mincher, B.J., Arbon, R.E., Knighton, W.B. and Meikrantz, D.H. (1994) Gamma-ray induced degradation of PCBs in neutral isopropanol using spent reactor fuel. *Appl. Radiat. Isot.* **45**, 879–193.

Mincher, B.J., Liekhus, K. and Arbon, R.E. (1996) PCB radiolysis in isooctane in the presence of ozone. *Appl. Radiat. Isot.* **47**, 713–715.

Mincher, B.J., Curry, R.D. and Brey, R. (2000) Method to simultaneously improve PCB radiolysis rates in oil and to close the chlorine mass balance. *Environ. Sci. Technol.* **24**, 3452–3455.

Nickelsen, M.G., Cooper, W.J., Kurucz, C.N. and Waite, T.D. (1991) Removal of benzene and selected alkyl-substituted benzenes from aqueous solution using continuous high-energy electron irradiation. *Environ. Sci. Technol.* **26**, 144–152.

Nickelsen, M.G., Cooper, W.J., Lin, K., Kurucz, C.N. and Waite, T.D. (1994) High energy electron beam generation of oxidants for the treatment of benzene and toluene in the presence of radical scavengers. *Water Res.* **28**, 1227–1237.

Nickelsen, M.G., Kajdi, D.C., Cooper, W.J., Kurucz, C.N., Waite, T.D., Gensel, F., Lorenzl, H. and Sparka, U. (1998) Field application of a mobile 20 kW electron beam treatment system on contaminated groundwater and industrial wastes. In: *Environmental Applications of Ionizing Radiation*, Cooper, W.J., Curry, R.D. and O'Shea, K.E. (eds). John Wiley and Sons, Inc., New York, pp. 451–466.

Pikaev, A.K. (1969) *Solvated Electron in Radiation Chemistry*. Nauka, Moscow, 457 pp. (in Russian).

Pikaev, A.K. (1998) Electron-beam purification of water and wastewater. In: *Environmental Applications of Ionizing Radiation*, Cooper, W.J., Curry, R.D. and O'Shea, K.E. (eds). New York, pp. 495–506.

Pikaev, A.K. (2000a) Current status of the application of ionizing radiation to environmental protection. I. Sources of ionizing radiation, natural and drinking water purification. *Khim. Vys. Energ.* **34**, 3–15.

Pikaev, A.K. (2000b) Current status of the application of ionizing radiation to environmental protection. II. Wastewater and other liquid wastes. *Khim. Vys. Energ.* **34**, 83–103.

Pikaev, A.K. (2000c) Current status of the application of ionizing radiation to environmental protection. III. Sewage sludge, gaseous and solid systems. *Khim. Vys. Energ.* **34**, 163–176.

Pikaev, A.K. and Kabakchi, S.A. (1982) *Reactivity of Primary Products of Water Radiolysis*. Reference Book, Moscow: Energoizdat, 201 pp. (in Russian).

Pikaev, A.A. and Podzorova, E.A. (1999) Radiation treatment of wastewater under heterogeneous conditions for removal of petroleum products. *Khim. Vys. Energ.* **33**, 233–234.

Pikaev, A.K., Bludenko, A.V., Makarov, I.E., Ponomarev, A.V., Minin, V.N., Ponomarev, V.I. and Linnik O.A. (1996) Electron-beam treatment of highly-coloured river water. *Radiat. Phys. Chem.* **48**, 75–80.

Pikaev, A.K., Kabakchi, S.A., Putilov, A.V. and Vanyushkin, B.M. (1997a) Combined electron-beam and biological purification of industrial wastewater from surfactant. *J. Adv. Oxid. Technol.* **2**, 217–222.

Pikaev, A.K., Kartasheva, L.I., Zhestkova, T.P., Yurik, T.K., Chulkov, V.N., Didenko, O.A., Kim, D.K., Kim, Y. and Han, B. (1997b) Removal of heavy metals from water by electron-beam method in the presence of an hydroxyl radical scavenger. *Mendeleev Commun.* 52–53.

Pikaev, A.K., Makarov, I.E., Ponomarev, A.V., Kim, Y., Han, B., Yang, Y.W. and Kang, H.J. (1997c) A combined electron-beam and coagulation method of purification of water from dyes. *Mendeleev Commun.* (5) 176–178.

Pikaev, A.K., Podzorova, E.A. and Bakhtin, O.A. (1997e) Combined electron-beam and ozone treatment of wastewater in the aerosol flow. *Radiat. Phys. Chem.* **49**, 155–157.

Pikaev, A.K., Poniomarev, A.V., Bludenko, A.V., Minin, V.N. and Elizar'eva, L.M. (2001) Combined electron-beam and coagulation purification of molasses distillery slops. Features of the method, technical and economic evaluation of large-scale facility. *Radiat. Phys. Chem.* **61**, 81–87.

Podzorova, E.A., Pikaev, A.K., Belyshev, V.A. and Lysenko, S.L. (1998) New data on electron-beam treatment of municipal wastewater in the aerosol flow. *Radiat. Phys. Chem.* **52**, 361–364.

Podzorova, E.A., Pikaev, A.A. and Pikaev, A.K. (1999a) New radiation-chemical method of the removal of petroleum products from water. *Mendeleev Commun.* (3) 127–128.

Podzorova, E.A., Pikaev, A.K., Belyshev, V.A. and Lysenko, S.I. (1999b) The removal of pollutants from municipal wastewater by electron-beam treatment in the aerosol flow. *Khim. Vys. Energ.* **33**, 332–335.

Podzorova, E.A., Pikaev, A.A., Buryak, A.K., Ul'yanov and Pikaev, A.K. (2001) Chromato-mass-spectrometric study of radiation-chemical purification of water from petroleum products. *Khim. Vys. Energ.* **35**(2), 83–90.

Ponomarev, A.V., Bludenko, A.V., Makarov, I.E., Pikaev, A.K., Kim, D.K., Kim, Y. and Han, B. (1997) Combined electron-beam and adsorption purification of water from chromium and mercury using materials of vegetable origin as sorbents. *Radiat. Phys. Chem.* **49**, 473–476.

Ponomarev, A.V., Makarov, I.E., Minin, V.N., Kim, D.K., Han, B. and Pikaev, A.K. (1999) Combined electron-beam and coagulation treatment of wastewater with high concentrations of organic substances. *Khim. Vys. Energ.* **33**, 177–182.

Proksch, E., Gehringer, P., Szinovatz, W. and Eschweiler, H. (1987) Radiation-induced decomposition of small amounts of perchloroethylene. *Appl. Radiat. Isot.* **38**, 911–919.

Ross, A.B. (1973) Selected specific rates of reaction of transients from water in aqueous solution. Hydrated electron, supplemental data. *Nat. Stand. Ref. Data Ser., Nat. Bur. Stand. (US)* **43**(Suppl), 43.

Salimov, R.A., Kuksanov, N.K., Petrov, S.E., Spiridonov, G.A., Ivanchenko, V.A., Sludkin, O.P., Kundo, N.N. and Vanyushkin, B.M. (1998) BINP's accelerators for environmental applications. In: *Environmental Application of Ionizing Radiation*, Cooper, W.J., Curry, R.D. and O'Shea, K.E. (eds). Wiley, New York, pp. 47–62.

Schwarz, H.A. (1962) Determination of rate constants for the radical processes in the radiation chemistry of water. *J. Phys. Chem.* **66**, 255–263.

Schmelling, D., Poster, D., Chaychian, M., Neta, P., McLaughlin, W.J., Silverman, J. and Al-Sheikhly, M. (1998) Applications of ionizing radiation to the remediation of materials contaminated with heavy metals and polychlorinated biphenyls. *Radiat. Phys. Chem.* **52**, 371–377.

Shin, H.-S., Kim, Y., Han, B., Makarov, I.E., Ponomarev, A.V. and Pikaev, A.K. (2002). Application of electron beam to treatment of wastewater from papermill. *Radiat. Phys. Chem.* (in press).

Taghipour, F. and Evans, G.J. (1996) Radiolytic elimination of organochlorine in pulp mill effluent. *Environ. Sci. Technol.* **30**, 1558–1564.

Tobien, T., Cooper, W.J. and Asmus, K.-D. (2000a) Removal Simulation for the radiation induced degradation of the disinfection by-product chloroform. In: *Natural Organic Matter and Disinfection By-products*, Barrett, S., Krasner, S. and Amy, G. (eds). ACS Symposium Series 761, Washington DC, pp. 270–281.

Tobien, T., Cooper, W.J., Nickelsen, M.G., Pernas, E., O'Shea, K.E. and Asmus, K.-D. (2000b) Odor control in wastewater treatment: the removal of thioanisole from water – a model case study by pulse radiolysis and electron beam treatment. *Environ. Sci. Technol.* **34**, 1286–1291.

USEPA (1997) High Voltage Environmental Applications, Inc. Electron Beam Superfund Innovative Technology Evaluation Report, EPA/540/R-96/504, August.

Wang, T., Waite, T.D., Kurucz, C.N. and Cooper, W.J. (1994) Oxidant reduction and biodegradability improvement of paper mill effluent by irradiation. *Water Res.* **28**, 237–241.

Westerhoff, P., Aiken, G., Amy, G. and DeBroux, J. (1999) Relationship between the structure of natural organic matter and its reactivity towards molecular ozone and hydroxyl radical. *Water Res.* **33**, 2265–2276.

Woods, R.J. and Pikaev, A.K. (1994) *Applied Radiation Chemistry: Radiation Processing.* Wiley, New York, 535 pp.

Yak, H.K., Mincher, B.J., Chiu K.-H. and Wai, C.M. (1999) Supercritical fluid extraction/γ-radiolysis of PCBs from contaminated soil. *J. Haz. Mat.* **B69**, 209–216.

Yurik, T.K. and Pikaev, A.K. (1999) Radiolysis of weakly acidic and neutral aqueous solutions of hexavalent chromium ions. *Khim. Vys. Energ.* **33**, 248–252.

Zele, S. Nickelsen, M.G., Cooper, W.J., Kurucz, C.N. and Waite, T.D. (1998) Modeling kinetics of benzene, phenol and toluene irradiation in water using the high energy electron-beam process. In: *Environmental Applications of Ionizing Radiation*, Cooper, W.J., Curry, R.D. and O'Shea, K.E. (eds). John Wiley and Sons, Inc., New York, pp. 395–415.

Zhestkova, T.P., Pikaev, A.A., Pikaev, A.K., Kim, D.K., Kim, Y. and Han, B. (1997) Removal of lead from aqueous solutions by electron-beam treatment in the presence of an OH radical scavenger. *Khim. Vys. Energ.* **31**, 252–255.

10

Wet air oxidation processes

L. Patria, C. Maugans, C. Ellis, M. Belkhodja, D. Cretenot, F. Luck and B. Copa

10.1 INTRODUCTION

Wet oxidation, also known as wet air oxidation (WAO) or hydrothermal treatment, refers to the process of oxidising materials suspended or dissolved in liquid water with dissolved oxygen at elevated temperature. WAO is an attractive treatment for waste streams, which are too dilute to incinerate and too concentrated for biological treatment.

The wet oxidation technology has been successfully commercialised for 50 years. Over 200 full scale wet oxidation systems have been constructed and operated for a variety of applications ranging from elimination of hazardous wastes to the production of vanilla flavouring agent. While wet oxidation chemistry is also used as a chemical production technology as well as in the manufacture of computer chips, the focus of this chapter is on waste elimination.

After a first patent claiming the oxidation of a sulphite liquor from pulp production by compressed air at 180°C (Strehlenert, 1915), most of the original WAO

 Advanced Oxidation Processes for Water and Wastewater Treatment. Edited by Simon Parsons. ISBN: 1 84339 017 5

development work took place in the US some 50 years ago. While WAO is considered a waste treatment technology today, it was first successfully commercialised in the 1930s and 1940s for the manufacture of artificial vanilla flavouring (vanillin). The WAO process was commercialised as the Zimmermann process, named after its developer, and the term Zimpro® process remains a common synonym for WAO (Intelecom, 1997).

Zimmermann did extensive testing of the WAO technology and its application in treating paper mill waste liquor, black liquor, for waste elimination and energy and soda recovery (Barton and Schoeffel, 1956; Mishra *et al.*, 1995; Schoeffel, 1954, 1956; Zimmermann, 1950, 1958). In the early 1960s, the Zimpro company built several large WAO plants for the treatment of municipal wastewater sludge, either to improve sludge dewatering by thermal conditioning, or to achieve complete oxidation of the sludge (Seiler, 1987). A number of years later, high-pressure sludge oxidation was commercially applied to paper-mill sludges for waste destruction and recovery of filler clay materials used in the paper making process (Ellis, 1980; Flynn, 1976).

In the 1970s, the application of WAO for the treatment of industrial wastewaters began to be commercialised (Mishra *et al.*, 1995). Other forms of catalytic and non-catalytic WAO for treatment of industrial wastewaters were also commercialised during this time frame (Ely *et al.*, 1975; Flynn and Flemington, 1979; Holzer *et al.*, 1992; Sadana and Katzer, 1979; Wilhelmi and Knopp, 1979). During the 1980s, the market activity for the low-pressure oxidation (LOPROX®) technology for the treatment of sludge began to wane at the same time that WAO began to see more use as an industrial waste treatment technology. WAO was commercialised for the treatment of sulphide laden industrial spent caustics and other industrial wastes (Brandenburg and Bauer, 1983; Brandenburg *et al.*, 1989; Chowdhury, 1982; Chowdhury and Whilhelmi, 1980; DeAngelo and Wilhelmi, 1983). The spent caustic and industrial market continues to be an active market for new commercial WAO applications (Carlos and Maugans, 2000; Copa, 1994; Ellis, 1994; Ellis *et al.*, 1994; Grover and Gooma, 1993).

From the late 1990s through the present, there has been speculation of a resurgence in the municipal sludge destruction/minimisation market, this time in Europe. WAO is being commercialised for this market under trade names, such as Athos® and the LPO (Holzer *et al.*, 1992; Lendormi *et al.*, 2001; Luck *et al.*, 1999; Teletzke, 1966).

Another recent chemical industry WAO application trend is the migration to becoming more integrated into the overall chemical production process. An example of this is in the area of improving product crystals, where the mother liquor that is purged from a crystalliser is now being wet air oxidised for the destruction of the concentrated organic build-up thus allowing the recycle of the inorganic product to the crystalliser (Giudici and Maugans, 2000; Lehmann, 1997; Lemmens *et al.*, 2001; Noboru *et al.*, 1994).

About 100 WAO plants are in operation today, mostly to treat waste streams from petrochemical, chemical and pharmaceutical industries as well as residual sludges from wastewater treatment plants. These plants aim either at a complete oxidative decomposition of pollutants or at a partial oxidation into low-molecular weight compounds which can then be treated further by conventional biological processes.

10.2 PROCESS FUNDAMENTALS

WAO can be defined as the oxidation of organic and inorganic substances in an aqueous solution or suspension by means of oxygen or air at elevated temperatures and pressures. WAO is an aqueous phase process where the water is an integral part of the reaction.

Typical conditions for wet oxidation range from 100°C to 372°C at elevated pressures to maintain water in the liquid phase and ensure that oxidation reactions take place in liquid phase. Residence times may range from 15 to 120 min, and the chemical oxygen demand (COD) removal may typically be about 75–90%. Insoluble organic matter is converted to simpler soluble organic compounds which are in turn oxidised and eventually converted to carbon dioxide and water, without emissions of NO_x, SO_2, HCl, dioxins, furans, fly ash, etc.

The temperature range can be further subdivided into low (100–200°C), medium (200–260°C), and high temperature (260–320°C) operation. Higher temperature (320–372°C) systems can be designed as well but are rarely used due to the high capital cost. Due to the necessity to maintain water in the liquid form, the system pressure is maintained above the steam pressure. As steam pressure increases with temperature, the terms such as high temperature and high pressure are used interchangeably.

A complete mineralisation of the waste stream is difficult by WAO, since some low molecular weight oxygenated compounds (especially acetic and propionic acids and methanol) are resistant to oxidation. For instance, removal of acetic acid is usually negligible at temperatures lower than 260°C. Organic nitrogen compounds are easily transformed into ammonia, which is also very stable in WAO conditions. WAO plants aim either at a complete oxidative decomposition of pollutants or at a partial oxidation into low-molecular weight compounds, which can then be treated further by conventional biological processes.

10.2.1 Non-catalytic WAO

Several studies have been devoted specifically to the elucidation of the mechanism of WAO, most of them using either carboxylic acids (Devlin and Harris, 1984) or phenol (Nikolaou *et al.*, 1994) as model compounds. The free radical chain mechanism involving the formation of hydroperoxides and oxyradicals allows the description of the gaseous and liquid oxidation products. As a first step, hydrogen abstraction from the α, β and γ-CH_2 groups of carboxylic acids by O_2 results in the formation of free radicals which react immediately with O_2 to form peroxy radicals. The peroxy radicals can lead to decarboxylation and formation of CO_2

by H abstraction from the —COOH group of another molecule of carboxylic acid. The predominance of acetic acid as a major by-product is explained by the relative stability of the ethoxyl radical (Devlin and Harris, 1984).

Many attempts have been made to study reaction mechanisms for wet oxidation of organics. Understanding the reaction mechanisms is essential for the development of kinetic models. For engineering purposes, it is sufficient to quantify the global reaction rate by identifying the major wet oxidation pathways. Models of organic compounds families have been used for that specific purpose. Ploos van Amstel and Rietema (1973) and Foussard *et al.* (1989) were important actors in the development of wet oxidation kinetic models. On some occasions, kinetic models applied to specific applications were proposed such as the oxygen dependence equilibrium kinetic WAO model that was developed for different organic sludge types, describing organic matter mineralisation and main by-products production (Belkhodja, 2000).

A generalised kinetic model based on a simplified reaction scheme with acetic acid as the rate-limiting intermediate (Figure 10.1) was proposed by Li *et al.* (1991) and is usually used to represent the kinetic data of WAO reactions with reasonably good fits (Li and Lin, 1997; Luck, 1996).

10.2.2 Catalytic wet air oxidation

The stringent operating conditions of WAO provided a strong incentive to investigate catalysts which would allow substantial reductions in temperature, pressure and residence time. Another major benefit of using catalysts in WAO is the oxidation of the refractory compounds, namely acetic acid and ammonia, at much lower temperatures than in the absence of catalysts. Catalytic wet air oxidation (CWAO) was investigated over the years both with heterogeneous and homogeneous catalysts.

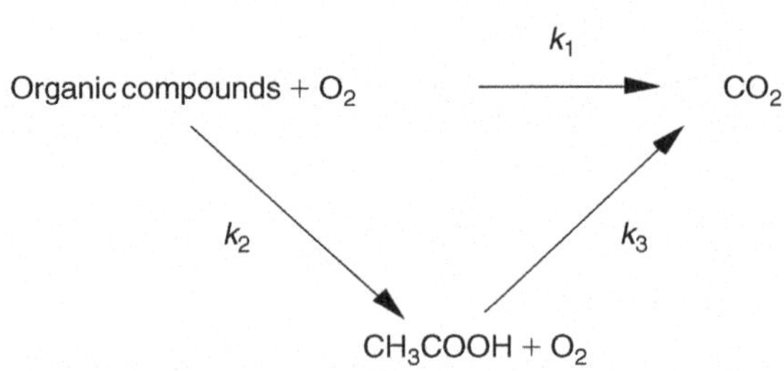

$$\frac{[OC + CH_3COOH]}{[OC + CH_3COOH]_0} = \frac{k_2}{k_1 + k_2 - k_3} e^{-k_3 t} + \frac{k_1 - k_3}{k_1 + k_2 - k_3} e^{-(k_1 + k_2)t}$$

Figure 10.1 Simplified kinetic model for WAO (Li *et al.*, 1991).

10.2.2.1 Heterogeneous CWAO

Several crucial issues have to be solved related to chemical and physical stability of the heterogeneous oxidation catalysts during WAO: leaching and sintering of the active phase and loss of surface area of the support material. Leaching can be controlled to a large extent by a proper choice of the catalytic metal or metal oxide according to the available solubility data, and by pH control during CWAO. The same parameters are also of importance to control the hydrothermal stability of the catalyst carrier. Oxides of Al, Hf, Zr and Ti have been shown to be stable in CWAO conditions and should be selected as catalyst supports. However, high surface area metastable oxides, such as γ-Al_2O_3, could undergo phase transitions detrimental to the durability of the supported catalytic metals and oxides. In addition to these deactivation factors associated with the physical properties of the liquid phase (temperature, pressure and pH), poisoning of the active sites by deposition of organic or inorganic compounds may also compromise the durability of the catalyst.

The first patent was filed by DuPont in 1950 to claim a catalytic composition based on Mn–Zn–Cr oxides to perform CWAO at temperatures in the range of 120–200°C, for the destructive catalytic oxidation of industrial wastes of an organic nature, to permit their economical disposal, to abate stream pollution, to inhibit stream contamination and to otherwise eliminate industrial and health problems associated therewith (DuPont patent, 1950). Several types of heterogeneous catalysts were studied in the last decade, based first on supported or unsupported base metal oxides and more recently on supported precious metals (Table 10.1). Supported precious metal catalysts are generally less prone to deactivation by leaching of the active phase, and present higher overall activities for the oxidation of various pollutants, especially acetic acid and ammonia. Additional information about academic studies of CWAO can be found in three recent reviews (Imamura, 1999; Matatov-Meytal and Sheintuch, 1998; Mishra *et al.*, 1995).

This research paved the way for the development of several commercial catalytic wet oxidation systems especially in Japan, where three CWAO technologies have been developed since the mid-1980s based on heterogeneous catalysts containing precious metals deposited on titania or titania–zirconia carriers. Compared to standard WAO, these processes are able to oxidise two refractory compounds, namely acetic acid and ammonia, thus allowing the treated water to be discharged directly or reused as process water.

10.2.2.2 Homogeneous CWAO

An alternative to insoluble precious metal catalysts is the use of homogeneous transition metal catalysts, which need however to be separated and recycled to the reactor or discarded. Several processes based on different variations of this concept have been developed in the last decade. The development took place mainly in Europe for the application on residual solid waste, such as sewage sludge. Many experiments were undertaken to evaluate the catalytic effect of different transition metal, one by one or in association as shown in Figure 10.2.

Table 10.1 Summary of reported heterogeneous CWAO research.

Catalyst		Application	Reference
Active phase	Carrier		
Cu	Alumina	Phenol	Sadana and Katzer (1974)
	Alumina	Phenol	Kim *et al.* (1991)
	Alumina	*p*-Cresol	Mishra *et al.* (1993)
Cu	Alumina, silica	Chlorophenols	Sanger *et al.* (1992)
Cu–Zn	Alumina, silica	Phenol compounds	Pintar and Levec (1992)
Cu–Mg–La	Zn aluminate	Acetic acid	Box and Fahra (1974) Levec *et al.* (1976)
Mn	Alumina	Phenol	Sadana and Katzer (1974)
Mn	SR 115	Chlorophenols	Sanger *et al.* (1992)
Mn–Ce	None	Poly(ethyleneglycol)	Imamura *et al.* (1986)
Mn–Zn–Cr	None	Industrial wastes	Moses and Smith (1954)
Cu–Co–Ti–Al	Cement	Phenol	Schmidt *et al.* (1990)
Co	None	Alcohols, amines, etc.	Ito *et al.* (1989)
Co–Bi	None	Acetic acid	Imamura *et al.* (1982)
Co–Ce	None	Ammonia	Imamura *et al.* (1985)
Fe	Silica	Chlorophenols	Sanger *et al.* (1992)
Ru	Cerium oxide	Alcohols, phenol, etc.	Imamura *et al.* (1988)
Ru	Cerium oxide	Acetic acid	Barbier Jr. *et al.* (1998)
Ru	Titania–zirconia	Industrial wastes	Harada and Yamasaki (1993)
Ru	Cerium oxide	Acetic acid	Gallezot *et al.* (1996, 1997)
Ru–Rh	Alumina	Wet oxidised sludge	Takahashi *et al.* (1991)
Pt	Titania	Phenol	Maugans and Akgerman (1997)
Pt	Cerium oxide, titania, zirconia	Acetic acid	Barbier *et al.* (1998)
Pt–Pd	Titania–zirconia	Industrial wastes	Ishii *et al.* (1991)
Pt–Pd–Ce	Alumina	Black liquor	Zhang and Chuang (1998)

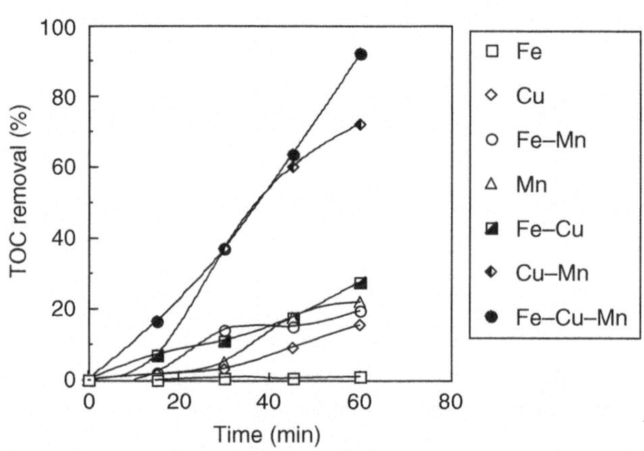

Figure 10.2 Determination of synergetic effects of metals (Fe, Cu and Mn) for oxidation of a mixture of oxalic, malonic, succinic and acetic acid (total organic carbon = $5\,g\,L^{-1}$) with H_2O_2 at 98°C, pH 3.5 and $[MnSO_4]$ = 5.4 mM (Striolo *et al.*, 1994).

10.2.3 Comparison of WAO and incineration

The common feature of the existing wet oxidation processes is that by contrast with incineration, they are intrinsically energy conservative. The only energy required is the difference in enthalpy between the incoming and outgoing streams, while for incineration, one has to supply not only the sensible heat and heat of vaporisation of the liquid, but also the energy required to heat the water vapour, combustion products and excess air to combustion temperatures, between 820°C and 1100°C, as shown in Figure 10.3.

Incineration and WAO differ in three main ways. Jung *et al.* (2002) demonstrated the key differences between incineration and WAO for the application on residual sludges. The main three key differences were as follows:

- oxidation mechanisms,
- modes of energy recovery,
- quantities and qualities of by-products.

Incineration is a thermal oxidation technology applied under dry conditions. Its main characteristic is a direct combustion with excess oxygen from air in the flame. The process is nearly instantaneous.

On the other hand, WAO is a thermal oxidation technology applied under wet conditions. The kinetics of the oxidation reaction are limited by oxygen solubilisation and diffusion in liquid phase. The kinetic is enhanced by temperature and

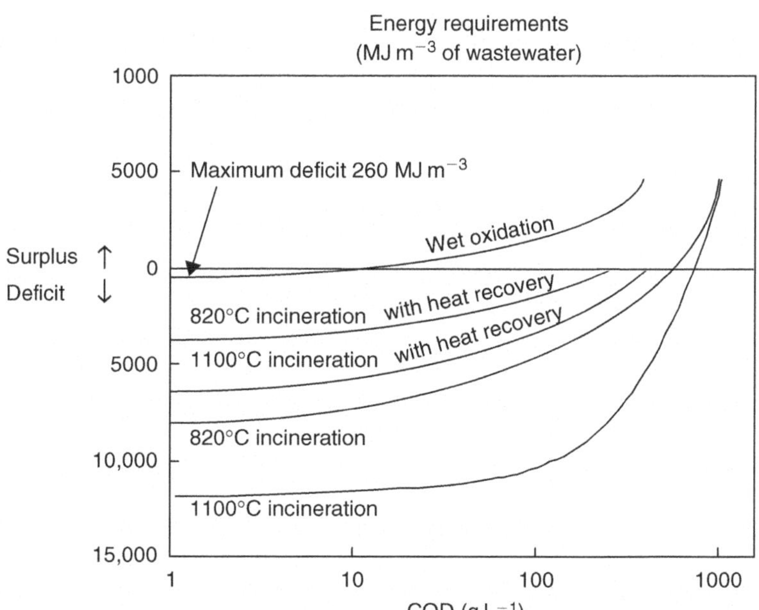

Figure 10.3 Comparison of wet oxidation and incineration processes for heat requirements as a function of the amount of organics in the wastewater (Chowdhury and Copa, 1986).

pressure, but otherwise the matrix as well as the reaction agents are compared to those of advanced oxidation processes.

Incineration and WAO are both thermal processes. In both cases, the objective is to operate the process under autothermal conditions. Therefore:

- Incineration is applied either to dewatered or dried sludge.
- WAO is applied directly to thickened sludge, without any further dewatering required. The oxidised effluent is cooled down by heat exchange with the feed before pressure release through a control valve. The heat released in the process can be used to produce steam or hot water and mechanical energy can be produced when the off-gas is expanded.

Incineration by-products are mainly in the gas (including NO_x, SO_2, HCl, dioxins, furans, etc.) and solid phases (bottom and fly ashes). Whereas, WAO by-products are in the gas (without emissions of NO_x, SO_2, HCl, dioxins, furans, fly ash, etc.), solid (different nature from ashes) and liquid phases (fatty acids and ammonia eventually recycled to a biological treatment system).

10.3 APPLICATIONS

WAO is an attractive treatment for waste streams which are too dilute to incinerate and too concentrated for biological treatment. About 100 plants are in operation today, mostly to treat waste streams from petrochemical, chemical and pharmaceutical industries as well as residual sludges from wastewater treatment facilities. These WAO plants aim either at a complete oxidative decomposition of pollutants or at a partial oxidation into low-molecular weight compounds which can then be treated further by conventional biological processes.

Due to the necessity to maintain water in liquid form, the system pressure is maintained above the steam pressure. Figure 10.4 is a diagram of the typical non-catalytic WAO waste treatment application spectrum.

Commercial applications of low-temperature oxidation (100–200°C) include the low-pressure thermal conditioning of sludge (i.e. LPO) and low-strength sulphidic spent caustic treatment. Other industrial wastes treated by low-temperature oxidation include cyanide and phosphorus wastes as well as non-chlorinated pesticides.

Medium temperature (200–260°C) oxidation is used for typical ethylene spent caustics, as well as for autothermal thermal sludge conditioning (TSC), and some industrial wastes. Ethylene spent caustics, derived from sulphidic acid gas washings in the ethylene industry are treated in the 200–220°C range. Autothermal TSC operations occur in the temperature range of 220–260°C as does wet air regeneration (WAR) treatment of powdered activated carbon (PAC) for use in wastewater biological treatment applications. Some other industrial wastes containing phenolic or cresylic compounds can be treated at the high end of this temperature spectrum.

High temperature (260–320°C) oxidation is used for refinery spent caustic, sludge destruction, and most WAO treated industrial wastewaters. Within the 260–280°C range, refinery spent caustics, containing significant concentrations of naphthenic and cresylic acids, are treated. Most organic industrial wastes are

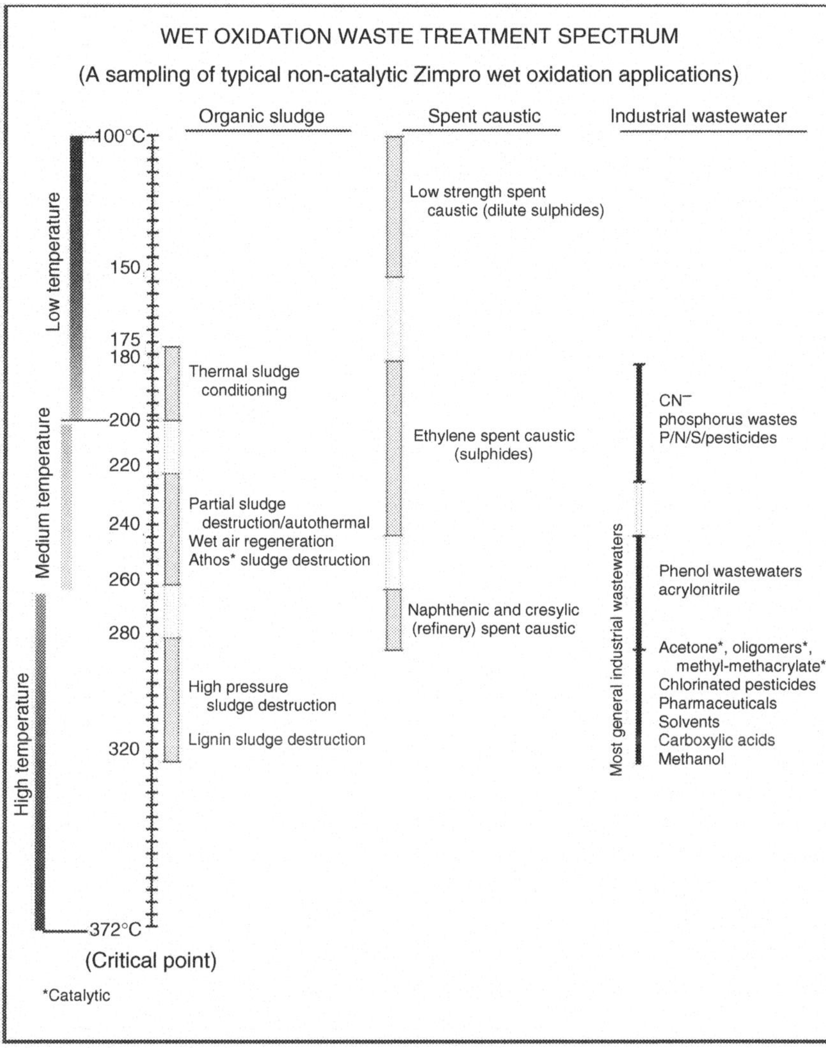

Figure 10.4 Spectrum of typical non-CWAO applications.

oxidised in this temperature range, including pesticides, solvents, pharmaceutical wastes and other dissolved organic chemical wastes. Sludge destruction systems are designed to operate within the higher end of this temperature range (280–320°C) and are employed for complete destruction of municipal, pulp and paper, or other organic sludges. Chemical recovery systems, which are used in the treatment of pulping liquors also operate within this higher temperature range.

CWAO is used when performance at reduced temperature, pressure and residence time is necessary. CWAO also allows for the effective oxidation of refractory compounds, such as acetic acid and ammonia. CWAO is used in the ATHOS® process for sludge and ammonia destruction and is also used in various industrial applications.

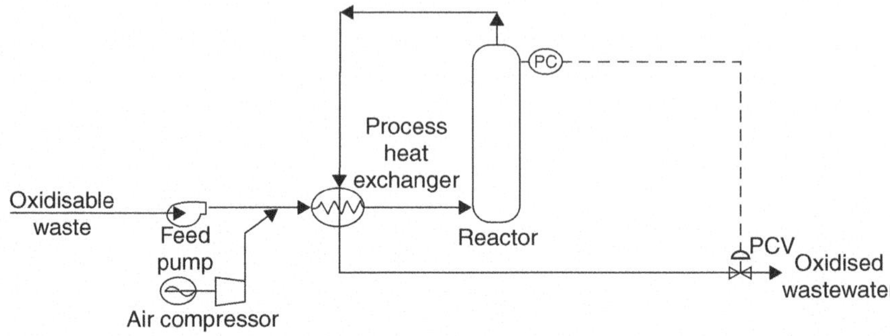

Figure 10.5 Flow diagram of a WAO process.

10.4 REACTOR DESIGN

After a first patent claiming the oxidation of a sulphite liquor from pulp production by compressed air at 180°C (Strehlenert, 1911, 1915), most of the original WAO development work took place in the US some 50 years ago. Zimmermann did extensive testing of WAO technology and its application in treating spent pulp mill liquor (Zimmermann, 1958). In the 1960s, the Zimpro company built several large WAO plants for the treatment of municipal wastewater sludge, either to improve sludge dewatering by thermal conditioning, or to achieve complete oxidation of the sludge (Seiler, 1987).

Figure 10.5 shows a basic flow diagram of a WAO plant, which consists mainly of a high-pressure pump, an air or oxygen compressor, a heat-exchanger as well as a reactor with a relief valve and a downstream separator. The waste is retained in the reactor for a sufficient period of time to achieve the desired chemical oxidation (30–120 min). The simplest reactor design is usually a cocurrent vertical bubble column with a height-to-diameter ratio in the range of 5–20. Some design aspects of WAO reactors have already been reported (Beyrich *et al.*, 1979; Joglekar *et al.*, 1991; Ploos van Amstel and Rietema, 1973).

10.5 COMMERCIAL PROCESSES

10.5.1 Non-catalytic WAO processes

10.5.1.1 The Zimpro TSC process

More than 130 Zimpro TSC units, the first widely successful commercial WAO process, were installed in the US and in Europe for the oxidation of sewage sludge, most of them for 'low-pressure' conditioning by partially oxidising the organic fraction – breaking up the structure of the sludge floc so that it would more easily release its bound water for improved dewatering (Figure 10.6). Some of these units were used for wet air carbon regeneration. After use in the PACT®

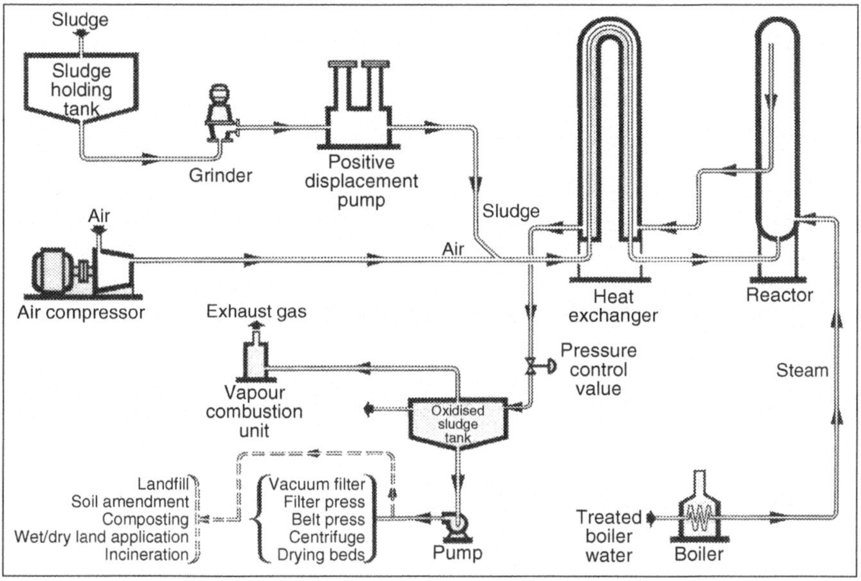

Figure 10.6 Flow diagram of the Zimpro® LPO process for TSC.

process, the mixture of spent PAC and biomass from the aeration basin was oxidised to destroy excess biological solids, oxidise organics that are adsorbed on the carbon, and reactivate the PAC for reuse in the biological/adsorption system.

Details of the installation and operation of Zimpro TSC units in wastewater treatment plants were already reported (Goldstein and Lokatz, 1963; McDonald *et al.*, 1974). Sludge destruction is carried out with air as oxidant, at 250–270°C and under pressures of 8.5–12 MPa, in reactors presenting a height-to-diameter ratio ranging between 5 and 20, while thermal conditioning is performed at 175–220°C.

The largest Zimpro TSC unit currently in service for sludge conditioning is located in New Jersey (Passaic Valley Sewerage Commissioners, 1996). It treats sludge thickened to 4% dry solids (DS) (100,000 t DS $year^{-1}$), which is disinfected and easily dewatered to 55% DS on a filter press. The process is used there for thermal conditioning, since the treated sludge is intended after lime treatment to reuse in land application and restoration of landfills (Passaic Valley Sewerage Commissioners, 1996).

Other Zimpro WAO units are in operation in chemical, petrochemical and pharmaceutical plants worldwide (Chowdhury and Copa, 1986). Some of these units use optionally a homogeneous copper catalyst to help achieving higher COD removal. In the 1990s, 25 Zimpro units have been sold for oxidation of ethylene plant spent caustic (USFilter/Zimpro, 1999). The gas from the hydrocarbon cracking furnace is scrubbed in a caustic wash tower leading to a purge which is laden with sulphides and organics, such as condensed oils and benzene. WAO achieves oxidation of reactive sulphide to soluble thiosulphate, sulphite and sulphate.

This treated stream is then suitable for biotreatment in the plant's wastewater treatment system.

10.5.1.2 The VerTech process

The VerTech aqueous phase oxidation (APO) process treats sludge using pure oxygen in a sealed below-ground oxidation vessel consisting of two concentric tubes (downcomer and upcomer) and a heat exchange system (Knops *et al.*, 1993). The underground portion of the APO system extends to a depth of approximately 1200 m with a diameter of 0.95 m, and uses drilling and casing technology developed in the natural gas and oil industries. The depth of the vessel results in high bottom pressures due to the hydraulic head without the need for aboveground, high-pressure pumps. This original system facilitates heat exchange and minimises the cost of pumping, but requires a weekly stop for descaling of the tubes with nitric acid, in order to preserve an effective heat exchange.

In the VerTech unit brought into service in 1994 at Apeldoorn (The Netherlands), the temperature reaches approximately 275°C at the bottom of the reactor, under a pressure of 100 bar. This unit requires 55 t day^{-1} of oxygen to eliminate 70% of the COD from the 80 dry tons of sludge it processes daily. The residual solids are landfilled after dewatering to 55% DS, while most of the nitrogen load in the oxidised filtrate is treated by biological nitrification–denitrification in the wastewater plant. The off-gas is treated by thermal incineration with a catalytic reactor as backup.

10.5.2 Heterogeneous CWAO processes

10.5.2.1 The NS-LC process

The NS-LC process introduces air into wastewater and performs oxidation, decomposition and purification in one stage using a catalyst, usually cylindrical pellet type, under WAO conditions (Nippon Shokubai).

The NS-LC process initially used to involve a Pt–Pd/TiO_2–ZrO_2 honeycomb catalyst two-phase flow in vertical monoliths gave in a broad range of gas–liquid velocities a very beneficial flow pattern, the slug flow (segmented gas–liquid flow). Since each liquid plug was sandwiched between two gas plugs, a recirculation pattern is developed within each liquid plug, which improved mass transfer significantly (Hatzantoniou *et al.*, 1986) and prevented solids deposition. In addition, a thin liquid film was formed between the gas and the channel wall which allowed high mass-transfer rates while keeping the catalyst continuously wetted.

Typical operating conditions of the NS-LC process are a temperature of 220°C, a pressure of 4 MPa and a space velocity of 2 h^{-1}. At these conditions the oxidation of compounds, such as phenol, formaldehyde, acetic acid, glucose, etc., reaches or exceeds 99% (Nippon Shokubai). In the absence of catalyst the removal efficiencies would be limited to 5–50% (Zhang and Chuang, 1998).

10.5.2.2 The Osaka Gas process

The Osaka Gas CWO process is based on a mixture of precious and base metals on titania or titania–zirconia carriers (honeycomb or spheres). It has been demonstrated to work on several industrial and urban wastes:

1. a coal gasifier effluent in a pilot plant at British Gas's London Research Station (Cooke and Taylor, 1993);
2. wastewater from coke ovens (Fukatawa *et al.*, 1993);
3. concentrated cyanide wastewater of the Tufftride process (a soft nitriding process for steel) (Harada and Yamasaki, 1993);
4. sewage sludge and residential wastes (Fukatawa *et al.*, 1993).

Table 10.2 shows some operational data on the treatment of wastewater from coke ovens in a test plant built in 1979 with several reactors (0.30 m inner diameter, 6.50 m height, total capacity 6 $m^3\,day^{-1}$). Highly concentrated COD and ammonia are decomposed to a level of 10 $mg\,L^{-1}$ or less after 24 min contact time (SV = 2.5 h^{-1}). Phenol and cyanide, the major pollutants present together with ammonia, are decomposed to levels below the detection limit. Furthermore, there is no detectable emission of NO_x or SO_x, and a catalyst lifetime longer than 8 years is claimed (Fukatawa *et al.*, 1993).

Without catalyst, there is no removal of ammonia and the COD decomposition rate is very low. With a two- to sixfold increase of contact time, a residual COD value of *ca.* 1000 $mg\,L^{-1}$ is left in the effluent (acetic acid).

10.5.2.3 The Kurita process

Ammonia can also be selectively oxidised into N_2 and N_2O by the NO_2^- ion at lower temperatures than in the presence of oxygen in conventional WAO.

Table 10.2 Typical coke oven wastewater wet oxidation results [Osaka Gas process (Fukatawa *et al.*, 1993)].

Treatment conditions				Raw waste water	Treated water	Exhaust gas composition	
Temperature (°C)	250	pH	–	10.5	6.4	O_2 (%)	9.9
Pressure (MPa)	6.86	T-N	($mg\,L^{-1}$)	3.750	160	N_2 (%)	83.1
Liquor volume ($L\,h^{-1}$)	200	NH_3-N	($mg\,L^{-1}$)	3.080	3	CO_2 (%)	7.0
Air volume ($Nm^3\,h^{-1}$)	14.4	COD_{Mn}	($mg\,L^{-1}$)	5.870	<10	NO_x ($mg\,Nm^{-3}$)	ND (<0.5)
Reaction time (min)	24	TOD	($mg\,L^{-1}$)	17.500	ND	SO_x ($mg\,Nm^{-3}$)	ND (<2.0)
Catalyst	Type A	T-CN	($mg\,L^{-1}$)	15	ND	NH_3 ($mg\,Nm^{-3}$)	ND

ND: not detected.

The Kurita company recently developed a process tailored to abate ammonia with nitrite at 170°C in the presence of a supported platinum catalyst (Kurita, 1997). With a catalyst close to those of the above-mentioned processes, the use of nitrite instead of oxygen allows operating temperatures 50–100° lower.

10.5.3 Homogeneous catalytic wet oxidation processes

10.5.3.1 The Ciba-Geigy process and MINERALIS process

The Ciba process involves the use of a copper salt which is separated as copper sulphide and recycled into the reactor which is lined with titanium to withstand corrosion. Compressed air was selected as the oxidant in the three units installed in the 1990s within the German and Swiss plants of the pharmaceutical company. They achieve high oxidation efficiencies (95–99%) on chemical and pharmaceutical wastes at elevated temperature (300°C). Such temperatures coupled with copper catalysis are essential for the oxidation of the rate-limiting acetic acid, but ammonia is still present in the treated waste. The unit located in Grenzach, Germany (near Basle) is currently highly reliable with 97% availability (Schrode, 1995). The off-gas from this unit is treated by catalytic oxidation.

Recently the Ciba-Geigy process was marketed for the treatment of industrial wastewaters and sewage sludge, in catalysed as well as uncatalysed versions, by the Swiss company Granit (Granit, 1998). The process is known as the MINERALIS process in its application for sewage sludge. The MINERALIS process operates at 300°C and does not require catalyst addition for sewage sludge treatment (Lendormi *et al.*, 2000).

10.5.3.2 The LOPROX process

The low pressure wet oxidation (LOPROX) process was developed by Bayer AG for the separate treatment of organic substances which degrade too slowly in normal biological plants or adversely affect the degradation of other substances. Oxidation takes place with oxygen in the acidic range in a multistage bubble column oxidation reactor, under relatively mild operating conditions (temperature below 200°C and pressures in the range 0.5–2.0 MPa) and is catalysed by a combination of Fe^{2+} ions and quinone-generating substances. To compensate for the mild reaction temperature, a residence time of *ca.* 2 h in the reactor is needed (range 1–3 h). At a reaction temperature of 150°C, a destruction efficiency of about 10 g COD L^{-1} effluent is sufficient in order to allow the process to take place autothermally (Holzer, 1991).

Several LOPROX plants are in operation at Bayer AG. The process serves also as an alternative in the handling of sludge from wastewater treatment plants (Holzer and Horak, 1994). About 85–90% of the organic matter will be eliminated in the overall process, with 60–65% oxidised to CO_2 directly in the oxidation reactor at approximately 190°C and 25% dissolved in the treated liquid phase as biodegradable COD to be recycled in the waste water treatment plant. The catalysts Fe^{2+}

and quinones are provided by the sludge. The operating costs for sludge treatment by the LOPROX process are estimated by Bayer in the range 125–175€ per ton of dry matter. A pilot unit with a capacity of 600 L h^{-1} is in operation in Switzerland, in the wastewater treatment plant Birs II, in an effort to optimise the process for the treatment of sludge.

10.5.3.3 The WPO process

The WPO process derived from the classical Fenton's reagent by IDE (France) and INSA, Toulouse (France) allows high-oxidation efficiencies (up to 98%) at mild conditions (90–130°C, 0.1–0.5 MPa), using hydrogen peroxide as the oxidant catalysed by Fe^{2+} in solution at pH 3. This process was successfully operated in Spain in 1992 on two different sites (aquifers) with a 5 $m^3 h^{-1}$ demonstration unit.

On both sites, the process was able to achieve a COD removal efficiency of at least 95% (Striolo *et al.*, 1994). Recently, a homogeneous Fe–Cu–Mn catalyst was developed to treat refractory oxidation by-products, such as acetic acid and bicarboxylic acids (oxalic, succinic, etc.). The catalyst is recovered after reaction by precipitation at pH 9 and filtration (Falcon *et al.*, 1993; Striolo *et al.*, 1994).

10.5.3.4 The ORCAN process

A modified WPO process has been recently developed in Switzerland based on R&D performed at ETH Zürich (Orcan web site). The oxidation reactor for carbon and nitrogen compounds (ORCAN) process pretreats refractory wastes at 120°C and 0.3 MPa, with air as the stoichiometric oxidant. The oxidation is catalysed by Fe^{2+} ions as well as minor amounts of hydrogen peroxide (H_2O_2/COD = 0.20). The decomposition of H_2O_2 is likely to initiate the free radical chain mechanism at the relatively low temperature, since with oxygen only and Fe^{2+} ions the oxidation reactions are initiated at *ca.* 150°C. A first commercial unit came into operation in 1996 in a textile factory (feed 4 $m^3 h^{-1}$, COD 7–10 g L^{-1}).

10.5.3.5 The ATHOS® process

The ATHOS® wet oxidation process is dedicated to WAO of residual sludges. It consists of four main steps: preparation of the sludge, oxidation (liquid and gas phases), heat recovery and dewatering of the residual solids (Figure 10.7).

The sludge preparation unit consists of a shredder designed to prevent objects of more than 3 mm and fibres from entering the oxidation reactor. The pH of the sludge is adjusted and optionally a soluble oxidation catalyst added. From the mixed feed tank, which also collects the recycled oxidised and thickened sludge, a high pressure pump is then used to convey the sludge through a heater (a group of shell-and-tube heat exchangers), which heats the sludge up to the required temperature in the oxidation reactor, which is intensively mixed by means of a recirculation pump.

Oxidation is carried out at a temperature of 235–250°C and a pressure level of about 44–52 bar. The oxidant is pure oxygen, stored on-site under liquid form. The residence time is 1 h. Once the process reaches a steady state, it is autothermal without any energy input, because the reaction is exothermic. The oxidised sludge is then withdrawn and cooled down in a second group of heat exchangers, the recovered heat is transmitted to the incoming sludge by means of a thermal fluid circulating between the two exchangers.

The off-gas produced is treated in a catalytic reactor, where CO and the other pollutants are readily oxidised. The excess vent is discharged to the atmosphere, with no plume, nor dust.

Separation of the liquid and solid phases is carried out in a lamellar settler at a pressure of 3 bar and a temperature of 110°C. The supernatant is cooled to ambient temperature then recycled to the head of the plant. The solid phase is pumped from the settling tank and easily dewatered in a filter press to 55% DS, without any additives.

The main equipment item in the pre-treatment section is the screening equipment, which is used only on high fibres content sludges, like primary sludges. In the heating section, a high-pressure pump is used to feed the sludge to the reactor via the heater exchanger. The heat exchangers are designed as dual sludge/hot water high pressure and high velocity shell and tube type heat exchangers. The sludge flow is inside the tubes and hot water is circulating in a closed loop through the shell in order to minimise consumption of soft water. The cooler and heater can be operated in opposite functions by using additional automatic valves. The design and autothermal operation characteristics of the ATHOS® heat exchangers have been made with success in order to minimise fouling and maintenance of the ATHOS®

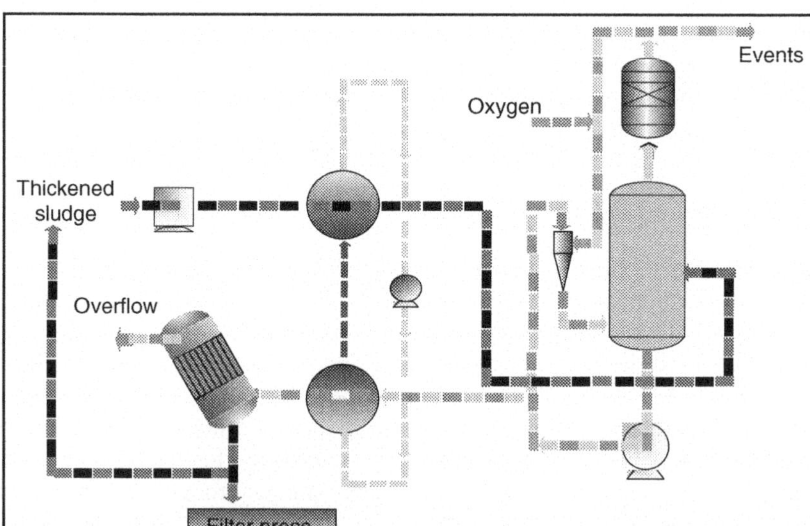

Figure 10.7 Flow diagram of the ATHOS® process.

heat exchangers. In the oxidation section, the WAO reactor is considered as a completely stirred reactor by using a recirculation pump at a appropriate recirculation ratio. Oxygen is introduced at a controlled flow-rate in the recirculation loop using a sludge-eductor. Other important equipment items are the catalytic reactor, the lamellar settler and the dewatering equipment. Special designed process control valves are used to withstand the abrasion characteristics of the concentrated sludge.

10.6 CASE STUDIES

10.6.1 Refinery Spent Caustic – RPDM, Rio de Janeiro, Brazil

In 1995, a WAO system was put into operation for the treatment of refinery spent caustic from gasoline sweetening, gasoline and LPG prewashing, and from gasoline and LPG mercaptans extraction at the Refinaria de Petroleos de Manguinhos, S.A. (RPDM) in Rio de Janeiro. RPDM is a producer of liquid fuels and is located near residential property within Rio de Janeiro. Prior to installing a WAO system, the refinery was dependent on off-site disposal of their spent caustic. In considering on-site treatment, it was desired to avoid odours since the refinery was in a populated area. The WAO system was installed in 1995 and operates at 2 gpm and 260°C with a 1 h residence time. Figure 10.8 is a photograph of the WAO system. After treatment by WAO, the oxidised spent caustic is sent to the refinery's biological treatment system for final treatment. System performance includes odour-free operation with complete removal of sulphides, mercaptans and thiosulphates as well as significant reductions in phenols and overall COD.

Figure 10.8 RPDM Refinery Spent Caustic WAO Unit.

Figure 10.9 ATOFINA Italia AMS Crystalliser Mother Liquor Catalytic WAO System.

10.6.2 Ammonium sulphate crystalliser mother liquor – ATOFINA, Rho, Italy

ATOFINA Italia has incorporated WAO into its approach for dealing with the large quantities of unavoidable by-product formed from the methyl methacrylate production process. Through utilising crystallisers, by-product ammonium sulphate (AMS) salt is recovered in high quality form. In order to ensure that the quality of the AMS is, and remains, acceptable for agricultural use, a purge stream is generated from the operation of the crystallisers. This purge stream, which contains the organic by-products, passes through a WAO system where the organic components are destroyed and the treated AMS returns to the crystallisation system. The exothermic WAO process produces surplus energy (steam), which is returned to the crystallisers. AMS in the WAO effluent is returned to the crystalliser system for recovery of AMS product. Large volumes of crystalliser blowdown liquor are no longer trucked off-site for third party disposal. The WAO reactor is operated at 280°C using a homogenous copper catalyst, which is recycled within the system. A photograph of the WAO system is shown in Figure 10.9.

10.6.3 Ethylene spent caustic – Chinese Petroleum Corporation, Kaohsiung, Taiwan

In the late 1980s, Chinese Petroleum Corporation (CPC) contracted with the M.W. Kellogg company in Houston, Texas, to build a 500,000 Mt year^{-1} grass roots ethylene plant. The refinery where this facility was to be located was in a heavily populated area and the whole project came under considerable public

Figure 10.10 CPC Ethylene Spent Caustic WAO System; Insert: Feed and Treated Samples.

scrutiny because of concerns over odour issues surrounding the spent caustic that the plant would produce. Public pressure was so great that the entire ethylene expansion project was being threatened. CPC evaluated a number of alternatives including using WAO as a means of treating their spent caustic. Ultimately CPC chose to construct three WAO systems, each with a capacity of 7.3 $m^3 h^{-1}$ to handle the spent caustic from the new ethylene facility, plus what was produced from the existing ethylene units at the refinery.

In the late fall of 1990, the 200°C WAO units were started up for testing to evaluate their performance. The effluent produced by the WAO systems easily met the guarantee requirements. The most important result was that the effluent quality and the operation of the WAO systems completely satisfied the local neighbours such that a positive solution to the spent caustic problem had been implemented and protest of the continued construction of the new ethylene plant subsided. A photograph of one of the units is shown in Figure 10.10.

Figure 10.11 Repsol Quimica PO/SM Wastewater Pure Oxygen WAO system.

10.6.4 Polyol ether/styrene monomer wastewater – Repsol Quimica, Tarragona, Spain

In 2000, a wastewater treatment system was constructed to accommodate a new Repsol facility in Tarragona, Spain. A WAO system operating at 295°C using pure oxygen and a 1-h retention time was built for the treatment of propylene oxide/styrene monomer (PO/SM) wastewaters. A photograph of the WAO system is shown in Figure 10.11. The impetus for WAO usage was to destroy the biologically refractory components in the PO/SM wastewater and to decrease the COD loading on the overall wastewater plant.

After WAO treatment, the effluent is mixed with all other facility wastewaters and is routed through a PACT-WAR system. The PACT sludge is regenerated in a 240°C WAR/WAO reactor. The WAR reactor and part of the PACT system is shown in Figure 10.12.

10.6.5 Municipal sewage sludge – Veolia Water, Toulouse, France

In 1998, an ATHOS® plant was constructed at the Toulouse wastewater treatment plant in France (Figure 10.13). From 1998 to 2001, the plant was operated as a demonstration unit with the support of the LIFE programme of the EU and several full-scale trials were undertaken. Since the beginning of 2002, the ATHOS® unit is in full operation as one of the sludge treatment processes on site.

The ATHOS® unit was set up as a modular facility (ground footprint 20 × 25 m, height 12 m). The WAO system is operating at 235°C, 44 bars using pure

Figure 10.12 Repsol Quimica WAR reactor (right) and PACT system (left).

Figure 10.13 View of the ATHOS® demonstration unit in operation at the Toulouse wastewater treatment plant.

oxygen and a 1-h residence time for the treatment of sludge with a 30–50 $g\,L^{-1}$ DS content. It was designed with a treatment capacity of 3 $m^3\,h^{-1}$.

Operating at 235°C, a 85–90% COD removal is routinely obtained together with a 45–70% ammonia removal. The main characteristics of the overflow recycled to the head of the plant were 7–9 $g\,L^{-1}$ COD (including 4–6 $g\,L^{-1}$ acetic acid or equivalent), 0.8–1.2 $g L^{-1}$ N-NH_4 and a suspended solids content below 500 $mg L^{-1}$. The residual solids contain a small amount of organic matter and are easily settled and dewatered to 50–60% DS by means of a filter-press.

10.6.6 Municipal sewage sludge – Aquiris, Brussels, Belgium

A new wastewater treatment plant is to be built in Brussels, Belgium, by the year 2006. The large capacity of the plant (1.2 million population equivalent) raised several issues to be handled in order to get a sustainable and cost effective sludge treatment and disposal or reuse system.

An innovative sludge treatment train was selected, including the ATHOS® process. The objective was to maximise waste sludge volume reduction using a thermal oxidation process with optimised overall energy recovery. Designing the plant with this in mind is a cost effective integration of the sludge treatment in the plant and the optimised volume and quality of the final residues to be disposed.

The sludge treatment train selected for the Brussels plant will include a thickening/dewatering step which will thicken the sludge up to 15–16% DS, followed by thermal hydrolysis using steam injection at 165–180°C and 10 bars, anaerobic digestion, WAO with the ATHOS® process and a final solid residue processing using a drying filter-press to 98% DS.

Two ATHOS® units will be operated in parallel. Their treatment capacity represents $2 \times 12\, m^3 h^{-1}$.

The 98% DS residue will be landfilled as inert residues. Several reuse options remain opened for consideration in the long term, such as use for secondary raw material in building and construction.

Energy recovery was considered at every step of the incorporation of thermal technologies.

10.7 CONCLUSIONS

Wet oxidation has been used for many years and has been applied to a wide variety of waste treatment projects. These applications range from the destruction of sludge to the destruction of waste liquors with extremely toxic components, such as cyanides and pesticides. Wet oxidation has also been used to enhance the biodegradability of certain wastewaters and for the general disruption of the COD of other wastewaters. At present, WAO is widely used for the treatment of spent caustic liquors in the ethylene and refining industries, and custom applications are made in the chemical process industry. With the high thermal efficiency of WAO, it offers an economical alternative to liquid waste incineration. Since the WAO system utilises a direct chemical oxidation process, it is not subject to toxic upsets as would be the case with biological treatment.

Adding catalyst to WAO offers lower energy requirements and much higher oxidation efficiencies. Commercial catalytic wet oxidation processes rely either on supported precious metal and/or base metal oxide catalysts or on homogeneous catalysts, such as Fe or Cu. WAO is today a cost-effective solution for the treatment of organic sludge and refractory industrial wastewater.

Besides the existing applications for the technology, either WAO or CWAO, additional applications will surely be realised as more stringent discharge criteria are applied by regulatory agencies.

REFERENCES

(2001) Planta de tratamiento de aquas residuales de las nuevas plantas de OP/SM y derivados de Repsol YPF en el complejo industrial de Tarragona. *Technologia Del Aqua* **217**, 66–71.

Andre, D. (1993) Quel avenir pour la valorisation agricole des boues des stations d'épuration? *L'Eau, L'Industrie, Les Nuisances* **163**, 58–60.

Barbier Jr., J., Delanoe, F., Jabouille, F., Blanchard, G. and Duprez, D. (1998) Total oxidation of acetic acid in aqueous solutions over noble metal catalysts. *J. Catal.* **177**, 378.

Barbier, J., Delanoe, F., Jabouille, F., Duprez, D., Blanchard, G. and Isnard, P. (1998) Total oxidation of acetic acid in aqueous solutions over noble metal catalysts. *J. Catal.* **177**(2), 378–385.

Barner, H.E., Huang, C.Y., Johnson, T., Jacobs, G., Martch, M.A. and Killilea, W.R. (1992) Supercritical water oxidation: an emerging technology. *J. Haz. Mater.* **31**, 1–17.

Barton, D.M. and Schoeffel, E.W. (1956) *Recovery of Cooking Liquor from Spent Soda Pulping Liquors*. US Patent 2774666.

Barton, D.M. and Schoeffel, E.W. (1956) *Ammonia–Sulfur Dioxide Cooking Acid Regeneration*. US Patent 2752243.

Bayer, A.G. (1994) LOPROX® Technical Leaflet K-RP LTK 19033e, May 1994.

Belkhodja, M. (2000) *Etude expérimentale et modélisation du procédé d'oxydation par voie humide. Application au traitement des boues d'épuration.* Thèse de doctorat, Ecole Centrale Paris.

Beula, D.A., Momont, J.A. and Copa, W.M. (1992) *Caustic Sulfide Wet Oxidation Process*. US Patent 5082571.

Beyrich, J., Gautschi, W., Regenass, W. and Wiedmann, W. (1979) Design of reactors for the wet air oxidation of industrial waste water by means of computer simulation. *Comp. Chem. Eng.* **3**, 161.

Beziat, J.C., Besson, M., Gallezot, P. and Durecu, J. (1999) Catalytic wet air oxidation of carboxylic acids on TiO_2-supported ruthenium catalysts. *J. Catal.* **182**(1), 129–135.

Box Jr., E.O. and Fahra, F. (1974) US Patent 3823088.

Brandenburg, B.L. and Bauer, G.L. (1983) *Process and Apparatus for Treatment of Thiosulfate Containing Liquors*. US Patent 4372940.

Brandenburg, B.L., Cooley, C.D. and Ellis, C.E. (1989) *Process for Treating Caustic Cyanide Metal Wastes*. US Patent 4812243.

Carlos, T.M. and Maugans, C.B. (2000) Wet air oxidation of refinery spent caustic: a refinery case study. In: *Proceedings of the NPRA Conference*, September.

Carlos, T.M. and Maugans, C.B. (2002) Manage refinery spent caustic efficiently. *Hydrocarb. Process.* **February**, 89–92.

Chowdhury, A.K. (1982) *Process for Treatment of Caustic Waste Liquors*. US Patent 4350599.

Chowdhury, A.K. and Wilhelmi, A.R. (1980) Treatment of spent caustic liquor by wet oxidation. In: *Proceedings of the 8th Annual Industrial Pollution Conference*. Sponsored by Water and Wastewater Equipment Manufacturers Association, June.

Chowdhury, A.K. and Copa, W.C. (1986) Wet air oxidation of toxic and hazardous organics in industrial wastewaters. *Ind. Chem. Eng.* **28**, 3.

Cooke, B.H. and Taylor, M.R. (1993) The environmental benefit of coal gasification using the BGL gasifier. *Fuel* **72**, 305.

Copa, W. (1994) Wet air oxidation of spent caustics. *Nat. Environ. J.* **November/December**.

Daun, M. (1996) Ver Tech-Naßoxidations-anlage Apeldoorn Erfahrungen aus dem Dauerbetrieb. *Abwassertechnik* **1**, 46–49.

DeAngelo, D.J. and Wilhelmi, A.R. (1983) Wet air oxidation of spent caustic liquors. *Chem. Eng. Prog.* **March**, 68–73.

Devlin, H.R. and Harris, I.J. (1984) *Ind. Eng. Chem. Fundam.* **23**, 391.

Ding, Z.Y., Frisch, M.A., Li, L. and Gloyna, E.F. (1996) Catalytic oxidation in supercritical water. *Ind. Eng. Chem. Res.* **35**(10), 3257–3279.

Djafer, M., Luck, F. and Bourbigot, M.M. (1994) Catalytic wet air oxidation of municipal sludge: effect of reaction parameters. In: *Proceedings of the 1st International Conference on Advanced Oxidation Technologies*, London, Ontario, June, p. 207.

Djafer, M., Luck, F., Wacheux, H. and Bourbigot, M.M. (1994) Wet air oxidation: study of the by-products and biodegradation tests. In: *Proceedings of the 1st International Congress on By-products of Water Treatment*, Poitiers, France, September 29–30.

Djafer, M. and Luck, F. (1997) Characterization of the residual solids from wet oxidation of sludge. In: *Proceedings of the 2nd International Congress on By-products of Water Treatment*, Rennes, France, March.

Earle, J.S. and Zimmermann, F.J. (1966) *Sewage Treatment Process*. US Patent 3272739.

Ellis, E.C. (1980) Disposing of paper mill sludge and recovering filler clay. *Indus. Wastes* **March/April**.

Ellis, C.E. (1994) Taiwan refineries adopt wet air oxidation to treat spent caustic liquors. *Hazmat World* **March**, 37.

Ellis, C.E., Lawson, R.L. and Brandenburg, B.L. (1994) Wet air oxidation of ethylene plant spent caustic. In: *Proceedings of the 6th Annual Ethylene Producers Conference*.

Ely, R.B., Pogainis, E.M. and Hoffman, C.A. (1975) *Catalyzed Process and Catalyst Recovery*. US Patent 3912626.

Falcon, M., Peyrille, B., Peilhac, P., Foussard, J.N. and Debellefontaine, H. (1993) Oxydation en voie humide de la pollution organique aqueuse par le peroxyde d'hydrgène. Procédé. Wet Peroxide Oxidation (WPO®). Etude de nouveaux catalyseurs. *Rev. Sci. Eau* **6**, 411.

Flynn, B.L. (1976) Swiss mill is trying wet air oxidation to get rid of sludge and recover filler. *Paper Trade J.* **May 1**.

Flynn, B.L. and Flemington, W.Va. (1979) Wet air oxidation of waste streams. *Chem. Eng. Prog.* **April**, 66–69.

Fortuny, A. *et al.* (1999) Water pollution abatement by catalytic wet air oxidation in a trickle bed reactor. *Catal. Today* **53**, 107–144.

Foussard, J.N., Debellefontaine H. and Besombes-Vailhe, J. (1989) Efficient elimination of organic liquid wastes: wet air oxidation. *J. Environ. Eng.* **115**, 367–385.

Fr. Pat. 2,726,761, 1995 (Appl. Date 27 October 1994), to OTV.

Fr. Pat. 2,750,976, 1998, to Rhône-Poulenc Chimie.

Fr. Pat. 2,752,534, 1998, to Rhône-Poulenc Chimie.

Fukatawa, M., Takahashi, H., Inoue, G. and Fujioka, T. (1993) Treatment of concentrated cyanide waste water. In: *Proceedings of the IDA/WRPC World Conference on Desalination and Water Treatment*, Yokohama, Vol. 1, 231.

Futakawa, M., Takahashi, H., Inoue, G. and Fujioka, T. (1993) Treatment of concentrated cyanide wastewater. In: *Proceedings of the IDA/WRPC World Conference on Desalination and Water Treatment*, Yokohama/Japan, November 3–6, Vol. II, pp. 505–512.

Gallezot, P., Laurain, N. and Isnard, P. (1996) Catalytic wet-air oxidation of carboxylic acids on carbon-supported platinum catalyts. *Appl. Catal. B. Environ.* **9**(1–4).

Gallezot, P., Chaumet, S., Perrard, A. and Isnard, P. (1997) Catalytic wet air oxidation of acetic acid on carbon-supported ruthenium catalysts. *J. Catal.* **168**(1), 104–109.

Gitchel, W. (2002) Former Zimpro Vice President of Research, Interview.

Gitchel, W.B. and Hoffman, C.A. (1966) *Sewage Sludge Treatment Process*. US Patent 3272740.

Gitchel, W.B., Hoffman, C.A. and Schoeffel, E.W. (1967) *Partial Wet Air Oxidation of Sewage Sludge*. US Patent 3359200.

Giudici, D. and Maugans, C.B. (2000) Improvement of industrial synthesis of methyl methacrylate application of a wet air oxidation process (WAO). *La Chemica e L'Industria* **September**, 873–878.

Giudici, D. and Maugans, C.B. (2000) Wet air oxidation recovers energy and waste. *Water Wastewater Int.* **October**, 13–15.

Gloyna, E.F., Li, L. and McBrayer, R.N. (1994) Engineering aspects of supercritical water oxidation. *Water Sci. Technol.* **30**(9), 1–10.

Goldstein, A. and Lokatz, S. (1963) Sewage sludge oxidation at Chicago. In: *Proceedings of the WPCF Conference*, Seattle, Washington, October.

Granit (1998) Commercial leaflet.

Grover, R. and Gomaa, H.M. (1993) Proven technologies manage olefin plant's spent caustic. *Hydrocarb. Process.* **September**, 61–69.

Harada, Y. and Yamasaki, K. (1993) Treatment of wastewater and sludge by a catalytic wet oxidation process. In: *Proceedings of IDA/WRPC World Conference on Desalination and Water Treatment*, Yokohama, Japan, November 3–6, Vol. II, pp. 231–242.

Hatzantoniou, V., Andersson, B. and Schöön, N.-H. (1986) *Ind. Eng. Chem. Prod. Des. Dev.* **25**, 964.

Helling, R.K. and Tester, J.W. (1988) Oxidation of simple compounds and mixtures in supercritical water: carbon monoxide, ammonia and ethanol. *Environ. Sci. Technol.* **22**(11), 1319–1324.

Holzer, K. (1991) Behandlung von Problemabwässern. *Umwelt* **21**, 179.

Holzer, K. and Horak, O. (1994) The BAYER-LOPROX®-System for oxidative treatment of wastes and sludges. In: *Proceedings of the Conference on ACHEMA 94*, Environmental Technology, DECHEMA e.V., Frankfurt-am-Main.

Holzer, K., Horak, O. and Lawson, J.F. (1992) LOPROX®: A flexible way to pretreat poorly biodegradable effluents. In: *Proceedings of the 46th Purdue Industrial Waste Conference.*

Hutton, D.G. and Robertaccio, F.L. (1978) *Industrial Waste Water Treatment Process.* US Patent 4069148.

Imamura, S. (1999) Catalytic and noncatalytic wet oxidation. *I&EC Res.* **38**, 1743–1753.

Imamura, S., Doi, A. and Ishida, S. (1985) *Ind. Eng. Chem. Prod. Res. Dev.* **24**, 75.

Imamura, S., Fukuda, I. and Ishida, S. (1988) Wet oxidation catalyzed by ruthenium supported on cerium (IV) oxides. *Ind. Eng. Chem. Res.* **27**, 718.

Imamura, S., Hirano, A. and Kawabata, N. (1982) *Ind. Eng. Chem. Prod. Res. Dev.* **21**, 570.

Imamura, S., Nakamura, M., Kawabata, N., Yoshida, J. and Ishida, S. (1986) *Ind. Eng. Chem. Prod. Res. Dev.* **25**, 34.

Intelecom (1997) *Preserving the Legacy: Thermal Treatment Technologies*. PTL 0401, Audiovisual VHS, INTELECOM, Pasadena CA.

Ishii, T., Mitsui, K., Sano, K. and Inoue, A. (1991) Eur. Pat. 431 932, to Nippon Shokubai.

Ishii, T., Mitsui, K., Sano, K. and Inoue, A. (1992) *Method for Treatment of Wastewater*. US Patent 5145587.

Ito, M.M., Akita, K. and Inoue, H. (1989) Wet oxidation of oxygen- and nitrogen-containing organic compounds catalysed by cobalt (III) oxide. *Ind. Eng. Chem. Res.* **28**, 894.

Jarosz, J. (1994) L'incinération des boues par four à lit fluidisé. Colloque 'Mieux gérer l'eau', Hydrotop 94, Marseille, France, April 12–15.

Joglekar, H.S., Samant, S.D. and Joshi, J.B. (1991) Kinetics of wet air oxidation of phenol and substituted phenols. *Water Res.* **25**, 135.

Jung, C.G., Fontana, A., Cretenot, D. and Belkhodja, M. (2002) Traitement des boues d'épuration: comparaison entre les procédés par oxydation par voie humide et par incinération, Eau, Industrie, Nuisances **249**, 49–53.

Kim, S., Shah, Y.T., Cerro, R.L. and Abraham, M.A. (1991) Aqueous phase oxidation of phenol in a monolithic reactor. In: *Proceedings of the AIChE Annual Meeting*, Pittsburgh, August, p. 46.

Knopp, P.V. and Gitchel, W.B. (1970) Wastewater treatment with powdered activated carbon regenerated by wet air oxidation. In: *Proceedings of the 25th Purdue Industrial Waste Conference*, May 5–7.

Knops, P.C.M., Bowers, D.L., Krielen, A. and Romano, F.J. (1993) The start-up and operation of a commercial below-ground aqueous phase oxidation system for processing sewage sludge. *VerTech Treatment Systems Technical Documentation.*

Kurita (1997) Personal communication.

Lehmann, R.W. (1997) *Wet Oxidation of High Strength Liquors with High Solids Content.* US Patent 5651897.

Lemmens, J.A.W., Lansu, A., Ma., Ekkelenkamp, G. and Debeij, J.E.P. (2001) Process for treating a mixture comprising an ammonium sulfate solution phase and an aqueous Lactam phase. *World Patent Application* WO 01/12549 A1.

Lendormi, T., Prévot, C., Poignant, A., Doppenberg, F. and Debellefontaine, H. (2000) Subcritical wet oxidation of municipal sewage sludge: comparison of pilot and full scale plant experiments. In: *Proceedings of the IWA (poster)*, Paris, France, July.

Lendormi, T., Prévot, C., Doppenberg, F., Spérandio, M. and Debellefontaine, H. (2001) Wet oxidation of domestic sludge and process integration: the Mineralis® process. *Water Sci. Technol.* **44**(10), 163–169.

Levec, J., Herskowitz, M. and Smith, J.M. (1976) An active catalyst for the oxidation of acetic acid solutions. *AIChE J.* **22**, 919.

Li, L., Chen, P. and Gloyna, E.F. (1991) Generalized kinetic model for wet oxidation of organic compounds. *AIChE J.* **37**(11), 1687–1697.

Li, S.H. and Lin, S.J. (1997) *Environ. Technol.* **123**, 852.

Luck, F. (1996) A review of industrial catalytic wet air oxidation processes. *Catal. Today* **27**(1–2), 195–202.

Luck, F. (1999) Wet air oxidation: past, present, and future. *Catal. Today* **53**, 81–91.

Luck, F., Bonnin, C., Niel, G. and Naud, G. (1995) By-products from sludge treatment by subcritical and supercritical water oxidation. *Revue des Sciences de l'Eau* **8**, 481.

Luck, F., Bonnin, C., Rose, J.P. and Cretenot, D. (2000) From sludge to inert solids with wet oxidation. In: *Proceedings of the 2nd International Conference on Oxidation Technologies for Water and Wastewater Treatment*, Clausthal-Zellerfeld, May 28–31.

Luck, F., Djafer, M. and Rose, J.P. (1999) ATHOS®: a novel process for sludge disposal. In: *Proceedings of the AWWA 18th Federal Convention*, Adelaide, Australia, April.

Mantzavinos, D., Sahibzada, M., Livingston, A.G., Metcalfe, I.S. and Hellgardt, K. (1999) Wastewater treatment: wet air oxidation as a precursor to biological treatment. *Catal. Today* **53**(1), 93–106.

Matatov-Meytal, Y.I. and Sheintuch, M. (1998) Catalytic abatement of water pollutants. *I&EC Res.* **37**, 309–326.

Maugans, C.B. and Akgerman, A. (1997) Catalytic wet oxidation of phenol over a Pt/TiO_2 catalyst. *Water Res.* **31**, 3116–3124.

McDonald, J.R., Page, V.W. and Belliss, R.M. (1974) Wet-air oxidation of sludge at Hockford sewage works. *Effluent Water Treat. J.* **1**, 13–21.

Meijerink, J. (1997) Oxydation par voie humide: trois années d'expérience en application industrielle. In: *Proc. of Séminaire Boues-Euroforum*, Paris, December 2–3.

Mishra, V.S., Joshi, J.B. and Mahajani, V.V. (1993) *Ind. Chem. Eng.* **34**, 211.

Mishra, V.S., Mahajani, V.V. and Joshi, J.B. (1995) Wet air oxidation. *Ind. Eng. Chem. Res.* **34**, 2–48.

Momont, J.A., Beula, D.A. and Copa, W.M. (1994) *Low Temperature Caustic Sulfide Wet Oxidation Process*. US Patent 5298174.

Moses, D.V. and Smith, E.A. (1954) US Patent 2690425.

Nikolaou, N., Abatzoglou, N., Gasso, S. and Chornet, E. (1994) The oxidation of valeric acid in aqueous solution. *Can. J. Chem. Eng.* **72**, 522 and references therein.

Nippon Shokubai. *NS-LC Wastewater Treatment System – Catalytic Oxidation*, Nippon Shokubai Technical Leaflet.

Nippon Shokubai Web site http://www.shokubai.co.jp

Noboru, I., Katsumi, M. and Nagaben, M. (1994) *Manufacturing Method for Decolorised Rearranged Ammonium Sulphate*. Chinese Patent CN 1023790C.

Ohta, H., Goto, S. and Teshima, H. (1980) Liquid-phase oxidation of phenol in a rotating catalytic basket reactor. *I&EC Fundam.* **19**, 180–185.

ORCAN Web site http://www.orcan.ch.

Passaic Valley Sewerage Commissioners (1996) Annual Report.

Patria, L. and Cretenot, D. (2002) Wet air oxidation: the ATHOS process. In: *Proceedings of the Advanced Oxidation Conference*, Cranfield, UK, April.

Patria, L., Bonnin, C., Cretenot, D. and Luck, F. (2001) Biosolids disposal: new stabilisation and oxidation processes. In: *Proceedings of the Leading Edge Waste Water Treatment Workshop, IWA Conference*, Berlin, Germany, October.

Pintar, A. and Levec, J. (1992) Catalytic liquid-phase oxidation of refractory organics in waste water. *Chem. Eng. Sci.* **47**, 2395–2400.

Ploos van Amstel, J.J.A. and Rietema, K. (1973) Wet-air oxidation of sewage sludge. Part II: The oxidation of real sludges. *Chemie-Ing.-Techn.* **45**, 1205–1211.

Reddy, K.J., Gloss, S.P. and Wang, L. (1994) Reaction of CO_2 with alkaline solid waste to reduce contaminant mobility. *Water Res.* **28**, 1377–1382.

Rose, J.P., Bonnin, C., Patria, L., Cretenot, D. and Luck, F. (2000) ATHOS: a new hydrothermal oxidation of sludge. In: *Proceedings of the Weftec 73rd Annual Conference & Exposition on Water Quality and Wastewater Treatment*, Anaheim, US, October.

Rose, J.P., Cretenot, D., Bonnin, C. and Luck, F. (2000) Industrial results from the demonstration plant using a novel WAO process for sludge and waste disposal. In: *Proceedings of the IWA World Water Congress*, Paris, France, July 3–7.

Sadana, A. and Katzer, J.R. (1979) An analysis of the heterogeneous catalyzed free radical oxidation of phenol in aqueous solution. *Ind. Eng. Chem. Process Des. Dev.* **18**, 50–56.

Sanger, A.R., Lee, T.T.K. and Chuang, K.T. (1992) In: *Progress in Catalysis*, Smith, K.J. and Sanford, E.C. (eds). Elsevier, p. 197.

Schoeffel, E.W. (1954) *Preparation of Calcium Bisulfite Cooking Acid from Waste Sulfite Liquor*. US Patent 2696424.

Schoeffel, E.W. (1956) *Recovery of Cooking Liquor from Spent Semi-chemical Pulping Liquors*. US Patent 2750290.

Schoeffel, E.W. (1962) *Sewage Treatment*. US Patent 3060118.

Schoeffel, E.W. and Zimmermann, F.J. (1968) *Wet Air Oxidation of Combustible Materials Adsorbed on Carbonaceous Adsorbent*. US Patent 3386922.

Schoeffel, E.W. and Zimmermann, F.J. (1969) *Oxidation and Reclamation Process*. US Patent 3442798.

Schrode, R. (1995) In: *Proceedings of the 2nd Conference Wastewater Treatment by Wet Oxidation*, Vogelpohl, A. (ed.), Clausthal, Germany, September, CUTEC-Reports Nr. 20.

Seiler, G.S. (1987) Twenty-five years of sludge management by wet oxidation. *Sludge Manag. Ser.* **17**, 100.

Shmidt, F.K., Kochetkova, R.P., Babikov, A.F., Shiverskaia, I.P., Shpilevskaia, L.I. and Eppel, C.A. (1990) In: *Proceedings of the 8th French-Soviet Meeting Catal.*, Novossibirsk, June, p. 140.

Sibony, J. and Guibelin, E. (1997) Traitement des boues: les techniques éprouvées, les besoins nouveaux, les développements en cours. In: *Proc. of Séminaire Boues-Euroforum*, Paris, France, December 2–3.

Strehlenert, R.W. (1911 and 1915) Swed. Patent 34 941, September 1911; US Patent 1,149,420, August 1915.

Striolo, Ph., Foussard, J.N. and Debellefontaine, H. (1994) Advanced method for the treatment of organic aqueous wastes: wet peroxide oxidation – WPO®. In: *Proceedings of the 1st International Conference on Advanced Oxidation Technologies for Water and Air Remediation*, London, Ontario, Canada, June 25–30, pp. 363–365.

Striolo, Ph., Pech, J.L. and Debellefontaine, H. (1994) Wet peroxide oxidation (WPO)™: a new oxidation technique for the remediation of organic aqueous wastes. In: *Proceedings of the 1st International Conference on Advanced Oxidation Technologies for Water and Air Remediation*, London, Ontario, Canada, June 25–30, pp. 184–186.

Suarez, F.J. (1996) Pluses and minuses of caustic treating. *Hydrocarb. Process.* **October**, 117–123.

Takahashi, Y., Takeda, N., Aoyagi, T. and Tanaka, K. (1991) Catalytic wet-oxidation of human waste produced in a space habitat: purification of the oxidized liquor for human drinking. In: *Proceedings of the 4th European Symposium on Space Environment and Control System*, Florence, ESA SP-324, Vol. 2, 643.

Teletzke, G.H. (1966) Wet air oxidation – industrial waste applications. In: *Proceedings of the 39th Annual Conference: Water Pollution Control Federation.*

Teletzke, G.H. (1966) Wet air oxidation of sewage sludge. In: The Transactions of the 16th Annual Conference of Sanitary Engineering. *Bull. Eng. Archit.* **56**, 25–30.

Teletzke, G.H. (1966) *Sewage Treatment Process*. US Patent 3256179.

Teletzke, G.H., Knopp, P.V. and Erickson, A.H. (1972) *Heat Treatment of Sewage Sludge*. US Patent 3697417.

US Patent 2 (1950), 690, 425, 1954 (Appl. Date 23 March 1950), to DuPont.

Wiart, J. and Reveillere, M. (1995) La teneur en éléments-traces métalliques des boues résiduaires des stations d'épuration urbaines françaises. T.S.M., 12, 913–922.

Wilhelmi, A.R. and Knopp, P.V. (1979) Wet air oxidation – an alternative to incineration. *Chem. Eng. Progr.* **August**, 46–52.

Zhang, Q. and Chuang, K.T. (1998) Kinetic of wet oxidation of black liquor over a Pt-Pd-Ce-Alumina catalyst. *Appl. Cat. B : Environ.* **17**, 321.

Zimmermann, F.J. (1943) *Process of Making Vanillin*. US Patent 2399607.

Zimmermann, F.J. (1945) *Process for Making Vanillin*. US Patent 2434626.

Zimmermann, F.J. (1950) *Waste Disposal*. US Patent 2665249.

Zimmermann, F.J. (1958) New waste disposal process. *Chem. Eng.* **August 25**, 117–120.

Zimpro (1999) WAO (Wet Air Oxidation) Units, Installation List, USFilter/Zimpro, February.

11

AOPs for VOCs and odour treatment

K. Van Craeynest, H. Van Langenhove and R.M. Stuetz

11.1 INTRODUCTION

In the European Solvent Directive, volatile organic compounds (VOCs) are defined as organic compounds which have a vapour pressure above 10 Pa at 25°C. Another, effect-oriented definition used by the United States Environmental Protection Agency defines VOCs as organic compounds that participate in atmospheric photochemical reactions. Both definitions cover a broad variety of organic compounds with diverse chemical properties. VOCs include hydrocarbons, olefins, aromatics and various oxygen-, nitrogen-, sulphur- and halogen-containing molecules. Their effects on human health vary from a simple nuisance to a serious hazard (Hunter and Oyama, 2000).

VOCs are among the most common air pollutants. VOCs emissions result from a broad range of human activities including (Hewitt, 1999):

- mining, treatment, storage, distribution and combustion of fossil fuels;
- solvent use (e.g. paints, thinners, printing inks, adhesives and degreasers);

- industrial processes (e.g. production of organic chemicals, chipboard, food and beverages);
- biological processes (e.g. agriculture, municipal waste and wastewater treatment).

In the atmosphere, VOCs contribute to several major environmental problems: photochemical ozone formation, global warming and stratospheric ozone depletion (Calvert, 1994). Furthermore, some of these compounds and their degradation products may be important in the epidemiology of respiratory disorders and cancer (Conroy and Forst, 1998; Larson and Weber, 1994). Next to this, several VOCs have very low odour thresholds and can cause severe odour nuisance. Although odour nuisance may appear of little importance compared to the other effects mentioned, it is often the most direct confrontation of the public with air pollution. Hence, complaints about environmental odours are among the most frequently registered official complaints (Van Langenhove *et al.*, 2000). Combined with a growing public awareness on health and environmental issues, these insights have led to strict legislation on VOCs emissions (European Solvent Directive, United States Clean Air Act).

In this context, there has been a growing interest in VOCs emission control measures. In fact, a survey among American chemical engineers (Moretti and Mukhopadhyay, 1993) has shown that regulatory compliance is the main driving force behind VOC abatement technology.

From an environmental point of view, process and equipment modifications (e.g. use of waterborne paints or powder paints, and ultrasonic cleaning) that prevent VOC emissions are the most desirable. In some processes however, end-of-pipe techniques remain necessary. Traditional treatment technologies include:

- condensation,
- membrane separation,
- adsorption,
- absorption,
- thermal oxidation,
- catalytic oxidation,
- biotechnological treatment.

All these techniques operate in a competitive way in a limited field of application (Dvorak *et al.*, 1996). If the recovered VOCs can be reused, recovery techniques are the most favourable from an environmental point of view. In most applications, however, it is economically more suitable to combine recovery techniques with disposal or destruction.

Next to this, there is a continued effort in the development of new abatement technologies. One possibility is the adaptation of advanced oxidation processes (AOPs), that were initially developed as water treatment techniques, for VOCs and odour abatement. AOPs which have been applied in air treatment include: direct photolysis, the ozone/ultraviolet (UV) process, heterogeneous photocatalysis, non-thermal plasma processing and advanced oxidative scrubbers (AOSs).

This chapter presents an introduction to the application of AOPs for VOCs and odour abatement. The fundamentals of the different treatment methods are presented as well as specific case studies, examples and some economics.

11.2 DIRECT UV PHOTOLYSIS

11.2.1 Principle

UV photons with energies from 4 to 7 eV (300–175 nm, respectively) can break molecular bonds through photodissociation (Wekhof, 1991). In this way, many VOCs can undergo direct photolysis in the presence of UV radiation. The efficiency of direct photolysis depends on a number of factors:

- molar absorptivity of the VOCs at the used wavelength (Figure 11.1),
- quantum yield of the reaction,
- intensity of the UV source.

The chemical and photochemical processes that take place in gas-phase photo-oxidations have been widely studied. Most of these studies have been directed primarily towards improving the understanding of atmospheric photochemistry. The conditions prevailing in gas treatment, however, are quite different from those applied in atmospheric studies. Photo-oxidation mechanisms at higher concentration levels frequently include radical chain reactions that result in higher rates and apparent quantum yields.

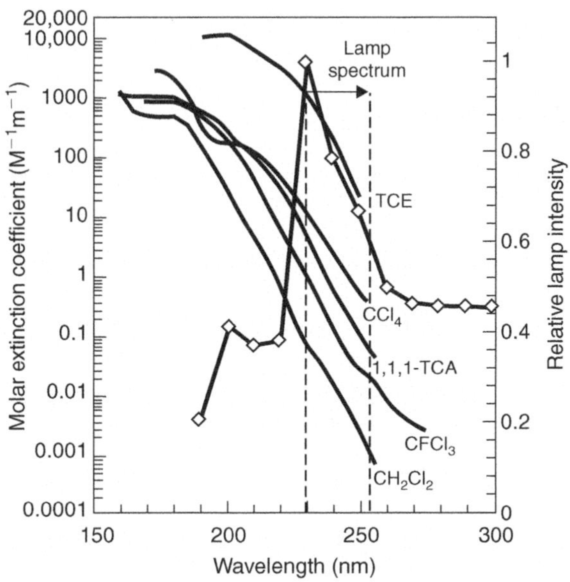

Figure 11.1 Emission spectrum of a UV flashlamp compared to absorption spectra of chlorinated VOCs (Johnson *et al.*, 1994).

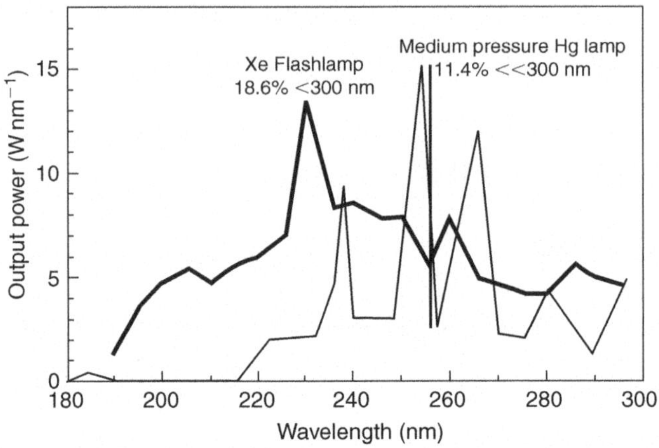

Figure 11.2 Emission spectra of flashlamps and medium-pressure mercury lamps (Johnson *et al.*, 1994).

11.2.2 UV sources

The low-pressure mercury lamps that are commonly used for disinfection purposes emit monochromatic light at 254 nm (a second emission line at 185 nm is absorbed by the encasing). This is effective for breaking the hydrogen bonds in DNA molecules that absorb light at 265 nm. However, for the cleaving of molecular bonds, more energetic deep UV photons (<250 nm) are more effective. Moreover, most VOCs (though not all, e.g. aromatics) have higher molar absorptivities at lower wavelengths (Glaze, 1993).

Several lamps that provide deep UV radiation are now commercially available (Figure 11.2):

- Low-pressure mercury lamps with high purity silica encasing that does not absorb the emission line at 185 nm.
- Medium- and high-pressure mercury lamps (second generation UV systems) with dopants to increase the number of UV lines in the deep UV region.
- Xenon flashlamps (third generation UV systems) that are based on pulsed corona discharge technology and emit a broad band of deep UV radiation.
- Xenon excimer lamps that emit monochromatic light at 172 nm (vacuum UV) that can dissociate water molecules.

11.2.3 Applications

11.2.3.1 VOCs

Most research in this field has focussed on chlorinated VOCs. While these compounds pose major difficulties to traditional treatment technologies, they are interesting target compounds in photolysis. Due to the low bonding energy of the C—Cl bond (compared to C—C and C—H bonds), they are more rapidly photolysed by UV photons. Furthermore, $Cl^{\bullet}$ atoms formed by this reaction can initiate chain reactions resulting in high apparent quantum yields (Johnson *et al.*, 1994):

$$HClC{=}CCl_2 + h\nu \rightarrow HClC{=}CCl^{\bullet} + Cl^{\bullet}$$

$$Cl^{\bullet} + HClC{=}CCl_2 \rightarrow HCl_2C{-}CCl_2^{\bullet}$$

$$HCl_2C{-}CCl_2^{\bullet} + O_2 \rightarrow HCl_2C{-}CCl_2OO^{\bullet}$$

$$2HCl_2C{-}CCl_2OO^{\bullet} \rightarrow 2HCl_2C{-}CCl_2O^{\bullet} + O_2$$

$$HCl_2C{-}CCl_2O^{\bullet} \rightarrow HCl_2C{-}CClO\ (DCAC) + Cl^{\bullet} \rightarrow$$

Bhowmick and Semmens (1994) compared the photo-oxidation of chloroform ($CHCl_3$), tetrachloromethane (CCl_4), trichloroethene/ethylene (TCE), tetrachloroethene/ethylene (PCE) and 1,1,2 trichloroethane (TCA) using a low-pressure mercury lamp with high purity silica encasing. They found that only the 185 nm emission line induced photolysis and that the reactivities of the VOCs decrease in the following order: TCE > PCE > CCl_4 > $CHCl_3$ > TCA.

Blystone *et al.* (1992) determined apparent quantum yields up to 31 for the photo-oxidation of TCE with a flashlamp. Wang and Ray (2000) observed an increase in photodegradation of VOCs with increasing moisture and $Cl^{\bullet}$ concentrations. Both phenomena point towards radical reaction mechanisms. Chlorine radicals generated in TCE photo-oxidation can sensitise degradation of other VOCs (1,1-dichloroethene) in gas mixtures (Haag *et al.*, 1996).

However, $Cl^{\bullet}$ chain reactions in TCE photo-oxidation yield products like dichloroacetyl chloride (DCAC), phosgene and trichloroacetyl chloride (TCAC) that are more toxic than TCE. Furthermore, DCAC and TCAC exhibit different UV absorption characteristics than TCE and require a longer exposure time to be fully mineralised (Figure 11.3), which can drastically increase the treatment costs (Haag *et al.*, 1996).

Johnson *et al.* (1994) constructed a pilot-scale photoreactor using Xenon flashlamps for the destruction of TCE (Figure 11.4). The system was operated at flows up to 300 scfm (260 ppmv TCE) and concentrations up to 10,600 ppmv (100 scfm). Residence times ranged from 5 to 75 s. In all cases, the observed TCE removal efficiencies were greater than 99%. However, photo-oxidation products DCAC and phosgene pose a major obstacle for full-scale operation and require further treatment.

Phosgene and DCAC can easily be removed in an alkaline scrubber, since they are rapidly hydrolysed into CO_2, HCl and dichloroacetic acid (DCAA) (Haag *et al.*, 1996; Yamaguchi *et al.*, 1998). To meet emission standards for DCAA in the scrubbing liquid, Haag *et al.* (1996) suggested a hydrothermal treatment of the effluent to achieve further dechlorination.

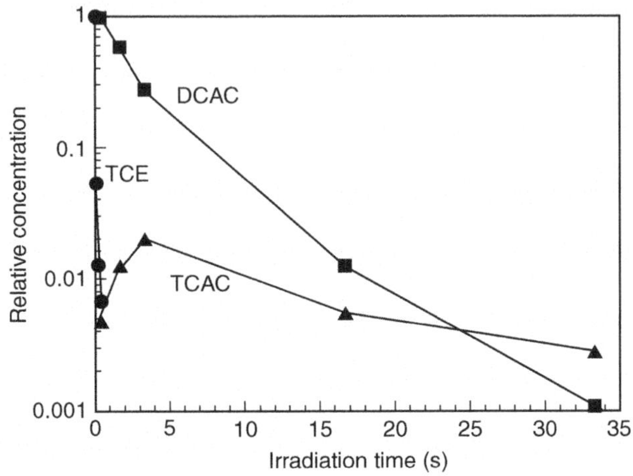

Figure 11.3 Product yields from the photolysis of TCE (Haag *et al.*, 1996). (Reprinted with permission from *Environ. Sci. Technol.* 1996, **30**, 414–421; © 1996 American Chemical Society.)

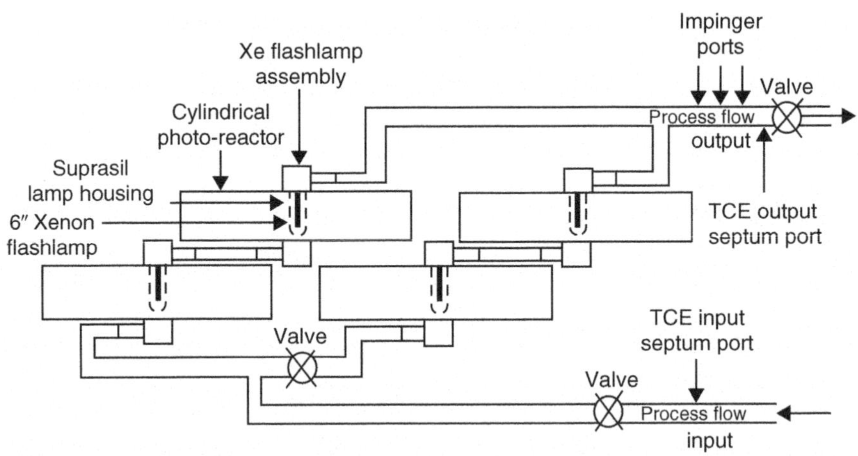

Figure 11.4 Pilot-scale photoreactors (Johnson *et al.*, 1994).

Gassiot *et al.* (1998) used a Xenon excimer lamp (172 nm) for the degradation of tetrahydrofurane (THF) in air. They found that THF degradation depended on oxidation by radicals formed in the photolysis of water vapour and that oxygen was needed for complete oxidation of THF even at high water concentrations.

A commercial VOCs photoxidation system is marketed by SolAirPur (www.SolAirPur.com) (Figure 11.5).

Figure 11.5 SolAirPur photo-oxidation reactor at coating plant (courtesy of SolAirPur, Belgium).

Figure 11.6 Hanovia UV photoreactor for odour abatement (courtesy of Hanovia, UK).

11.2.3.2 Odour

Photolysis of sulphurous compounds is initiated by the photodissociation of the H—S bond (McClean *et al.*, 1999):

$$H_2S + h\nu \rightarrow HS^{\bullet} + H^{\bullet}$$

$$RSH + h\nu \rightarrow RS^{\bullet} + H^{\bullet}$$

The radicals formed are effectively degraded further by reaction with oxygen to odourless compounds, such as sulphates and sulphonates, which are acceptable for release into the environment.

The UK company Hanovia (www.hanovia.co.uk) markets a commercial UV system for odour control in sewage systems and food processing plants (Figure 11.6).

11.2.4 Prospects

The prospects of direct photolysis for gaseous VOCs treatment are closely related to the evolutions in UV lamp technology, offering increased spectral and energetic efficiency. UV systems appear particularly interesting for the abatement of chlorinated VOCs. Furthermore, they offer a high degree of flexibility due to modular design and they require little space.

The formation of noxious photostable by-products forms a major obstacle for full-scale application. This means additional treatment is required to achieve complete mineralisation.

11.3 O_3/UV PROCESS

11.3.1 Principle

In this process, ozone is fed into a UV photoreactor. In the presence of water, UV-sensitised ozone decomposition yields $OH^•$ radicals that oxidise organic compounds (as low wavelength UV light generates ozone, it could be said that gas-phase direct UV photolysis processes are essentially UV/O_3 processes). Since ozone exhibits an absorption maximum at 254 nm, traditional UV sources can be applied in the O_3/UV process.

In the O_3/UV process, gaseous VOCs are degraded via a combination of reaction mechanisms including (Shen and Ku, 1998):

- direct UV photolysis,
- direct ozonation,
- indirect $OH^•$ radical oxidation,
- $Cl^•$ chain reactions.

The relative contribution of these mechanisms to VOCs degradation depends on UV radiation wavelength and intensity, ozone concentration, VOCs concentration and type, and the presence of moisture (Bhowmick and Semmens, 1994).

11.3.2 Applications

Again, most of the research in this field has focussed on chlorinated VOCs. McGregor *et al.* (1988) evaluated the use of an air-phase UV/O_3 oxidation process to treat the VOC laden off-gas from an air stripper. Bhowmick and Semmens (1994) investigated UV/O_3 oxidation of several chlorinated VOCs ($CHCl_3$, CCl_4, TCE, PCE, 1,1,2-TCA) using a low-pressure mercury lamp with high purity silica encasing. They found that direct ozonation contributed to the oxidation of chloroethenes and that $OH^•$ reactions contributed to the oxidation of $CHCl_3$, 1,1,2-TCA and to a lesser extend TCE. Neither ozone nor $OH^•$ significantly enhanced the oxidation of CCl_4. Shen and Ku (1997, 1998, 1999) applied UV/O_3 oxidation on several chlorinated compounds using a 254 nm low-pressure mercury lamp. They concluded that adding ozone slightly improved the oxidation of chlorinated

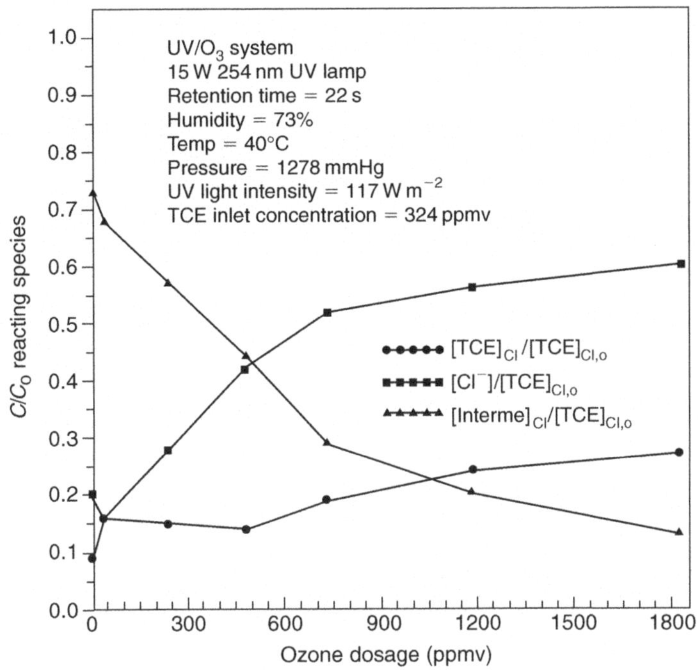

Figure 11.7 UV photo-oxidation of TCE in the presence of ozone. (Reprinted with permission from Shen and Ku. *J. Hazard. Mater.* 1997, **54**, 189–200; © 1997 Elsevier Science.)

VOCs, promoting further dechlorination and mineralisation of photo-oxidation products (Figure 11.7).

Striebig *et al.* (1997) used a UV/O_3 photolytic reactor in a pilot-scale hybrid air pollution control system. They observed significant degradation of TCE, but little to no effects on non-chlorinated VOCs (ethanol, ethylbenzene and methyl ethyl ketone).

11.3.3 Prospects

The UV/O_3 process faces similar obstacles as direct photolysis. Although ozone promotes further dechlorination, complete mineralisation is difficult. Thus, additional treatment remains necessary. Furthermore, ozone needs to be removed from the gas stream before discharge.

11.4 HETEROGENEOUS PHOTOCATALYSIS

11.4.1 Principle

Semiconductors can act as photocatalysts, sensitising the oxidation of organic compounds. If a semiconductor is irradiated, photons with an energy higher than the band gap energy of the semiconductor can excite electrons from the valence band into the conduction band, leaving a positively charged vacancy or hole in the valence band. The combination of both is referred to as an electron–hole pair. Many of the electron–hole pairs will recombine as the electron returns to its original state, emitting light or heat. However, some pairs may migrate to the catalyst surface, where they can participate in redox reactions with compounds adsorbed at the surface. The reaction products desorb from the surface back into the bulk medium (Chapter 5).

Sometimes, ozone is added to enhance the rate of the surface reduction step (Gilbert, 2002; Shen and Ku, 2002).

Many semiconductors (e.g. TiO_2, ZnO, Fe_2O_3, CdS and ZnS) can act as photocatalysts. However, TiO_2 is most commonly applied in environmental applications for a number of reasons (Fu *et al.*, 1996):

- It has high absorptivities for UV radiation. The band gap energy is in the range of 3.0–3.2 eV, corresponding with near UV radiation (300–370 nm) produced by UV black light lamps.
- It exhibits good adsorption characteristics for many contaminants.
- The redox potential of its valence and conduction band edges are suitable for oxidising organic contaminants.
- It displays high photocatalytic activity, while being highly resistant to photocorrosion.
- It is inexpensive and non-toxic.

The efficiency of near UV in photocatalytic processes enables the use of sunlight as an energy source. Only 5% of the solar spectrum is of use for the TiO_2 band gap (Romero *et al.*, 1999). However, the catalyst efficiency may be enhanced by doping TiO_2 with metals to expand the absorption spectrum into the visible light (Dvoranova *et al.*, 2002).

In water-phase applications, the catalyst surface is saturated with water molecules. Photocatalytic reactions yield hydroxyl radicals that can oxidise organic contaminants. In gas-phase applications, direct reactions of adsorbed contaminants are more important. In general, photo-oxidation rates obtained in gas–solid photoreactors are much higher than those typically reported for liquid–solid reactors. In fact, quantum yields higher than one may occur in gas-phase photooxidation (Lichtin and Avudaithai, 1996).

11.4.2 Reactor configurations

A large number of reactor configurations (e.g. powder layer, transport reactor, fluidised bed, honeycomb monolith, etc.) have been applied in laboratory-scale studies of photocatalytic oxidation processes (Peral *et al.*, 1997).

Next to more traditional reactor aspects (e.g. pressure drop, mixing, etc.), photocatalyst reactors should exhibit a large catalyst surface area and an efficient irradiation of this surface.

Since TiO_2 may be coated or synthesised on almost any surface, photocatalyst reactors could be produced in any geometry. Catalyst coated optical fibres have been tested for liquid and gas treatment (Peral *et al.*, 1997). In another approach, TiO_2 is coated on adsorbents to increase contaminant concentrations at the catalyst surface (Nozawa *et al.*, 2001). However, it is believed that with maturation of this technology, large-scale commercial applications will be limited to conventional heterogeneous reactor configurations (e.g. fixed and fluidised beds, transport reactors and monoliths) (Peral *et al.*, 1997).

Another interesting aspect is the application of sunlight as an energy source, which will require the development of innovative reactor configurations (Romero *et al.*, 1999).

11.4.3 Applications

Heterogeneous photocatalysis has been a hot topic during the last decade. Although the focus has been mainly on water-phase applications, there is a rapidly increasing amount of work on the oxidation of gaseous VOCs and odours. At the United States National Renewable Energy Laboratory (NREL), attempts are made to keep track of the work in this field. The NREL has published a series of bibliographies (Blake, 1994, 1995, 1997, 1999, 2001) listing over 6200 citations.

11.4.3.1 VOCs

Due to the large number of publications in this field, the focus will be on pilot-scale applications. References to laboratory-scale research can be found in NREL bibliographies (Blake, 1994, 1995, 1997, 1999, 2001).

Watt *et al.* (1999) conducted a pilot-scale study with a non-tracking solar-powered photocatalytic oxidation unit. The system effectively treated off-gases from an air stripper, achieving 95% destruction of TCE for a 10 scfm air stream loaded with 10–15 ppmv TCE. Dichloroethenes and possibly benzene (C_6H_6) were also photocatalytically oxidised.

Read *et al.* (1996) field tested a pilot-scale TiO_2 pellet-based reactor using VOCs-contaminated air (5000 ppmv) from a soil vapour extraction well. The primary contaminants were TCE and PCE. The reactor was operated at a flow rate ranging from 5.0×10^{-4} to 6.0×10^{-3} scmm; UV light was provided by fluorescent UV black light lamps. Removal efficiencies exceeding 97% were achieved when the reactor was operated with the undiluted air stream. However, many undesirable by-products (phosgene, $CHCl_3$, CCl_4, pentachloroethane and hexachloroethane) were yielded under these conditions. Diluting the feed stream to concentrations below 1000 ppmv reduced by-product concentrations to below 1 ppmv (except hexachloroethane).

Two more pilot systems were evaluated in a large-scale test of different off-gas technologies for abatement of chlorinated VOCs (Jarosch *et al.*, 1994). Off-gases containing a few thousand ppm chlorinated VOCs (mainly TCE and PCE) were treated. A reactor from the University of Wisconsin operated at flow rates up to 51pm achieved degradation efficiencies over 99%. For a second system from NuTech, operated at flow rates from 12 to 55 scfm, degradation efficiencies

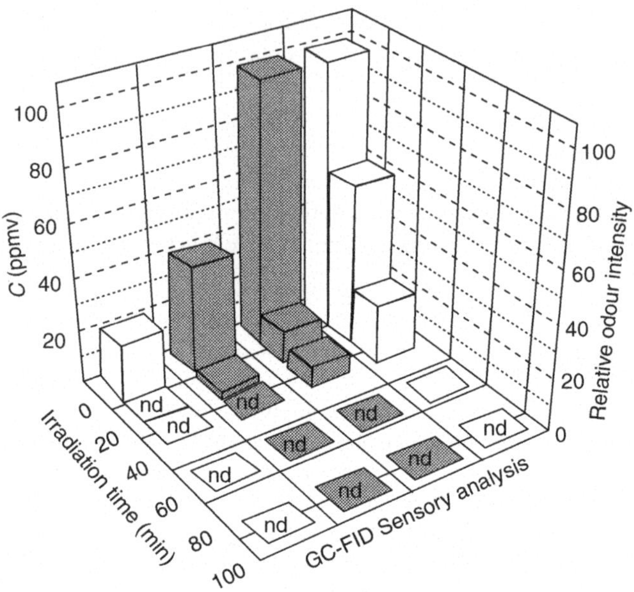

Figure 11.8 Confirmation of photocatalytic oxidation of propylene sulphide (white) and trimethylene sulphide (grey) by sensory analysis (Canela, 1999). (Reprinted with permission from *Environ. Sci. Technol.* 1999, **33**, 2788–2792; © 1999 American Chemical Society.)

ranged from 50% to 99%. In both systems, hexachloroethane was found to be an important by-product.

Commercial-scale photocatalytic oxidation systems developed by KSE, Inc. and Matrix Photocatalytic, Inc. (formerly NuTech) have been tested in the Superfund Innovative Technology Evaluation Program of the United States Environmental Protection Agency.

11.4.3.2 Odour

Canela *et al.* (1998, 1999) have evaluated the use of heterogeneous photocatalysis for the destruction of odorous sulphur compounds (H_2S, trimethylene sulphide, propylene sulphide, thiophene and dimethyl disulphide). Degradation efficiencies of 99% were observed. Odour removal was confirmed by sensory analysis (Figures 11.8 and 11.9).

The main degradation products were sulphate and CO_2. SO and SO_2 were detected as partial oxidation products from thiophene and dimethyl disulphide (Figure 11.9).

At high H_2S concentrations (>600 ppm), catalyst deactivation occurred, experiments with organic sulphur compounds were performed at lower concentrations (<100 ppm). Vorontsov *et al.* (2001) also observed catalyst deactivation in the photocatalytic oxidation of gaseous diethyl sulphide at a concentration in the order of hundreds of ppm. They did not observe a high degree of mineralisation.

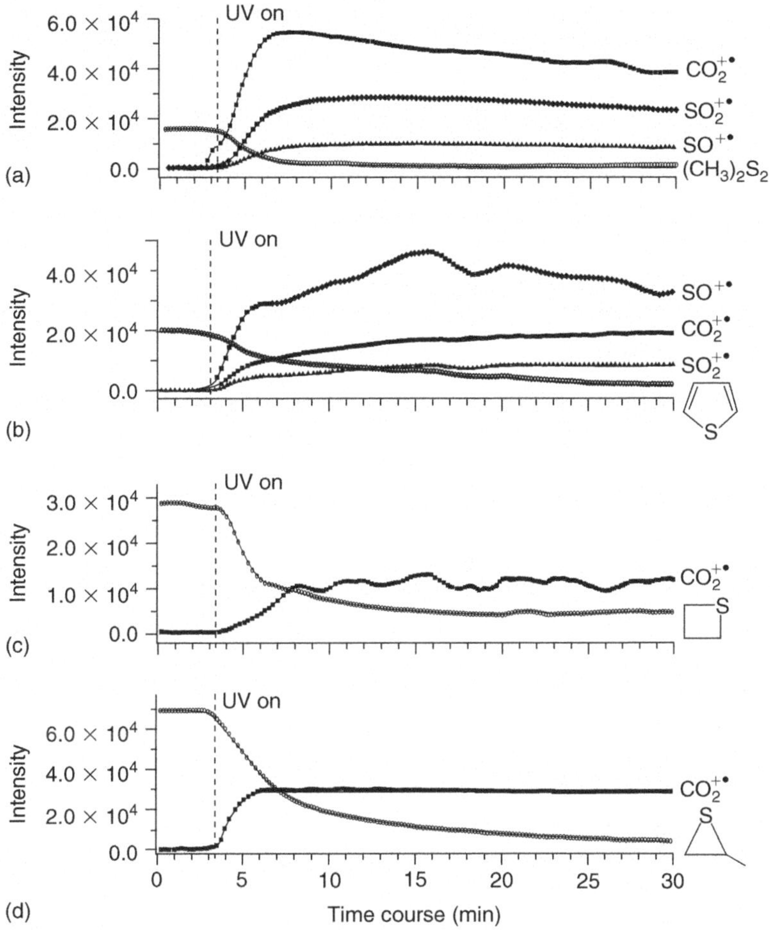

Figure 11.9 Degradation of dimethyl disulphide (Canela *et al.*, 1998). (Reprinted with permission from *Environ. Sci. Technol.* 1999, **33**, 2788–2792; © 1999 American Chemical Society.)

Nozawa *et al.* (2001) used a TiO_2 catalyst supported on fibre activated carbon for the deodorisation of ammonia, methyl mercaptan and H_2S. Pichat *et al.* (2000) constructed and tested a prototype air purifier/deodoriser for the treatment of indoor air.

11.4.4 Prospects

Heterogeneous photocatalysis provides a promising approach for the treatment of VOC loaded air streams. However, complete mineralisation appears difficult.

By-products formed during VOCs degradation may require further treatment and may deactivate the catalyst.

For full-scale application, more work is needed in the field of reactor design and optimisation of heterogeneous photoreactors. Also, catalyst efficiencies and relative reaction rates remain low, resulting in high energy requirements. Catalyst improvement and the use of solar power may resolve these problems.

11.5 NON-THERMAL PLASMA

11.5.1 Principle

Non-thermal plasmas are characterised by the electrons in the plasma having a higher mean energy ($T_e \gg 1$ eV) than the other constituents of the gas (ions or atoms and molecules) ($T_g < 0.1$ eV). This is compared to a thermal plasma, where the constituents of the gas are in thermal equilibrium (Urashima and Chang, 2000).

Through electron-impact dissociation and ionisation of the background gas molecules, the energetic electrons produce free radicals (oxygen and hydroxyl radicals) and additional electrons which, in turn, oxidise or decompose the VOC molecules (Penetrante *et al.*, 1996). Electrons may also react directly with the pollutants, which allows the treatment of perhalogenated saturated compounds (e.g. freons).

11.5.2 Plasma sources

Non-thermal plasmas can be generated by electron-beam irradiation or electrical discharges.

In the electron-beam process, high energy electrons are generated outside the reactor chamber under high vacuum in an electron accelerator. The electron beam generated is injected into the reaction chamber through a thin foil titanium window.

In the electrical discharge process, high energy electrons are produced by an electric field generated between two high-voltage electrodes within the reactor chamber. The plasma is generated by accelerating electrons in the field. Under atmospheric-pressure conditions, collisions with other gas constituents will transfer energy from the electrons to the rest of the gas. Local heating of the gas will result in a spark discharge in which the plasma is confined to a small filament. However, different reactor configurations have been developed that extend non-thermal discharge plasma processing to atmospheric-pressure conditions (Figure 11.10) (Urashima and Chang, 2000):

- In dielectric barrier discharges (e.g. silent discharge, surface discharge and dielectric packed bed), sparks are prevented by a dielectric barrier between the metal electrodes.
- In pulsed corona discharges, the applied voltage time during pulses is too short to generate spark discharges.

- In flow stabilised discharges (e.g. corona torch, capillary tube, corona radical injection and shower), a fast gas flow cools the electrodes, preventing spark discharges.

The characteristics of different reactor configurations are listed in Table 11.1.

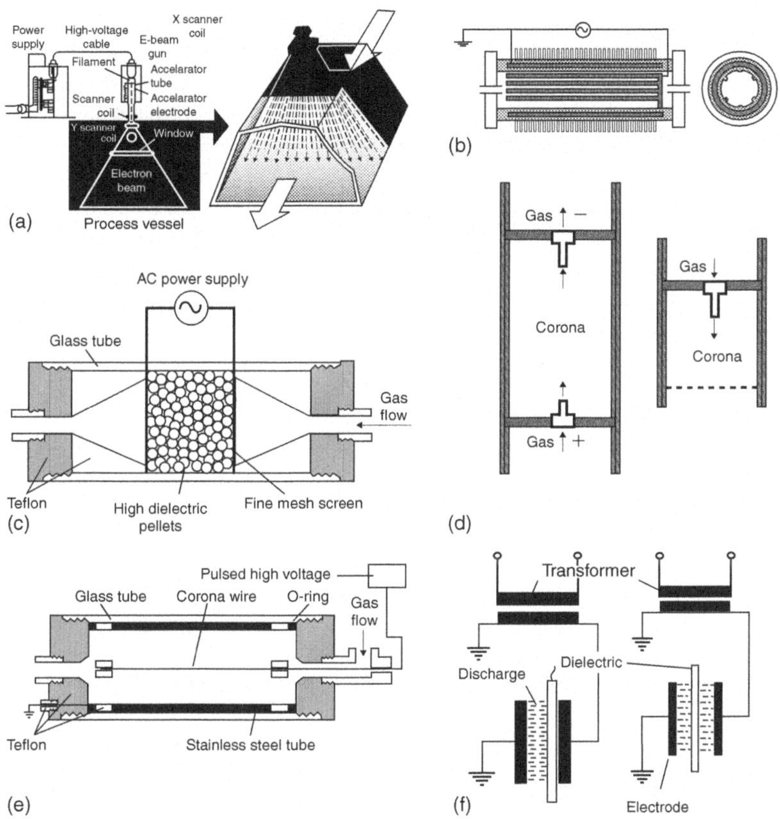

Figure 11.10 Non-thermal plasma reactor configurations: (a) electron beam (Urashima and Chang, 2000); (b) surface discharge (Oda *et al.*, 1996); (c) dielectric packed bed (Urashima and Chang, 2000); (d) corona torch (Urashima and Chang, 2000); (e) pulsed corona (Yamamoto *et al.*, 1992); (f) silent discharge (Urashima and Chang, 2000) (© 1992, 1996, 2000 IEEE).

Table 11.1 Characteristics of plasma sources (Urashima and Chang, 2000).

Plasma	Voltage (V)	Current (A)	Efficiency (%)	Pressure drop	Flow (m^3/h)
Electron beam	10^5–2×10^8	10^{-3}–10 (dc)	80–95	Small	10^3–10^5
Barrier discharge	5×10^3–2×10^4	10^{-3}–10	30–80	Large	10^{-2}–1
Pulsed corona	3×10^4–2×10^5	10^{-2}–1	20–70	Middle	1–10^2
Flow stabilised	10^4–10^5	10^{-5}–10^{-1}	90–95	Middle	1–10^2

11.5.3 Applications

Several reactor types have been applied for the abatement of a variety of VOCs. This research includes some pilot-scale test, other researchers have focussed on non-thermal plasma processing of odorous compounds.

11.5.3.1 VOCs

One of the first laboratory-scale studies on the application of non-thermal plasmas for the removal of VOCs from contaminated air streams was made by Slater and Hamilton (1981). They used a plasma generated by electron-beam irradiation for the destruction of vinyl chloride. In 1992, Yamamoto *et al.* applied discharge plasmas for VOCs abatement. They achieved complete destruction of toluene (TOL), 95% degradation of dichloromethane ($MeCl_2$) and 67% degradation of trichlorotrifluoroethane using a pulsed corona discharge reactor and a dielectric packed bed reactor. Since then, a number of papers have been published on degradation of VOCs in several discharge reactors and electron-beam reactors (Table 11.2).

Table 11.2 Non-thermal plasma processing of VOCs.

Reactor type[1]	VOCs	Reference
EB	Vinyl chloride	Slater and Hamilton (1981)
PCD, DPB	TOL, $MeCl_2$, trichlorotrifluoroethane	Yamamoto *et al.* (1992)
EB	CCl_4	Bromberg *et al.* (1993)
SD	TCE	Evans *et al.* (1993)
SD	Formaldehyde	Chang and Lee (1995)
PCD, SD	Methanol (MeOH), TCE	Hsiao *et al.* (1995)
EB	CCl_4, TCE	Koch *et al.* (1995)
EB, PCD	CCl_4	Penetrante *et al.* (1995)
SUD	Freon 113, TCE	Oda *et al.* (1996)
DPB	CCl_4	Tonkyn *et al.* (1996)
EB	1,1,1-TCA	Vitale *et al.* (1996)
EB, PCD	CCl_4, $MeCl_2$, TCE, C_6H_6, acetone, MeOH	Penetrante *et al.* (1996)
DPB, PCD	TCE	Futamura and Yamamoto (1997)
PCD, SD	TCE, methyl ethyl ketone, chloromethane	Korzekwa *et al.* (1997)
SD	TOL, TCE	Urashima *et al.* (1997)
PCD	Styrene, ethane (C_2H_4), TOL, pentane, TCA, propane, butane	Van Heesch *et al.* (1997)
EB	Ethyl chloride, vinyl chloride	Vitale *et al.* (1997)
CT	TOL, TCA, TCE	Kohno *et al.* (1998)
DPB	TCE	Oda *et al.* (1999)
SD	Freon 21, freon 142B	Spiess *et al.* (2000)
DPB, PCD, SD	TCE, tetrafluoromethane, bromomethane	Futamura *et al.* (2001)
DPB	TCE	Oda *et al.* (2001)
DPB	TCE	Oda *et al.* (2002)

[1]PCD: pulsed corona discharge; DPB: dielectric packed bed; SD: silent discharge; EB: electron beam; SUD: surface discharge; CT: capillary tube.

In a series of studies, Penetrante *et al.* (1995, 1996, 1997) have compared the energy efficiencies of discharge reactors (silent discharge and pulsed corona discharge) and an electron-beam reactor for the VOCs abatement. It was found that the electron-beam process required much lower energy densities to achieve similar removal efficiencies as the discharge processes (Figure 11.11).

Several authors have suggested hybrid or superimposed systems combining non-thermal plasma processing with adsorbents, catalysts and photocatalysts (Urashima and Chang, 2000). Kang *et al.* (2002) evaluated a dielectric packed bed reactor, packed with TiO_2 coated glass beads. They found that the combined plasma/TiO_2 system performed better than a simple plasma or photocatalytic system (Figure 11.12).

Anshumali *et al.* (1997) constructed a pilot-scale electron-beam reactor based on commercially available technology (Zapit, Inc.). Dry air contaminated with various VOCs (freon 112, propane, 1,1,1-TCA, vinyl chloride, TCE and PCE) in the concentration range of 50–1000 ppm was treated in the pilot system (flow rates were not specified). Reaction by-products were treated using an alkaline scrubber and an activated carbon adsorber.

Prager *et al.* (1998) field tested a mobile pilot-scale electron-beam system using air-stripper off-gas from groundwater remediation. The stripper off-gas contained low concentrations of dichloroethenes, 1,1,1-TCA, TCE and PCE with a total VOC concentration of about 3 ppmv. The system was operated at flow rate of $1000\,m^3\,h^{-1}$ and a mean residence time of 0.2 s. At the applied conditions, the system achieved about 90% removal of total VOCs. Reaction by-products (chloroacetyl chlorides and phosgene) were removed using a dry scrubbing process and ozone formed was removed by an activated carbon filter.

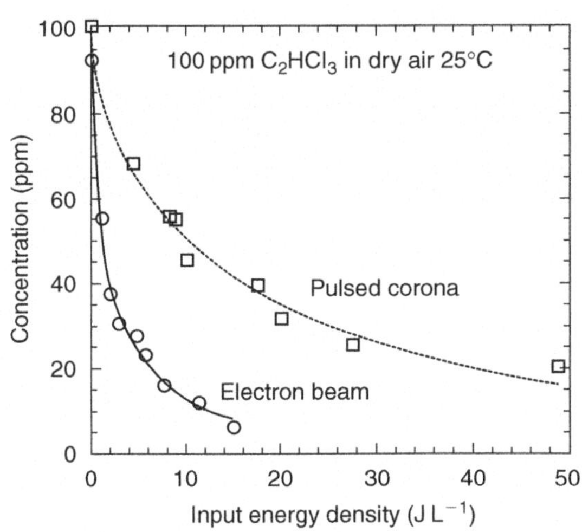

Figure 11.11 Comparison of electron beam and pulsed corona discharge process for TCE degradation. (Reprinted with permission from Penetrante *et al. Pure Appl. Chem.* 1996, **68**, 1083–1087; © 1996 IUPAC.)

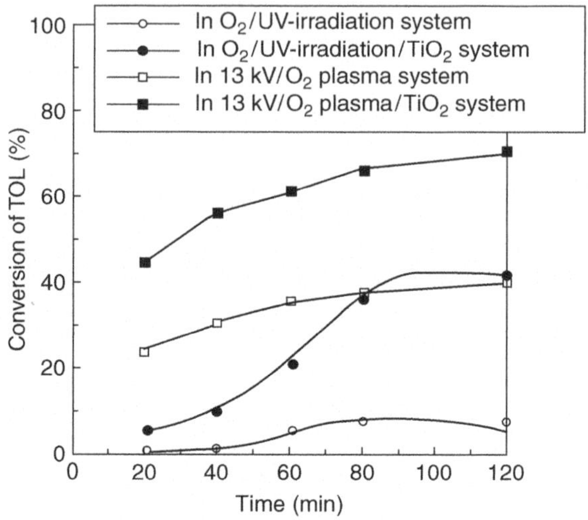

Figure 11.12 Comparison of toluene decomposition under plasma system and photocatalytic system. (Reprinted with permission from Kang *et al. J. Mol. Catal. A-Chem.* 2002, **180**, 125–132; © 2002 Elsevier Science.)

Figure 11.13 Non-thermal discharge reactor operating at an animal house (courtesy of Ruan R.).

Tipaz, Inc. (formerly Zapit, Inc.) markets commercial electron-beam systems for VOC abatement (www.zetron.net). Another US company (PlasmaSol) markets a capillary discharge non-thermal plasma system for the same purpose (www.plasmasol.com). An example plant can be seen in Figure 11.13.

11.5.3.2 Odour

Okubo *et al.* (2001a,b) developed an indoor air cleaner composed of a dielectric packed bed reactor and an electrostatic precipitator. The study focussed on two important odorous components in cigarette smoke: acetaldehyde and NH_3 as well as cigarette smoke aerosol particles. With the proposed air cleaner, complete removal of NH_3 and over 95% removal of acetaldehyde were achieved.

Chang and Tseng (1996) evaluated the effectiveness of a silent discharge process for the removal of H_2S and NH_3. Simultaneous removal efficiencies for H_2S (50 ppmv) and NH_3 (100 ppmv) of 100% and 88%, respectively, were observed.

Several studies have been directed towards odour control in livestock keeping. Zhang *et al.* (1996) used a dielectric packed bed reactor in decomposing NH_3 and odours in the air taken from a swine house. Another research group tested a silent discharge reactor for removal of H_2S and NH_3 (Ma *et al.*, 2001; Ruan *et al.*, 1999).

Tamon *et al.* (1995, 1996) proposed a gas-purification system based on electron attachment for the abatement of sulphur compounds (SF_6, H_2S, CH_3SH, $(CH_3)_2S$, CS_2, COS and SO_2). In the proposed system, the sulphur compounds are ionised by the attachment of electrons produced in a corona discharge reactor. The negative ions formed drift to the anode. Removal of the sulphur compounds was due to a combination of non-thermal plasma oxidation and electron attachment.

11.5.4 Prospects

At this stage, attempts are still made to establish the capabilities of non-thermal plasma processing and to identify where it is a sensible alternative to existing technologies. However, it is clear that the technology shows some distinct advantages (McAdams, 2001):

- Simultaneous destruction of different pollutants (VOCs, NO_x and SO_x).
- Plasma systems are electrically driven and require little warm-up (compared to catalytic systems), so they can be operated in batch mode.
- Pulsed corona systems are compact and low in mass. Their open structure results in little back pressure. This offers potential for indoor applications or vehicle exhaust treatment.
- Modularity.

Although non-thermal plasma systems are capable of destroying most VOCs, a significant amount of unwanted by-products (e.g. O_3, NO_x, CO and phosgene) may be formed. Therefore, one should consider an integrated environmental approach combining plasma systems with catalysts, adsorbents or scrubbers.

11.6 AOSs

11.6.1 Principle

Scrubbers are widely used for the removal of pollutants from contaminated air streams (McGinley, 1998). The absorption process is driven by the non-equilibrium concentration in the liquid phase. Therefore, the application of physical scrubbers is limited to highly soluble contaminants. In reactive scrubbers (alkaline, acid or oxidative scrubbers), reagents are added to the scrubbing liquid (Smet *et al.*, 1998). In this way, the contaminants react upon absorption and the driving force is maintained. Very fast reactions can even enhance the mass transfer (Lawson and Adams, 1999).

Traditional oxidative scrubbers using O_3, Cl_2, NaOCl and H_2O_2 have been very successful in the treatment of reduced malodorous compounds (e.g. mercaptans, organic sulphides and amines) (VDI Richtlinie, 1995). In an AOS, hydroxyl radicals generated via an AOPs are used as oxidants instead of traditional oxidants. Due to the high reactivity of hydroxyl radicals, AOSs can also be used for less reactive compounds. Theoretically, any AOP used in water treatment could be applied in AOSs. Since the constitution of liquid phase can be freely adjusted, typical problems reducing oxidation rates in AOP water treatment (e.g. scavengers and pH) are easily avoided. Due to their high reactivity, hydroxyl radicals are very short lived. Therefore, hydroxyl radicals should not be dosed in the scrubbing liquid storage tank (which is done in traditional oxidative scrubbers), but initiation of the AOP should take place in the scrubber reactor itself.

11.6.2 Applications

Two AOPs have been applied in AOSs, UV/H_2O_2 and peroxone (O_3/H_2O_2). Alibegic *et al.* (2001) used an AOS based on the UV/H_2O_2 process for the removal of PCE (70–530 ppmv). H_2O_2 was dosed via the scrubbing liquid and contacted with the waste-gas stream in a 254 nm UV-bubble column reactor. Depending on the applied conditions, degradation rates were in the range of 50–99%, no mass transfer enhancement could be observed. Small amounts of $CHCl_3$ could be detected in runs with high inlet concentrations. These were probably due to gas-phase reactions.

Lawson and Adams (1999) treated VOC-contaminated air streams with the peroxone process in a packed column. Tested contaminants include C_6H_6, TOL, xylenes, PCE, TCE and 1,1,1-TCA. They dosed H_2O_2 via the scrubbing liquid and ozone via the gas and could observe significant mass transfer enhancement.

Dewulf *et al.* (2000) and Van Craeynest *et al.* (2002) applied the same process for the abatement of TCE in a bubble column. Removal efficiencies up to 98% could be observed.

11.6.3 Prospects

Although little research has been performed in this field, AOSs could rapidly move up to full-scale application. Knowledge gained in the widespread application of oxidative scrubbers could be adapted to AOSs. Thus, technical questions like those facing photocatalytic and non-thermal plasma reactor configurations are avoided. Also, the reactions taking place in AOSs are essentially water-phase reactions that are well known from water treatment applications and yield less noxious by-products than their gas-phase counterparts.

11.7 ECONOMICS

As all of the discussed AOPs are still in the development stage, commercial systems are rare and reliable economic data on full-scale operation is hard to find. Although several studies include cost estimates, these extrapolations are often highly speculative. Predicting capital costs and operating reliability of full-scale systems is hard, nonetheless, both can significantly contribute to the total cost. For these reasons, some researchers limit cost estimates to energy requirements. Since most of the discussed technologies have not been optimised with respect to energy use, even these estimates should be handled with reservation.

Another problem is that, for a number of reasons, cost estimates from different researchers are often difficult to compare. Not all estimates include capital costs. Numerous assumptions have to be made in the extrapolations, which may give rise to significant differences in the final results. Costs can be expressed per year, per pound VOCs removed, per cubic metre air treated, etc. Since research papers often do not feature complete calculations and assumptions, it is hard to establish if cost differences are due to the extrapolation or to technical issues.

In one study performed for the United States Department of Energy, Jarosch *et al.* (1994) field tested several off-gas treatment technologies for the abatement of chlorinated VOCs. Tests were conducted with a soil vapour extraction system at a site contaminated mainly with PCE, TCE and 1,1,1-TCA. Among others, the field test included four AOP systems: direct photolysis, photocatalytic oxidation (2) and non-thermal plasma. For emerging technologies, the economic analysis was restricted to a comparison of energy requirements (a full analysis was performed for thermal catalytic oxidation) (Table 11.3).

The website of United States Naval Facilities Engineering Service Center features a VOC off-gas treatment technologies database (www.nfesc.navy.mil) containing cost estimates for AOP systems and mature technologies (Table 11.4).

Several researchers have compared cost estimates for AOP systems to treatment costs associated with traditional VOCs control technologies. Haag *et al.* (1996) calculated a cost of \$3.94 h^{-1} for a combined photolytic treatment of an air-stripper off-gas (100 ppmv TCE at 1000 cfm), including hydrothermal treatment of the scrubbed photoproducts. They concluded that such a system could be competitive with a catalytic oxidation system at concentrations below 50 ppm.

Table 11.3 Comparison of energy requirements for emerging technologies and thermal catalytic treatment (95% conversion efficiency at 100 scfm) (Jarosch *et al.*, 1994).

Technology	Power use (kW)
Thermal catalysis	16–19
Direct photolysis	5
Heterogeneous photocatalysis	5–23
Silent discharge plasma	7–324

Table 11.4 Cost estimates for AOP systems for VOC off-gas treatment (NFESC).

Technology	Cost ($ per lb)	Inlet concentration	Capacity (cfm)
AOPs			
Heterogeneous photocatalysis	2.04–8.86	Up to 10,000 ppm	500
Direct photolysis	3.16–9.36	Up to 5000 ppm	20,000
Silent discharge plasma	2.15–10.63	–	–
Other			
Thermal oxidation	4.14–12.55	Up to 25% of LEL	60,000
Catalytic oxidation	2.91–11.08	Up to 25% of LEL	80,000
Adsorption	1.20–6.70	Below 5000 ppm	100,000

The data only applies to the systems that were included in the NFESC evaluation.
LEL: lower explosive limit.

Henschel (1998) made an elaborate cost analysis of a photocatalytic oxidation system for removing organic compounds from indoor air (0.27 mg VOC/m^3 at 1 m^3/s). He concluded that photocatalytic system would have an installation cost more than 10 times greater than a similar activated carbon system and an annual cost almost seven times higher. Watt *et al.* (1999) concluded that annual costs for off-gas treatment with a solar-powered photocatalytic oxidation system ($33,000) were competitive with the costs for thermal oxidation ($37,000) or activated carbon adsorption ($32,000).

Anshumali *et al.* (1997) performed a cost analysis of an electron-beam system with activated carbon post-treatment. They concluded that such a system could be competitive to single activated carbon treatment.

11.8 CONCLUSIONS

Although AOPs have been applied mainly in water treatment, these processes show potential in the treatment of VOCs and odours in air. Of the discussed AOPs, heterogeneous photocatalysis and non-thermal plasma are the most documented.

AOPs show some advantages compared to traditional treatment techniques:

- AOP systems generally have a small footprint.
- Modular design of AOP systems offers high flexibility.
- Oxidants are generated electrically on site, thus no storage room for chemicals is required.
- Furthermore, electrical systems can be operated in batch mode.

At this stage, most of the technology is still in the development phase and several issues will need to be solved for full-scale operation:

- The removal of noxious by-products formed in gas-phase radical chain reactions.
- There is a lack of transparent cost estimates.

- The reliability of the technologies need to be evaluated.
- Reactor configurations need to be optimised.

Despite these unresolved issues, the first commercial applications of AOPs for VOCs and odour treatment are already on their way.

REFERENCES

Alibegic, D., Tsuneda, S. and Hirata, A. (2001) Kinetics of tetrachloroethylene (PCE) gas degradation and byproducts formation during UV/H_2O_2 treatment in UV-bubble column reactor. *Chem. Eng. Sci.* **56**, 6195–6203.

Anshumali, Winkleman, B.C. and Sheth, A. (1997) Destruction of low levels of volatile organic compounds in dry air by an electron-beam generated plasma. *J. Air Waste Manage. Assoc.* **47**, 1276–1283.

Bhowmick, M. and Semmens, M.J. (1994) Ultraviolet photooxidation for the destruction of VOCs in air. *Water Res.* **28**, 2407–2415.

Blake, D.M. (1994) *Bibliography of Work on the Heterogeneous Photocatalytic Removal of Hazardous Compounds from Water and Air*. Technical Report, National Renewable Energy Laboratory, Colorado.

Blake, D.M. (1995) *Bibliography of Work on the Heterogeneous Photocatalytic Removal of Hazardous Compounds from Water and Air (Update Number 1 to June 1995)*. Technical Report, National Renewable Energy Laboratory, Colorado.

Blake, D.M. (1996) *Bibliography of Work on the Heterogeneous Photocatalytic Removal of Hazardous Compounds from Water and Air (Update Number 2 to October 1996)*. Technical Report, National Renewable Energy Laboratory, Colorado.

Blake, D.M. (1999) *Bibliography of Work on the Heterogeneous Photocatalytic Removal of Hazardous Compounds from Water and Air (Update Number 3 to January 1999)*. Technical Report, National Renewable Energy Laboratory, Colorado.

Blake, D.M. (2001) *Bibliography of Work on the Heterogeneous Photocatalytic Removal of Hazardous Compounds from Water and Air (Update Number 4 to October 2001)*. Technical Report, National Renewable Energy Laboratory, Colorado.

Blystone, P.G., Johnson, M.D., Haag, W.R. and Daley, P.F. (1992) Advanced ultraviolet flashlamps for the destruction of organic contaminants in air. *ACS Symposium Series* **518**, 380–392.

Bromberg, L., Cohn, D.R., Koch, M., Patrick, C.M. and Thomas, P. (1993) Decomposition of dilute concentrations of carbon tetrachloride in air by an electron-beam generated plasma. *Phys. Lett. A* **173**, 293–299.

Calvert, J.G. (1994) *Chemistry for the 21st Century. The Chemistry of the Atmosphere: Its Impact on Global Change*. Blackwell Scientific Publications, Oxford.

Canela, M.C., Alberici, R.A. and Jardim, W.F. (1998) Gas-phase destruction of H_2S using TiO_2/UV–VIS. *J. Photoch. Photobio. A* **112**, 73–80.

Canela, M.C., Alberici, R.A., Sofia, R.C.R., Eberlin, M.N. and Jardim, W.F. (1999) Destruction of malodorous compounds using heterogeneous photocatalysis. *Environ. Sci. Technol.* **33**, 2788–2792.

Chang, M.B. and Lee, C.C. (1996) Destruction of formaldehyde with dielectric barrier discharge plasmas. *Environ. Sci. Technol.* **29**, 181–186.

Chang, M.B. and Tseng, T.D. (1996) Gas-phase removal of H_2S and NH_3 with dielectric barrier discharges. *J. Environ. Eng. ASCE* **121**, 41–46.

Conroy, L.M. and Forst, L. (1998) Health effects and exposure assessments of VOCs. In: *Odor and VOC Control Handbook*, Rafson H.J. (ed.). McGraw-Hill, New York.

Dewulf, J., Van Langenhove, H., De Smedt, E. and Geuens, S. (2000) Combination of advanced oxidation processes and gas absorption for the treatment of chlorinated solvents in waste gases. *Water Sci. Technol.* **44**(9), 173–180.

Dvorak, B.I., Herbeck, C.J., Meurer, C.P., Lawler, D.F. and Speitel, G.E. (1996) Selection among aqueous and off-gas treatment technologies for synthetic organic chemicals. *J. Environ. Eng. ASCE* **122**, 571–580.

Dvoranova, D., Brezova, V., Mazur, M. and Malati, M.A. (2002) Investigations of metal-doped titanium dioxide photocatalysts. *Appl. Catl. B: Environ.* **37**, 91–105.

Evans, D., Rosocha, L.A., Anderson, G.K., Coogan, J.L. and Kushner, M.J. (1993) Plasma remediation of trichloroethylene in silent discharge plasmas. *J. Appl. Phys.* **74**, 5378–5386.

Fu, X., Zeltner, W.A. and Anderson, M.A. (1996) Applications in photocatalytic purification of air. *Stud. Surf. Sci. Catal.* **103**, 445–461.

Futamura, S. and Yamamoto, T. (1997) Byproduct identification and mechanism determination in plasma chemical decomposition of trichloroethylene. *IEEE Trans. Ind. Appl.* **33**, 447–453.

Futamura, S., Einaga, H. and Zhang, A. (2001) Comparison of reactor performance in the nonthermal plasma chemical processing of hazardous air pollutants. *IEEE Trans. Ind. Appl.* **37**, 978–985.

Gassiot, I., Baus, C., Schaber, K. and Braun, A.M. (1998) VUV gas-phase photooxidation: a new tool for the degradation of VOCs in air. *J. Inf. Record.* **21**, 129–132.

Gilbert, Y.S. (2002) Influence of ozone on the photocatalytic oxidation of organic compounds. *Ozone Sci. Eng.* **24**, 75–82.

Glaze, W.H. (1993) An overview of advanced oxidation processes: current status and kinetic models. In: *Proceedings of the International Chemical Oxidation Association Symposium 'Chemical Oxidation: Technology for the Nineties'*, Nashville, USA, February 17–19.

Haag, W.H., Johnson, M.D. and Scofield, R. (1996) Direct photolysis of trichloroethene in air: effect of contaminants, toxicity of products, and hydrothermal treatment of products. *Environ. Sci. Technol.* **30**, 414–421.

Henschel, D.B. (1998) Cost analysis of activated carbon versus photocatalytic oxidation for removing organic compounds from indoor air. *J. Air Waste Manage.* **48**, 985–994.

Hewitt, C.N. (1999) *Reactive Hydrocarbons in the Atmosphere*. Academic Press, San Diego.

Hsiao, M.C., Merritt, B.T., Penetrante, B.M. and Vogtlin, G.E. (1995) Plasma-assisted decomposition of methanol and trichloroethylene in atmospheric pressure air streams by electrical discharge processing. *J. Appl. Phys.* **78**, 3451–3456.

Hunter, P. and Oyama, S.T. (2000) *Control of Volatile Organic Compound Emissions. Conventional and Emerging Technologies*. John Wiley & Sons Inc, New York.

Jarosch, T.R., Haselow, J.S., Rossabi, J., Burdick, S.A., Raymond, R., Young, J.E. and Lombard, K.H. (1994) *Final Report on Testing Off-Gas Treatment Technologies for Abatement of Atmospheric Emissions of Chlorinated Volatile Organic Compounds.* US Department of Energy.

Johnson, M.D., Haag, W., Blystone, P.G. and Daley, P.F. (1994) *Destruction of Organic Contaminants in Air using Advanced Ultraviolet Flashlamps*. Emerging Technology Report, Risk Reduction Engineering Laboratory Office of Research and Development, Environmental Protection Agency, Cincinnati, United States.

Kang, M., Kim, B.J., Cho, S.M., Chung, C.H., Kim, B.W., Han, G.Y. and Yoon, K.J. (2002) Decomposition of toluene using an atmospheric pressure plasma/TiO_2 catalytic system. *J. Mol. Catal. A-Chem.* **180**, 125–132.

Koch, M., Cohn, D.R., Patrick, R.M., Shuetze, M.P., Bromberg, L., Reilly, D., Hadidi, K., Thomas, P. and Falkos, P. (1995) Electron beam atmospheric pressure cold plasma decomposition of carbon tetrachloride and trichloroethylene. *Environ. Sci. Technol.* **29**, 2946–2952.

Kohno, H., Berezin, A.A., Chang, J.S., Tamura, M., Yamamoto, T., Shibuya, A. and Honda, S. (1998) Destruction of volatile organic compounds used in a semiconductor industry by a capillary tube discharge reactor. *IEEE T. Ind. Appl.* **34**, 953–966.

Korzekwa, R., Rosocha, L. and Falkenstein, Z. (1997) Experimental results comparing pulsed corona and dielectric barrier discharges for pollution control. In: *Proceedings of the 11th IEEE International Pulsed Power Conference*, Baltimore, USA, June 29–July 2.

Larson, R.B. and Weber, E.J. (1994) *Reaction Mechanisms in Environmental Organic Chemistry*. Lewiss Publishers, Boca Raton.

Lawson, R.B. and Adams, C.D. (1999) Enhanced VOC absorption using the ozone/hydrogen peroxide advanced oxidation process. *J. Air Waste Manage.* **49**, 1315–1323.

Ma, H.B., Chen, P. and Ruan, R. (2001) H_2S and NH_3 removal by silent discharge plasma and ozone-combo system. *Plasma Chem. Plasma P.* **21**, 611–624.

McAdams, R. (2001) Prospects for non-thermal atmospheric plasmas for pollution abatement. *J. Phys. D: Appl. Phys.* **34**, 2810–2821.

McClean, J.C., Hamilton, D.J. and Clark, N. (1999) Odour abatement using enhanced UV photolysis: a new application for a proven technology. In: *Proceedings of the CIWEM and IWAQ Conference on Control of Odoriferous Pollutants in Process Industries*, Vigneron, S., Hermia, J. and Chaouki, J. (eds). Elsevier Science, Amsterdam, pp. 471–496.

McGinley, C.M. (1998) Chemical scrubbing: packed columns. In: *Odor and VOC Control Handbook*, Rafson H.J. (ed.). McGraw-Hill, New York.

McGregor, F.R., Piscaer, P.J. and Aieta, E.M. (1988) Economics of treating waste gases from an air stripping tower using photochemically generated ozone. *Ozone Sci. Eng.* **10**, 339–351.

Moretti, E.C. and Mukhopadhyay, N. (1993) VOC control. Current practices and future trends. *Chem. Eng. Prog.* **89**(7), 20–26.

Nozawa, M., Tanigawa, K., Hosomi, M., Chikusa, T. and Kawada, E. (2001) Removal and decomposition of malodorants by using titanium dioxide photocatalyst supported on fiber activated carbon. *Water Sci. Technol.* **44**(9), 127–133.

Oda, T., Takahashi, T. and Kohzuma, S. (1999) Decompostition of trichloroethylene by non-thermal plasma with catalyst. In: *Proceedings of the IEEE Industry Applications Society Annual Meeting*, Phoenix, USA, October 3–7.

Oda, T., Takahashi, T. and Kohzuma, S. (2001) Decomposition of dilute trichloroethylene by using nonthermal plasma processing – frequency and catalyst effects. *IEEE Trans. Ind. Appl.* **37**, 965–970.

Oda, T., Takahashi, T. and Yamaji, K. (2002) Nonthermal plasma processing for dilute VOCs decomposition. *IEEE Trans. Ind. Appl.* **38**, 873–878.

Oda, T., Yamashita, R., Takahashi, T. and Masuda, S. (1996) Atmospheric pressure discharge plasma decomposition for gaseous air contaminants – trichlorofluoroethane and trichloroethylene. *IEEE Trans. Ind. Appl.* **32**, 227–232.

Okubo, M., Kuroki, T., Kametaka, H. and Yamamoto, T. (2001a) Odor control using the ac barrier-type plasma reactors. *IEEE Trans. Ind. Appl.* **37**, 1447–1455.

Okubo, M., Yamamoto, T., Kuroki, T. and Fukumoto, H. (2001b) Electric air cleaner composed of nonthermal plasma reactor and electrostatic precipitator. *IEEE Trans. Ind. Appl.* **37**, 1505–1511.

Penetrante, B.M., Hsiao, M.C., Bardsley, J.N., Merritt, B.T., Vogtlin, G.E., Wallman, P.H., Kuthi, A., Burkhart, C.P. and Bayless, J.R. (1995) Electron-beam and pulsed corona processing of carbon-tetrachloride in atmospheric-pressure gas streams. *Phys. Lett. A* **209**, 69–77.

Penetrante, B.M., Hsiao, M.C., Bardsley, J.N., Merritt, B.T., Vogtlin, G.E. and Wallman, P.H. (1996) Electron beam and pulsed corona processing of volatile organic compounds in gas streams. *Pure Appl. Chem.* **68**, 1083–1087.

Penetrante, B.M., Hsiao, M.C., Bardsley, J.N., Merritt, B.T., Vogtlin, G.E., Kuthi, A., Burkhart, C.P. and Bayless, J.R. (1997) Decomposition of methylene chloride by electron beam and pulsed corona processing. *Phys. Lett. A* **235**, 76–82.

Peral, J., Domenech, X. and Ollis, D.F. (1997) Heterogeneous photocatalysis for purification, decontamination and deodorization of air. *J. Chem. Technol. Biot.* **70**(2), 117–140.

Pichat, P., Disdier, J., Hoang-Van, C., Mas, D., Goutailler, G. and Gaysse, C. (2000) Purification/deodorization of indoor air and gaseous effluents by TiO_2 photocatalysis. *Catal. Today* **63**, 363–369.

Prager, L.R., Mehnert, A., Sobottka, H., Langguth, W., Baumann, H., Matzing, H.R., Paur, J., Schubert, R., Rashid, K.M., Taba, H.P., Schuchmann and Von Sonntag, C. (1998) Electron beam degradation of chlorinated hydrocarbons air-stripped from polluted ground water: a laboratory and field study. *J. Adv. Oxid. Technol.* **3**, 87–97.

Read, H.W., Fu, X.Z., Clark, L.A., Anderson, M.A. and Jarosch, T. (1996) Field trials of a TiO_2 pellet-based photocatalytic reactor for off-gas treatment at a soil vapor extraction well. *J. Soil Contam.* **5**, 187–202.

Romero, M., Blanco, J., Sanchez, B., Vidal, A., Malato, S., Cardona, A.I. and Garcia, E. (1999) Solar photocatalytic degradation of water and air pollutants: challenges and perspectives. *Sol. Energ.* **66**, 169–182.

Ruan, R., Han, W., Ning, A.R., Deng, S.B., Chen, P.L. and Goodrich, P. (1999) Effects of design parameters of planar, silent discharge, plasma reactors on gaseous ammonia reduction. *Trans. ASAE* **42**, 1841–1845.

Shen, Y.S. and Ku, Y. (1997) Treatment of gas-phase trichloroethene in air by the UV/O_3 process. *J. Hazard. Mater.* **54**, 189–200.

Shen, Y.S. and Ku, Y. (1998) Decomposition of gas-phase chloroethenes by the UV/O_3 process. *Water Res.* **32**, 2669–2679.

Shen, Y.S. and Ku, Y. (1999) Treatment of gas-phase volatile organic compounds (VOCs) by the UV/O_3 process. *Chemosphere* **38**, 1855–1866.

Shen, Y.S. and Ku, Y. (2002) Decomposition of gas-phase trichloroethene by the UV/TiO_2 process in the presence of ozone. *Chemosphere* **46**, 101–107.

Slater, R.C. and Douglas-Hamilton, D.H. (1981) Electron-beam-initiated destruction of low concentrations of vinyl chloride in carrier gases. *J. Appl. Phys.* **52**, 5820–5828.

Smet, E., Lens, P. and Van Langenhove, H. (1998) Treatment of waste gases contaminated with odorous sulfur compounds. *Crit. Rev. Env. Sci. Tec.* **28**, 89–117.

Spiess, F.J., Chen, X., Brock, S.L., Suib, S.L., Hayashi, Y. and Matsumoto, H. (2000) Destruction of freons by the use of high-voltage glow plasmas. *J. Phys. Chem. A* **104**, 11111–11120.

Striebig, B.A., Schneider, J.M., Spaeder, T.A., Mallery, M.R., Heinsohn, R.J. and Cannon, F.S. (1996) Analysis of advanced oxidation processes in a hybrid air pollution control system. In: *Proceedings of the International Chemical Oxidation Association Symposium 'Chemical Oxidation: Technology for the Nineties'*, Nashville, USA, April 14–17.

Tamon, H., Mizota, H., Sano, N., Schulze, S. and Okazaki, M. (1995) New concept of gas purification by electron attachment. *AIChE J.* **41**, 1701–1711.

Tamon, H., Sano, N. and Okazaki, M. (1996) Influence of oxygen and water vapor on removal of sulfur compounds by electron attachment. *AIChE J.* **42**, 1481–1486.

Tonkyn, R.G., Barlow, S.E. and Orlando, T.M. (1996) Destruction of carbon tetrachloride in a dielectric barrier/packed-bed corona reactor. *J. Appl. Phys.* **80**, 4877–4886.

Urashima, K. and Chang, J.S. (2000) Removal of volatile organic compounds from air streams and industrial flue gases by non-thermal plasma technology. *IEEE T. Dielect. El. In.* **7**, 602–614.

Urashima, K., Chang, J.S. and Ito, T. (1997) Destruction of volatile organic compounds in air by a superimposed barrier discharge plasma reactor and activated carbon filter hybrid system. In: *Proceedings of the IEEE Industry Apllications Society Annual Meeting*, New Orleans, USA, October 5–9.

Van Craeynest, K., Dewulf, J., Vandeburie, S. and Van Langenhove, H. (2002) Removal of trichloroethylene from waste gases via the peroxone process. In: *Proceedings of the 1st IWA Young Researchers Conference*, Cranfield, UK, September 9–10.

Van Heesch, E.J.M., Smulders, H.W.M., van Paasen, S.V.B., Blom, P.P.M., van Gompel, F.M., Staring, A.J.P.M. and Ptasinski, K.J. (1997) Pulsed corona for gas and water treatment. In: *Proceedings of the 11th IEEE International Pulsed Power Conference*, Baltimore, USA, June 29–July 2.

Van Langenhove, H., De Roo, K. and Overloop, S. (2000) Stank. In: *Mira S 2000. Milieu- en Natuurrapport Vlaanderen: Senario's*, Van Steertegem M. (ed.). Garant, Leuven.

VDI Richtlinie (1995) *Abgasreinigung durch oxidierende Gaswäsche. Waste-Gas Purification by Oxidative Gas Scrubbing.* VDI 2443, Düsseldorf, Germany.

Vitale, S.A., Hadidi, K., Cohn, D.R. and Falkos, P. (1996) Electron beam generated plasma decomposition of 1,1,1-trichloroethane. *Plasma Chem. Plasma P.* **16**, 651–668.

Vitale, S.A., Hadidi, K., Cohn, D.R. and Bromberg, L. (1997) Decomposition of ethyl chloride and vinyl chloride in an electron beam generated plasma reactor. *Radiat. Phys. Chem.* **49**, 421–428.

Vorontsov, A.V., Savinov, E.V. and Davydov, L. (2001) Photocatalytic destruction of gaseous diethyl sulfide over TiO_2. *Appl. Catal. B Environ.* **32**, 11–20.

Wang, J.H. and Ray, M.B. (2000) Application of ultraviolet photooxidation to remove organic pollutants in the gas phase. *Sep. Purif. Technol.* **19**, 11–20.

Watt, A.S., Magrini, K.A., Carlson, L.E., Wolfrum, E.J., Larson, S.A., Roth, C. and Glatzmaier, G.C. (1999) Pilot-scale demonstration of an innovative treatment for vapor emissions. *J. Air Waste Manage.* **49**, 1368–1373.

Wekhof, A. (1991) Treatment of contaminated water, air and soil with UV flashlamps. *Environ. Prog.* **10**(4), 241–247.

Yamaguchi, S.I., Iwaizumi, T. and Tsuda, S. (1998) Decomposition of gaseous trichloroethylene (TCE) by ultraviolet irradiation. *Combust. Sci. Technol.* **133**, 107–115.

Yamamoto, T., Ramanathan, K., Lawless, P.A., Ensor, D.S., Newsome, J.R., Plaks, N. and Ramsey, G.H. (1992) Control of volatile organic compounds by an ac energized ferroelectric pellet reactor and a pulsed corona reactor. *IEEE Trans. Ind. Appl.* **28**, 528–534.

Zhang, R.H., Yamamoto, T. and Bundy, D.S. (1996) Control of ammonia and odors in animal houses by a ferroelectric plasma reactor. *IEEE Trans. Ind. Appl.* **32**, 113–117.

12

Advanced oxidation of textile industry dyes

I. Arslan-Alaton

12.1 INTRODUCTION

Textile waste minimisation and remediation programs around the world have demonstrated that the colour problem is becoming more and more apparent as one of the major environmental tasks associated with this sector, and hence much research effort has transferred to this area. In the last two decades, there has been an immense research interest in the treatment of textile industry dyes with so-called 'advanced oxidation processes (AOPs)', not only as a consequence of growing environmental concerns tackled by legislative aspects, but also due to the rather 'visual' success of advanced oxidation systems in overcoming the colour and pollutant 'phase transfer' problem. Advanced oxidants formed by a variety of chemical (O_3/OH^-, O_3/H_2O_2, Fenton's reaction, modified Fenton or Fenton-like oxidation, hydrothermal oxidation processes in sub- and supercritical

 Advanced Oxidation Processes for Water and Wastewater Treatment. Edited by Simon Parsons. ISBN: 1 84339 017 5

water, power ultrasound), photochemical (O_3/UVC, H_2O_2/UVC, Photo-Fenton, modified Photo-Fenton's oxidation) as well as near-UV–Vis or solar light induced TiO_2-mediated heterogeneous photocatalytic oxidation systems appear to have the ability to completely decolourise and at least partially mineralise textile industry dyes in short reaction periods without producing any exhaust gases or harmful sludge. Besides, no toxic and/or carcinogenic by-products have usually been observed when applying AOPs for textile dye decolourisation. Herein, the major advanced oxidation application ranges for textile dye treatment have been reported following a brief description of available textile industry dyes and their application-based categorisation. Aside, the need for investigating and sizing the impact of dye auxiliary chemicals that in the case of fibre reactive dyes are potential hydroxyl radical scavengers has been further discussed.

12.2 TEXTILE DYES

Dyes are generally small molecules comprising two key components: the chromophore, responsible for the colour, and the functional group that allows dye fixation onto the fibres. In terms of chemical structure, the azo compounds represent the largest group (around 50%) of all textile dyestuffs produced. They derive their name from their characteristic azo group consisting of two nitrogen atoms (—N=N—). In accordance with the number of such groups, the dyes are described as monoazo, diazo, triazo, etc. dyes (Ryes and Zollinger, 1972). The second classification of dyes which is based on their mode of application distinguish acid, reactive, metal complex, disperse, vat, mordant, direct, basic and sulphur dyes. Most commonly in use today are the reactive and direct types for cotton, viscose-rayon dyeing and disperse types for polyester dyeing (Snowdon-Swan, 1995). There are more than 100,000 commercially available textile industry dyes with over 7×10^5 t of annual dyestuff production. Textile dyeing is done in a great variety of ways, and the continuous development of new dyes and auxiliary chemicals add to the complexity of the operation. General characteristics of major types of textile dyes together with their fixation degree for different dye and fibre combinations are presented in Table 12.1.

From the above table, it is obvious that fibre reactive dyes possess rather low fixation rates, while the highest fixation is achieved by basic dyes. It has been reported that after the reactive dyeing process is complete, up to $800\,mg\,L^{-1}$ of unfixed and hydrolysed dye may remain in the spent textile bath and cannot be reused (O'Neill *et al.*, 1999). Therefore, up to 50% of the applied reactive dye is usually discharged as a deeply coloured effluent. This rate is closely followed by disperse and sulphur dyes whose popularity can be verified by examining their appearance in textile wastewater (Easton, 1995).

Table 12.1 Classification of dyestuffs, their associated fibres and fixation rates (Buckley, 1992).

Dyestuff	Characteristics	Attachment mechanism and associated fibres	Loss to dyehouse effluent (%)
Acid	Anionic, highly water soluble, poor wet fastness	Ionic bond, nylon, wool	5–20
Metal (Cr, Co, Cu) complexed acid	Anionic, low water solubility, good wet fastness	Ionic bond, nylon, wool	2–10
Direct	Anionic, highly water soluble, poor wet fastness	Ionic bond, cotton, viscose	5–30
Basic and cationic	Cationic, highly water soluble	Ionic bond, acrylics	0–5
Disperse	Colloidal dispersion, very low water solubility, good wet fastness	Colloidal impregnation and adsorption, polyester, nylon, acrylic, cellulose acetate	0–10
Reactive	Anionic, highly water soluble, good wet fastness	Covalent bonds, cotton, viscose, wool	10–50
Sulphur	Colloidal after reaction in fibre, insoluble	Dye precipitated *in situ* in fibre, cotton, viscose	10–40
Vat	Colloidal after reaction in fibre, insoluble	Dye precipitated *in situ* in fibre, cotton, viscose	5–20
Azoic	Colloidal after reaction in fibre, insoluble	Dye precipitated *in situ* in fibre, cotton, viscose	2–3

12.3 THE PROBLEM OF RESISTANT COLOUR IN DYEHOUSE EFFLUENT

The textile industry is strongly associated with strong, persistent colour (Correa *et al.*, 1994), since dyes are intentionally designed to withstand microbial, chemical and photolytic degradation to satisfy the consumer's requirements (Reife, 1993). Many fibre acid, direct and reactive dyes are hardly biodegradable and hence pass untreated through the sewage works due to their high water solubility and relatively low molecular weight (Reife and Freeman, 1996). An important factor that has directed research towards 'destructive' AOPs is the fact that conventional physicochemical 'phase transfer' methods, such as activated carbon adsorption, coagulation, flocculation or precipitation, no longer give satisfactory treatment results in terms of colour consents. Besides sludge formation creates another source of waste problem. Moreover, among the oxidation techniques, chlorination, hydrogen peroxide treatment, direct short UV (UVC) photolysis, and in some instances even ozonation are insufficiently powerful to decolourise most of the commercial textile dyes at ambient reaction conditions (Arslan, 2001; Marechal *et al.*, 1997).

Research on textile effluent decolourisation has focused on fibre reactive dyes for the following reasons. First, reactive dyes represent an integral market share (almost 45% of all textile dyes produced annually are belonging to the fibre reactive class) due to the appreciable use of woollen, cotton and viscose-rayon fibres. Second, these dyes have low fixation rates, which results in highly coloured spent dye-baths whose treatment is inadequate in conventional wastewater treatment plants (Easton, 1995). Third, of special concern is the reactive dyeing process, where in the average 10 times more water is consumed for the preparation, dyeing, washing and rinsing stages than for dyeing with other dye types. Thus, special emphasis has been placed on the advanced chemical and photochemical oxidation of reactive dyestuffs in the forthcoming sections.

12.4 APPLICATION OF AOPs FOR TEXTILE INDUSTRY DYE TREATMENT

Ozone reacts with water and wastewater compounds in two different ways, namely direct molecular and indirect free radical-type reactions. Both pathways are important for colour removal. Application of ozone to decolourise different classes of textile dyes in dyehouse effluents is a well-known, relatively older process and has been reviewed two decades ago (Beszedits, 1980; Gould and Groff, 1987). Pilot-scale ozone treatment (calculated for 100,000 Mt day^{-1}) costs accompanied with activated sludge treatment (=1.57 US$ per tonne) is about twice the cost of traditional stand-alone activated sludge process (Reife and Freeman, 1996). Ozonation has been applied to dye waste to reduce its acute toxicity and biodegradability (Law *et al.*, 1996).

AOPs applied to textile industry dyes and dyehouse effluents have been mentioned in excellent review papers addressing the classification and reuse alternatives for textile dyes (Buckley, 1992), the environmental chemistry of dyes and pigments (Reife, 1993; Reife *et al.*, 1998), the traditional and emerging technologies for the treatment of textile wastewater (Vandevivere *et al.*, 1998), colour removal from dyehouse effluent (Slokar and Le Marechal, 1998), the problem of colour in textile wastewater (O'Neill *et al.*, 1999), technical and legislative aspects of pollution prevention in the textile industry sector (Reife and Freeman, 2000) and ozonation of textile industry effluents for their biodegradability improvement (Alvares *et al.*, 2001). Below photochemical and non-photochemical AOPs including pressurised oxidation and sonochemical methods applied to treat textile industry dyes are described in detail. Special emphasis was placed on reports examining the effect of various types of $OH^{\bullet}$ scavengers (particularly inorganic anionic compounds).

One might notice that Acid Orange 7 (also termed as Orange II) has been investigated as an index pollutant representing the recalcitrance of textile azo dyes and the fact that the more recently published works are almost entirely devoted to fibre reactive dyestuffs. Disperse dyes are hydrophobic and depending upon the pH tend to precipitate out of solution in the presence of transition metal salts added as catalysts and are barely oxidised by AOPs (Arslan, 2001; Beszedits, 1980). Thus, they will not be further evaluated in this chapter.

12.4.1 Non-photochemical AOPs

12.4.1.1 O_3/OH^- and O_3/H_2O_2 treatment

The principal advantage of combinations with O_3 is the enhanced mass transfer in solution arising as a logical consequence of accelerated O_3 decomposition (Glaze and Kang, 1989; Glaze *et al.*, 1987). It has been reported that trials with H_2O_2 and elevated pH (>11) have been carried out, but no appreciable improvements or cost benefits were usually determined over ozone alone (Arslan, 2000).

Chu and Ma (2000) quantitatively estimated the direct ozonation and free radical oxidation of direct, disperse and reactive dyes with assorted chromophores (mono, polyazo and anthraquinone). Ozone doses, reaction pH and dye concentrations were adjusted to observe different reaction mechanisms and kinetics for colour removal. Ozonation rate and efficiency was a combined function of dye solubility and structure.

12.4.1.2 Sonochemical degradation

Vinodgopal *et al.* (1998) studied the degradation of Reactive Black 5 solution using a high-frequency ultrasound generator; 65% mineralisation was achieved after 6 h treatment and remaining compounds were detected as oxalate, sulphate and nitrate ions.

Poon *et al.* (1998) studied the kinetic aspects of Cuprophenyl Yellow RL treatment by the UV/US/H_2O_2 process. The first-order rate constants of decolourisation and TOC removal were reduced significantly in the absence of H_2O_2 and also at acidic pH.

Stock *et al.* (2001) investigated the merits of combining the AOPs high-frequency sonolysis and TiO_2-mediated heterogeneous photocatalysis for Naphtol Blue Black degradation. Experiments were carried out in a sequential and combined manner. The enhancement in dye degradation was an additive effect and both sequential and simultaneous methods yielded similar pseudo-first-order decolourisation rate constants. In terms of TOC removal, however, photocatalysis appeared to be superior to sonolysis that was on the other hand more effective in colour removal. The simultaneous approach accelerated mineralisation rates drastically and up to 80% TOC removal could be achieved after 12 h treatment.

Ince and Tezcanli (2001) combined sonolysis and ozonation for hydrolysed Reactive Black 5 degradation. They found that the joint action of US and ozone induced a synergistic effect on both colour and TOC abatement kinetics.

12.4.1.3 *Wet-air oxidation or pressurised oxidation systems*

Shende and Mahajani (1994) examined the wet-air oxidation (WAO) of the reactive dyes Chemictive Brilliant Blue (Reactive Blue 19) and Cibacron Turquoise Blue G (Reactive Blue 21) in the temperature and pressure range of 150–250°C and 0.345–1.38 MPa, respectively. Chemical oxygen demand (COD) removal was more than 90% and 100% colour removal could be achieved in 4 h. They modelled dye degradation in terms of COD removal using the empirical Arrhenius equation and assuming a two-step oxidation process. The anthraquinone class dye was found to require more severe oxidation conditions (O_2 partial pressure and particularly temperature) than the phythalocyanine dye.

Donlagic and Levec (1997) investigated the reaction kinetics of Orange II WAO by modelling total carbon and parent dye disappearance in terms of lumped azo dye intermediates. Thermal decomposition of Orange II dye was also found to proceed as a first-order reaction with respect to Orange II. They also compared catalysed (a hybrid mixture of Al_2O_3, CuO and Zn was employed as the heterogeneous catalyst) and uncatalysed WAO of Orange II (1998). They evidenced a serious enhancement of the degradation rate in the presence of the catalyst, the oxidation route, however, remained unchanged as they have concluded from the nature of oxidation intermediates. Dye solutions treated catalytically were found to be more amenable to aerobic biodegradation as smaller quantities of oxidation intermediates were formed in the catalytic process.

Orange II degradation was also studied by Raffanier and von Rohr (2001) in the presence and absence of 100 mg L^{-1} $FeSO_{4s} \cdot 7H_2O$ catalyst. They observed that the activation energy decreased in the presence of the catalyst and their reaction shifted from zero to first-order speaking for a change in the reaction pathway/mechanism for azo dye degradation.

Chen *et al.* (1999) used WAO and wet-peroxide oxidation of five reactive dyes at extremely high concentration (50 g L^{-1} each). The effect of pH, temperature, H_2O_2 dose, initial dye concentration and copper catalyst dose was examined. Colour and TOC removal were most sensitive to the employed reaction temperature. Significant TOC reduction was not observed until the H_2O_2 dosage was 100% of the stoichiometric amount. TOC and colour removal were enhanced by using Cu^{2+}-ion salt (homogenous catalyst) and Cu metal supported on activated carbon (heterogeneous catalyst system). Up to 80% TOC was removed in less than 15 min under optimised process conditions. Again a two-step consecutive reaction scheme has been proposed, where first large parent dye molecules are broken down and a slower oxidation of the dye intermediates to low molecular weight carboxylic acids (acetic and formic acid) usually being the final refractory products of WAO and other AOPs.

Arslan-Alaton and Ferry (2002) have also evidenced that thermal dye decomposition through —N=N— cleavage played an important role in Orange II WAO and the catalyst has a minor effect on colour removal rates. However, the activation energy for TOC has been reduced considerably and the reaction pathway seemed to change in the presence of the heteropoly acids (silicadodecatungstic and phosphotungstic acid catalyst). The addition of polyoxometalates as selective oxidation catalysts shifted the reaction mechanism from a free radical to an electron transfer reaction by using organic and inorganic $OH^{\bullet}$ probe compounds.

12.4.2 Photochemical AOPs

12.4.2.1 O_3/UVC

Shu and Huang (1995) degraded eight commercial azo dyes in water using ozonation and UVC enhanced ozonation. UVC light did not significantly enhance the degradation rates under the examined conditions simply because the dyes competed with O_3 for UV Absorbance. Their experiments have also revealed that diazo dyes are more resistant to ozonation than monoazo dyes. Colour removal via ozonation appeared to be a kinetically fast, mass transfer limited oxidation process.

Of the major AOPs involving ozonation (O_3/OH^-, O_3/H_2O_2 and O_3/UVC), the O_3/H_2O_2 process should be preferred for large treatment plants when substrates of concern are not strong UV absorbers, since the O_3/UVC process will be more difficult to adopt on a large scale (Masten and Davies, 1993).

12.4.2.2 H_2O_2/UVC

Namboodri and Walsh (1996) investigated the H_2O_2/UVC treatment of simulated reactive dye-bath wastewater (hydrolysed Reactive Blue 21 and Reactive Red 195 solutions) by using both batch operation and continuous circulation methods. A pilot plant study was also conducted using a high-intensity reactor. For both reactive dyes under similar dye concentration, radiation dosage and H_2O_2 concentration levels, the continuous circulation method gave higher rates of decolourisation. Though compared to the low intensity lamp experiments in laboratory scale, the high-intensity lamps required a higher energy input per volume of wastewater, they were more compact and the number of bulbs was fewer becoming an important criterion for full-scale application.

Ince and Gönenç (1997) studied the H_2O_2/UVC oxidation of Reactive Black 5 in aqueous solution. Effects of the initial H_2O_2 concentration, initial dye concentration and alkalinity at neutral pH were tested by calculating pseudo-first-order reaction rate constants and electrical energy requirements per order of pollutant removal (EE/O values) in terms of colour removal.

Majcen-Le Marechal *et al.* (1997) described the decolourisation of the monochlorotiazine dye Reactive Red 120 in the absence and presence of dye-bath constituents. It was found that dye auxiliaries reduced the efficiency of the colour, COD and TOC removal process significantly. The decolourisation rate was proportional to the UV output of the light source.

Uygur and Kök (1999) examined the H_2O_2/UVC treatment of a synthetic reactive dye-bath liquor containing dichlorotriazine reactive systems. Colour removal rates increased at higher H_2O_2 concentrations accompanied with an obvious increase in the Biological oxygen demand (BOD) values. It was also concluded that during advanced oxidation TOC, COD and AOX decreased, while sulphate ions remain unchanged.

A detailed investigation was undertaken by Galindo and Kalt (1999) who studied the structure–degradability relationship for 28 azo dyes in aqueous media. They postulated that the dye solubility, size and hence dye structure affects their degradation rate significantly.

Colonna *et al.* (1999) reported first-order decolourisation kinetics and partial mineralisation for the H_2O_2/UVC oxidation of some azo-type acid and anthraquinone-type reactive dyes. Similar results were observed by Georgiou *et al.* (2002), who studied the H_2O_2/UVC degradation of several reactive azo dyes with vinylsulphone, halotriazine and chloropyridine anchor groups. After 2 h of irradiation, 80% TOC was destroyed, whereas complete colour removal was achieved in less than 20 min.

Imamura *et al.* (2000) modelled the H_2O_2/UVC oxidation of several azo dyes (including Orange II) in a flat surface–thin-film photoreactor. Shen and Wang (2002) modelled the treatment of Direct Yellow 84 in a continuous plug flow annular photoreactor under varying UV light intensities, dye and peroxide doses.

12.4.2.3 Fenton/Photo-Fenton oxidation systems

From the practical point of view, the Fenton process could be used for the removal of hydrophobic disperse as well as hydrophilic acid dyes at their natural acidic dyeing conditions by the dual action of oxidation and ferrous iron coagulation and precipitation. In fact, disperse dyes tend to precipitate upon ferrous iron addition even in the absence of H_2O_2. Although a pH adjustment would be required, fibre reactive dyes are also most effectively decolourised via Fenton's oxidation. However, Fenton's processes produce sludge during the course of after-treatment/pH adjustment from acidic to neutral pH.

Kuo (1991) was the first to study Fenton's treatment of mixtures of different types (acid, basic, direct, disperse and reactive) of commercial dye series. He examined and optimised several critical operating parameters, such as reaction pH, Fe(II) iron dose, initial H_2O_2 concentration and temperature and found similar treatment efficiencies for actual textile mill effluents. Similar treatability studies involving Fenton, Fenton-like processes and their light induced versions followed.

Spadaro *et al.* (1994) examined the degradation of ^{14}C labelled Disperse Yellow azo dye by $OH^{\bullet}$ generated in the Fenton-like reaction. ^{14}C formation was followed to study the extent of mineralisation by adding up the C mass balance. Disperse dye degaration yielded nitrobenzene and benzene, that is two products that are known to be more harmful and toxic than the intact phenylazo dye molecules.

Namboodri and Walsh (1995) studied the decolourisation of simulated direct and reactive dye-baths (hydrolysed azo, anthraquinone and phthalocyanine dye solutions) with hot peroxide (50% w/w) solution, in the absence and presence of cupric, ferric and ferrous ion acting as catalysts in the pH range of 3–7 and temperature near boiling (95–98°C). The presence of the copper phythalocyanine dye catalysed the decolourisation process by the catalytic effect of released (free) copper ion.

Solozhenko *et al.* (1995) studied the Fenton's oxidation of the azo dye Yellow Lightfast 2KT in aqueous solution; 95–97% decolourisation could be obtained at relatively low H_2O_2 dosages. Reaction temperature and the introduction of solar radiation intensified the decolourisation process drastically.

Herrera *et al.* (1999) carried out experiments to decolourise the anthraquinone dyes Remazol Brilliant Blue and Remazol Uniblue, that is its hydrolysed form (both are Reactive Blue 19) in the presence of Fe(III) iron and H_2O_2 under light irradiation at different wavelengths. Laser flash photolysis carried out at different wavelengths suggested that the electron transfer between the excited dye and Fe^{3+} (bimolecular charge transfer reaction) is the initiating step and the photoreduction of Fe(III)-ion to Fe(II)-ion is a meaningful acceleration route of radical chain reactions. In the same manner, they proposed that the photodissociation of dye–Fe(III) complex might also contribute to the initiation step of the free radical oxidation process.

Ince and Tezcanlý (1999) investigated the treatability of hydrolysed solution of the bifunctional Procion Red HE 7B by the photo-assisted and dark Fenton's oxidation. Under optimum reaction conditions (pH, temperature, Fe(II) : H_2O_2 ratio) complete colour and 79% TOC removal could be achieved after 20 min treatment. The authors concluded that a subsequent coagulation step combined with a membrane process was required to bring the wastewater to reuse quality. As for other reactive dyestuffs, the Fenton's reaction system appeared to be a rapid and effective oxidation method.

Nam *et al.* (2001) investigated the substituent effects of commercial textile azo dyes oxidation with the $Fe^{2+}/EDTA/H_2O_2$ system. They also determined the production rate of $OH^{\bullet}$ and H_2O_2 concentrations throughout the reaction by a unique analytical deoxyribose procedure. They detected the degradation products 4-hydroxy benzene sulphonic acid and 1,2,naphthaquinone for the azo dye Acid Orange 7 and proposed a free radical reaction pathway.

12.4.2.4 TiO_2-mediated heterogeneous photocatalysis

In a study by Hustert and Zepp (1992), four azo dyes were photocatalytically degraded. Both colour and TOC abatement followed first-order kinetics. Formation of aminophenols pointed out the involvement of an additional reductive cleavage pathway for the azo moiety by e^- produced on the conduction band of the semiconductor. Product studies verified the formation of the hydroxylated by-products benzoquinone and hydroquinone confirmed a free radical pathway.

Tang *et al.* (1997) studied the structure–degradation relationship for the TiO_2/UVA oxidation for 11 azo dyes using a Rayonet (merry-go-round) photoreactor. They observed that the photocatalytic degradation rate and pathway were significantly influenced by the electrical nature (charge), number of azo groups and the type of auxiliary groups attached to the dye chromophores. Hole oxidation appeared to be dominant at acidic pH (3) for neutral or negatively charged dyes, whereas $OH^{\bullet}$ should be the major oxidant at pH >7 for the rest.

Zhan and Tian (1998a,b) tested the effect of H/K acid-based acid azo dye substituents on the photocatalytic degradation in colloidal TiO_2 suspensions. They concluded that TiO_2 acts as a strong photosensitiser for $OH^{\bullet}$, $HO_2^{\bullet}$ and $O_2^{\bullet -}$ production (photosensitised degradation). Decolourisation of azo dyes is speculatively mainly due to the reaction of their hydrazone tautomer with $OH^{\bullet}$ and $O_2^{\bullet -}$. Their experimental results have also indicated that the pH values of the aqueous azo dye solution can change the dye structure, but only limitedly affect their reactivity of TiO_2. That the hydrazone form of Acid Orange 7 triggers and undergoes photocatalytic oxidation has also been confirmed via FT-IR spectroscopy in a recent paper of Bauer *et al.* (2001).

Shun and Yizhong (1999) studied the photodegradation and biodegradability changes for five azo dyes, three of which were commercial reactive dyes with monochlorotriazine reactive systems using TiO_2 suspensions irradiated with a medium-pressure mercury lamp. Experimental results demonstrated that the biodegradability of the wastewater increased considerably with a parallel decrease in TOC and COD.

Goncalves *et al.* (1999) examined the photocatalytic degradation of eight commercial azo dyes using different UV light sources. Conventional environmental sum parameters (COD and TOC) and UV–Vis spectrophotometry were assessed to compare treatment performances.

The reactive azo dyes Orange 16 (monoazo; Poulios and Aetopoulou, 1999) and Black 5 (diazo; Poulios and Tsachpinis, 1999) were treated by the TiO_2/UV process using artificial and solar light. Decolourisation kinetics could be fitted to the empirical Langmuir–Hinshelwood (L–H) model. The positive effects of the electron scavengers H_2O_2 and $S_2O_8^{2-}$ on the decolourisation rates were also studied.

The effect of Ca^{2+} and W^{6+} doping on TiO_2-mediated Acid Orange 7 degradation was investigated by Kiriakidou *et al.* (1999). For that purpose, conventional operation parameters (pH, initial dye concentration, catalyst lading, incident light flux) and dopant concentration was examined. The enhanced photoactivity of the W^{6+} dopant could be safely attributed to the formation of new energy levels and enhanced electron conductivity in the bulk semiconductor arising from the increased electron concentration in the conduction band of TiO_2 induced by higher valenced cations.

The solar photocatalytic degradation of eight commercial dyes with different substituent groups and structures was also investigated by Wang (2000). The dye degradation was followed up by TOC and decolourisation rates as well as inorganic ion production (extent of mineralisation).

Tanaka *et al.* (2000) degraded seven commercial azo dyes (including the biorecalcitrant index pollutant Acid Orange 7) in aqueous titania suspensions. According to their results, diazo dyes were less degradable than monoazo dyes. Major intermediate products were identified as phenolic compounds, amines and carboxylic acids. Experiments conducted under nitrogen gas sparging to prevent oxidation indicated that decolourisation could also be achieved via reductive electron transfer processes, and that under those conditions, no mineralisation was observed.

Prevot *et al.* (2001) also observed an appreciable acceleration of TOC and colour removal of Acid Blue 80 (an anthraquinoid dye) photocatalysis upon the addition of $S_2O_8^{2-}$. Identification of initial, unstable dye intermediates and mineralisation end products (NO_3^-, NH_4^+, SO_4^{2-} and CO_2) revealed that C—N bond breakage and mother compound hydroxylation were the early steps of degradation.

Chun *et al.* (2001) prepared a TiO_2 bond conjugated silica gel support (TiO_2/SiO_2) by an impregnation method and studied its activity for azo dye Reactive Red 15 degradation. They observed an abrupt decrease in the zeta potential of TiO_2 to pH 3, implying the presence of surface acidity that was expected to affect the adsorption–degradation behaviour of the semiconductor enormously. Although the fixed film (supported) catalyst had a lower reactivity than the more well-known, traditional Degussa P25, the use of a stable support would eliminate a costly and problematic separation procedure.

The photocatalytic oxidation of the monoazo dye Procion MX-5B (Reactive Red 2) was investigated by So *et al.* (2002). Again, the mineralisation rate was found to be relatively slow as compared to decolourisation.

Diaper *et al.* (2002) studied TiO_2/UVC (short UV) treatment of three commercial textile dyes (one acid, two reactive dyes). They obtained complete mineralisation in 24 h accompanied by an initial dye adsorption on the semiconductor surface at acidic pH. They concluded from their experimental results that good adsorption did not necessarily mean high removal rates. Treatment costs associated with photocatalysis were slightly lower and comparable to those established for conventional methods (activated carbon adsorption, coagulation, flocculation and chemical precipitation).

12.4.3 Comparison of AOPs

Morrison *et al.* (1996) comparatively studied the sunlight-induced decolourisation–degradation rates of Acid Orange 7 by heterogeneous (TiO_2/sunlight/solar simulator) and homogenous ($Fe^{3+}/Cu^{2+}/H_2O_2$) AOPs. The UV component of natural sunlight was sufficient to promote Orange II dye degradation in the presence of H_2O_2. The turnover number observed in the solar simulator was seven times lower than that observed for solar irradiation. The oxidation efficiency of the dark Fenton-like system was improved by the addition of Cu^{2+}, however, using Cu^{2+} catalyst alone colour removal kinetics were found insufficient for higher concentrations.

Peralta-Zamora *et al.* (1998) compared the degradation of Reactive Blue 19 by ozonation, enzymatic and heterogeneous photocatalytic processes. They observed the fastest colour removal for ozonation and the slowest for the enzymatic reaction. The use of ZnO and TiO_2 photocatalysts gave the highest mineralisation rate in 60 min.

Arslan and Akmehmet Balcioglu (1999) comparatively evaluated the advanced oxidation of the copper phythalocyanine dye Remazol Turquoise Blue G133 (Reactive Blue 21) in aqueous solution via heterogeneous photocatalysis, Fenton, Fenton-like, Photo-Fenton and Photo-Fenton-like processes. The highest colour and COD removal efficiencies were achieved by the Fenton-like process as a consequence of its dual oxidation + coagulation effect, whereas the Fenton and Photo-Fenton reaction proceeded faster due to the higher rate of $OH^{\bullet}$ formation. None of the selected AOPs appeared to be very effective in the degradation of the copper phythalocyanine dye that was partially attributable to the adsorption on the photocatalyst surface or the coagulating effect of iron applied in the homogenous AOPs.

Aplin and Waite (2000) comparatively studied three different AOPs (ozonation at high pH, Fenton and the modified Fenton-like process, Fe^{3+}/oxalate/UV treatment) for the degradation of the azo dye Reactive Red 235. The effect of initial dye concentration, pH and NaCl were explored for each AOP. Optimum operating conditions differed significantly for each AOP and chloride ions had a negative effect on dye oxidation rates. Inhibition in the presence of chloride was attributable to the formation of poor UV absorbing $FeCl_2^-$ and $FeCl^+$ complexes as well as its $OH^{\bullet}$ scavenging effect decreasing with increasing pH.

Figure 12.1a–c comparatively displays the degradation of hydrolysed Reactive Black 5 (Remazol Black SB) by several optimised photochemical (◇ H_2O_2/UVC; △ Photo-Fenton; □ Photo-Fenton-like and ○ TiO_2/UVA) and non-photochemical (◆ O_3/OH^-; ▲ O_3/H_2O_2; ■ Fenton and ● Fenton-like) AOPs in terms of the environmental sum parameters (a) colour or peak absorbance of Reactive Black 5 in the visible spectrum at $\lambda = 597$ nm (pH < 11.5) to 600 nm (pH $\geqslant 11.5$), (b) COD and (c) TOC. Reactive Black 5 is one of the oldest and most frequently sold vinysulphone reactive dyes in the market known for its unusually low substantivity, high solubility and overwhelming high fixation efficiency (Shore, 1995).

Figure 12.1a–c reveals that compared to the H_2O_2/UVC process, the O_3/OH^- and O_3/H_2O_2 processes are appreciably faster for Reactive Black 5 degradation. As has been already established in former studies, it seemed that the introduction of H_2O_2 did not enhance oxidation efficiency of the ozonation process due to fact that ozone decomposition to free radicals is of a rather pH-dependent nature and thus very dominant at the studied reaction pH (=11 for the ozonation reactions). Under these circumstances, the positive effect of H_2O_2 addition is partially suppressed. The superiority of the selected alkaline ozonation process at pH 12 to other AOPs, for example H_2O_2/UV, sonolysis and TiO_2/near-UV (Arslan and Akmehmet Balcioglu, 1999; Ince *et al.*, 1997; Vinodgopal *et al.*, 1998) applied for the decolourisation and partial mineralisation of aqueous Reactive Black 5 solution thus could be clearly demonstrated.

On the other hand, colour, COD and TOC reduction was accomplished within a few minutes during the Fenton and Photo-Fenton reactions and the difference between Fenton and Photo-Fenton treatment efficiencies was rather insignificant. Dark and UVA light-assisted Fenton-like (Fe^{3+}/H_2O_2) experiments were also conducted in parallel to the Fenton experiments. For the Fenton-like process, a significant initial reduction in colour (25.3%) and COD (30.7%) that took place by introducing the required amount of Fe^{3+} salt to the reaction solution was evident. The effect of UVA light introduction was more significant than in the case of Fenton and Photo-Fenton processes resulting in a two-fold acceleration of the pseudo-first-order decolourisation rate constant. This difference is attributable to

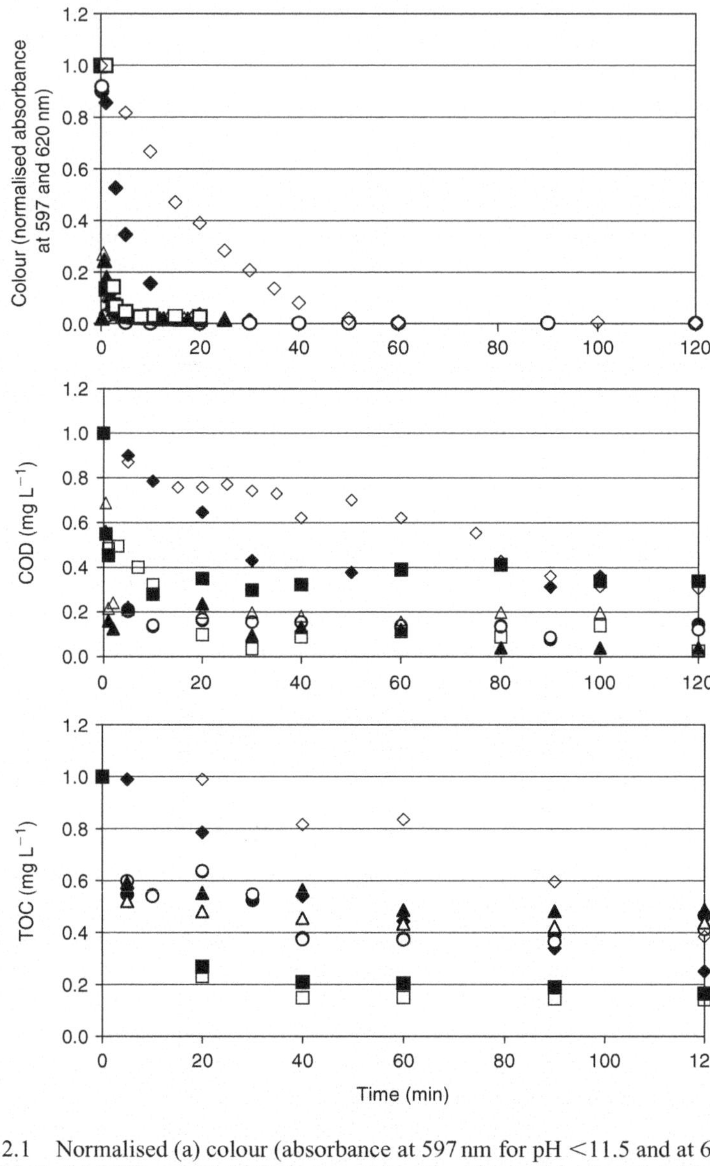

Figure 12.1 Normalised (a) colour (absorbance at 597 nm for pH <11.5 and at 620 nm at pH >11.5), (b) COD and (c) TOC abatement kinetics obtained for Reactive Black 5 degradation by different optimised photochemical (◇ H_2O_2/UVC; △ Photo-Fenton; □ Photo-Fenton-like and ○ TiO_2/UVA) and non-photochemical (◆ O_3/OH^-; ▲ O_3/H_2O_2; ■ Fenton and ● Fenton-like) AOPs. Experimental conditions: 75 mg L^{-1} Remazol Black SB; A_o = 162 L m^{-1}; COD_o = 57 mg L^{-1}; TOC = 12 mg L^{-1}. The reduced initial colour, COD and TOC values at t = 0 min for Fenton, Fenton-like and TiO_2/UVA experiments are the effects of ferrous/ferric iron coagulation and adsorption on the photocatalyst surface, respectively (adopted from Arslan, 2000; for details see also Arslan and Akmehmet Balcioglu, 1999, 2000, 2001; Arslan-Alaton and Akmehmet Balcioglu, 2001).

the indirect and hence delayed formation of $OH^•$. However, the presence of ferri-hyroxo complexes as in Fenton-like reactions is reported to cause a more complete destruction of organic pollutants. From the above results, one may conclude that due to the narrow emission band of the UVA lamp, only limited light quanta are available for the photo-reduction of ferric iron to ferrous iron and hence direct Fenton's oxidation plays a major role in the degradation of the dye molecules. Higher overall COD removal was obtained for the Photo-Fenton-like process indicating a remarkable enhancement of the Fenton-like process in the presence of UVA light irradiation. It should be mentioned that although the initial rate of mineralisation is found faster with the Fenton's reagent due to the force of more direct $OH^•$ production, the Fenton-like reagent is needed because the regeneration of Fe^{2+}-ion by organic radicals increases the length of the chain reaction that may result in a more complete destruction of the organic compounds, as was obvious from the present study.

Preliminary removal of colour, COD and TOC was also observed for TiO_2-mediated heterogeneous photocatalytic oxidation at acidic pH. For instance, the reduction of TOC by 23% at $t = 0$ min accounted for initial dark adsorption of the anionic dye at pH 4 on the positively charged semiconductor surface.

Generally speaking, besides the outstandingly fast colour removal rate, ozonation at alkaline pH provided relatively good treatment efficiency among the selected AOPs. Further, ozonation seemed to be the most feasible treatment option considering that it results in highest achieved oxidation rates at the natural high, alkaline pH of reactive dye-bath effluents that eliminates a costly pH adjustment prior to application of the selected AOP. However, overall treatment efficiency of the investigated photochemical AOPs (H_2O_2/UVC, TiO_2/UVA, etc.) was found higher than that of the non-photochemical AOPs.

Table 12.2 summarises optimised treatment conditions, process performance and electrical energy requirements per treated volume of wastewater for the selected AOPs for the degradation of hydrolysed Reactive Black 5. The EE/O and EE/M concept will be explained in Section 12.5.

12.4.4 Commercial-scale applications of AOPs for the treatment of textile industry wastewater

Commercial-scale photochemical advanced oxidation systems for drinking water, contaminated groundwater and landfill leachate include the Calgon peroxpure™ and Rayox® UV/H_2O_2 systems, the CAVOX® UV/H_2O_2, Wedeco UV/O_3 and the Rayox® Enox Photo-Fenton systems (EPA, 1998). However, no commercial-scale photochemical AOP is available for the treatment of textile industry effluents.

Fenton's oxidation of textile wastewater is practiced in full scale in some regions of South Africa (Vandevivere, 1998). Full-scale applications of ozone for textile industry wastewater treatment were briefly reviewed by Rice (1997). Many

Table 12.2 Comparison of different optimised photochemical and non-photochemical AOPs for hydrolysed Reactive Black 5 dye degradation. Experimental conditions: 75 mg L^{-1} Remazol Black SB; A_o at 597 nm = 157 L m^{-1}; A_o at 620 nm = 162 L m^{-1}; COD_o = 57 mg L^{-1}; TOC_o = 12 mg L^{-1} (adopted from Arslan, 2000; see also Arslan and Akmehmet Balcioglu, 1999, 2000, 2001; Arslan-Alaton and Akmehmet Balcioglu, 2001).

AOP type	k_d (L min^{-1})	Removal (%)		EE/O (kWh m^{-3} $order^{-1}$)	EE/M (kWh kg^{-1} TOC)
		COD	TOC		
10 mM H_2O_2/UVC (pH 7)	0.16	62	75	2.4	2000
48.75 mM O_3 (pH 12)	3.58	87	62	–	–
48.75 mM O_3/5 mM H_2O_2 (pH 12)	3.62	86		–	–
0.5 mM Fe^{2+}/ 5 mM H_2O_2 (pH 3)*	1.23	100	51	–	–
0.5 mM Fe^{2+}/ 5 mM H_2O_2)/UVC (pH 3)	1.27	98	55	0.59	660
0.5 mM Fe^{3+}/ 5 mM H_2O_2 (pH 3)**	0.33	77	64	–	–
0.5 mM Fe^{3+}/ 5 mM H_2O_2/ UVC (pH 3)**	0.66	98	86	1.16	812
TiO_2 (0.1%)/UVA (pH 4)	0.09	80	70	8.53	2424

* Dark-Fenton process.
** Dark-Fenton-like process.

of the early textile wastewater plants known in Japan use physicochemical treatment methods such a ozonation or Fenton's oxidation mainly due to space limitations. More recently, a number of Italian textile dyeing plants have opted for ozone treatment for decolourisation and destruction of non-ionic surfactants. At least 60 Italian plants are known to have installed ozone as a tertiary treatment stage after activated sludge treatment, sedimentation and sand filtration. In 1992, a UK wastewater treatment plant receiving municipal sewage and textile wastewater installed ozone in a process, which is essentially the same as the Italian procedure, with great benefit. The capital costs are likely to be very high, in the region of £500,000–1,000,000, with running costs of the order of 0.15–0.30 £ m^{-3}. However, although capital costs for alternative colour removal methods (application of biosorbents, eletrolysis, adsorption with activated carbon or clays) are lower in capital cost (in the range of £250,000–500,000) running costs are at least 0.4 £ m^{-3} (Churchley, 1995).

12.5 THE EFFECTS OF COMMON DYE AUXILIARY CHEMICALS AS OH• SCAVENGERS

As has been mentioned previously, colour in dyehouse effluent has become particularly identified with the use of reactive dyes, since up to 50% of these dyes may be lost to the effluent and cannot be recovered or reused. Hence, a central objective of textile research has been the improvement into novel reactive dyes and dyeing processes of higher fixation efficiency and minimisation of the proportion of applied dye lost by hydrolysis in the dye-bath (Shore, 1995). For technical and economical reasons, it is desirable to decolourise the spent dye-bath liquor at source and to concentrate wasted dye residual (e.g. after post-scouring) into the smallest volume possible. Separation of low load washing off stages makes water treatment more feasible. Hydroxide and carbonate alkalinity in the dyehouse effluent is another source of serious environmental concern. Reactive dyes are generally used in the range of moderately to highly alkaline pH conditions. Since alkali is introduced to cause the reaction between the dye and fibre, any reaction between the dye and alkalinity in the water before the dyeing process will result in significant loss of dyestuff due to so-called 'premature hydrolysis'.

Apart from contamination by unfixed reactive dye residues, the most serious ecological problem arising from reactive dye-baths is the high salt load in the effluent. A fundamental approach to minimise excessive salt content in effluents is to adjust dyeing parameters in the way that only the lowest possible salt concentration needs to be added at the dyeing stage. Cool dyeing implies the preference for dyes with high reactivity, which is not the ideal case when the substrate is difficult to penetrate. However, even under the most favourable circumstances, a concentration of at least $20\,g\,L^{-1}$ ordinary salt (NaCl) is needed in pale depths at relatively low liquor ratios (Shore, 1995).

A drawback resulting from the high reactivity and non-selectivity of $OH^\bullet$ is that it reacts with non-target compounds present in the water, such as CH_3COO^-, HCO_3^-, CO_3^{2-} and Cl^- ions, which are referred to as effective $OH^\bullet$ scavengers. Particularly the last two scavengers exist in textile dye-baths at very high concentrations that will result in a significant oxidant 'wastage' accompanied by a higher advanced oxidant demand to accomplish the desired degree of organic compound removal. In reactive dye-baths salt (NaCl) concentrations up to $100\,g\,L^{-1}$ are applied to enhance the dyestuff migration onto the fabric, and soda-ash (anhydrous Na_2CO_3, pH buffer and fixing agent) concentrations up to $20\,g\,L^{-1}$ are not unusual. Almost surprisingly, only very few papers have addressed the effect of free radical scavengers on textile industry dyes. In the present section the effect of some important $OH^\bullet$ scavengers (alkalinity and inorganic salts) on the advanced oxidation of textile industry (fibre reactive) dyes is described.

12.5.1 Effect of chloride

Pulse radiolysis studies have confirmed that chloride ions react with $OH^\bullet$ to form $ClOH^{\bullet-}$ at acidic pH, which in turn protonates and is converted to $Cl^\bullet$ depending upon the following reactions (Anbar and Neta, 1967; Kiwi *et al.*, 2000):

$$Cl^- + OH^\bullet \rightarrow ClOH^{\bullet -} \quad k_1 = 4.3 \times 10^9\,L\,(mol \times s)^{-1}\ (pH = 2) \qquad (12.1)$$

$$ClOH^{\bullet -} + H^+ \rightleftharpoons Cl^\bullet + H_2O \quad k_2 = 2.1 \times 10^{10}\,L\,(mol \times s)^{-1} \qquad (12.2)$$

$$ClOH^{\bullet -} \rightleftharpoons Cl^- + OH^\bullet \quad k_3 = 6.1 \times 10^9\,L\,(mol \times s)^{-1} \qquad (12.3)$$

$$Cl^\bullet + Cl^- \rightleftharpoons Cl_2^{\bullet -} \quad k_4 = 2.0 \times 10^{10}\,L\,(mol \times s)^{-1} \qquad (12.4)$$

$$Cl_2^{\bullet -} + OH^\bullet \rightleftharpoons HOCl^\bullet + Cl^- \quad K = 1 \times 10^9\ (2 < pH < 5) \qquad (12.5)$$

Acid dye Orange II reacts with $Cl_2^{\bullet -}$ at a rate of $3.7 \times 10^7\,L\,(mol \times s)^{-1}$, approximately 100 times more slowly than with $OH^\bullet$.

Kiwi *et al.* (2000) intensively studied the underlying mechanism and kinetics of $OH^\bullet$ intervention during Fenton's oxidation of Orange II in the presence of chloride ions added as NaCl. Near-UV light irradiation decreased the amount of organo-chloride complexes and AOX that were formed in the presence of background chloride. Under $OH^\bullet$ scavenging conditions $Cl_2^{\bullet -}$ significantly contributed to dye oxidation instead of $OH^\bullet$ at a (bimolecular) rate being similar in magnitude.

Gaehr *et al.* (1994) did not observe an inhibition during ozonation of reactive dyes in the presence of NaCl at the natural high pH of spent reactive dyes. This is consistent with the findings of Arslan and Akmehmet Balcioglu (2000) that did not evidence a decrease in COD and decolourisation rate constants established for the ozonation of Reactive Black 5 (vinylsulphone anchor group) and Reactive Yellow 84 (aminochlorotriazine anchor group) hydrolysates. Decrease in the COD removal rate (1–$70\,g\,L^{-1}$) was only observed in the presence of Na_2CO_3 (1–$20\,g\,L^{-1}$) at alkaline pH. In other words, at the natural high pH (=11.8) of spent vinylsulphone dye-baths, no significant inhibiting effect of NaCl ($70\,g\,L^{-1}$) on colour, UV_{254} and COD removal would be expected for the treatment with ozone.

12.5.2 Effects of HCO_3^-/CO_3^{2-} alkalinity

According to several former investigations, a possible mechanism for the reaction between the reactive dyestuffs and ozone in the presence of HCO_3^-/CO_3^{2-} alkalinity can be proposed as follows (Glaze and Kang, 1989; Reife and Freeman, 1996; Staehelin and Hoigné, 1984):

Initiation

$$O_3 + OH^- \rightarrow HO_2^- + O_2 \quad k_1 = 70\,M^{-1}s^{-1} \qquad (12.6)$$

$$O_3 + D \rightarrow Products \quad k_{D,O_3} = 10^8\text{–}10^9\,s^{-1} \qquad (12.7)$$

Propagation

$$HO_2^- + O_3 \rightarrow O_3^{\bullet -} + HO_2^\bullet \quad k_2 = 2.8 \times 10^6\,M^{-1}s^{-1} \qquad (12.8)$$

$$HO_2^\bullet \rightarrow O_2^{\bullet -} + H^+ \quad pK_3 = 4.8 \qquad (12.9)$$

$$O_2^{\bullet -} + O_3 \rightarrow O_3^{\bullet -} + O_2 \quad k_4 = 1.6 \times 10^9\,M^{-1}s^{-1} \qquad (12.10)$$

$$O_3^{\bullet -} + H^+ \rightarrow HO_3^{\bullet} \quad k_5 = 5 \times 10^{10} M^{-1}s^{-1} \tag{12.11}$$

$$HO_3^{\bullet} \rightarrow OH^{\bullet} + O_2 \quad k_6 = 1.4 \times 10^5 M^{-1}s^{-1} \tag{12.12}$$

Textile dye advanced oxidation

$$OH^{\bullet} + D \rightarrow DOH^{\bullet} \rightarrow DO_2^{\bullet} \rightarrow \text{Products} \tag{12.13}$$

where D stands for the dye molecules ($k_{D,OH^{\bullet}} \approx 10^9$–$10^{10} M^{-1}s^{-1}$) and

$$HO_2^{\bullet} + O_2^{\bullet -} \rightarrow HO_2^- + O_2 \quad k_8 = 8.7 \times 10^7 M^{-1}s^{-1} \tag{12.14}$$

$$HCO_3^- + OH^{\bullet} \rightarrow HCO_3^{\bullet} + OH^- \quad k_9 = 1.5 \times 10^7 M^{-1}s^{-1} \tag{12.15}$$

$$CO_3^{2-} + OH^{\bullet} \rightarrow CO_3^{\bullet} + OH^- \quad k_{10} = 4.2 \times 10^8 M^{-1}s^{-1} \tag{12.16}$$

Concerning Equations (12.15) and (12.16), the distribution of both species in dependence on the pH has to be considered. Similar to reactions (12.15) and (12.16), $OH^{\bullet}$ may be consumed by other constituents of the specific water system. Due to its extremely high reactivity, $OH^{\bullet}$ is not a very effective oxidising agent in the presence of alkalinity, however, $OH^{\bullet}$ and CO_3^{2-} may further initiate organic-radical-type chain reactions. For instance, carbonate radials ($HCO_3^{\bullet}$ and $CO_3^{\bullet -}$) formed under high alkalinity conditions are comparatively stable (long lived) radical species and possibly reactive enough to cleave the chromophoric groups of the dyestuffs as well as single chain aromatic moieties (Buxton *et al.*, 1988). The formed carbonate radicals may undergo self-reactions or reactions with aromatic compounds, such as dye molecules (Acero and Van Gunten, 2000; Westerhoff *et al.*, 1997):

$$HCO_3^{\bullet} \leftrightharpoons CO_3^{\bullet -} + H^+ \quad pK_a = 8.0 \tag{12.17}$$

$$CO_3^{\bullet -} + D \rightarrow CO_3^{2-} + \text{Products} \quad k_{D,CO_3^{\bullet -}} = 10^7\text{–}10^9 M^{-1}s^{-1} \tag{12.18}$$

Accordingly, dye decomposition may not be inhibited by the carbonate/bicarbonate radical scavenger system as a consequence of the bicarbonate radical formation at pH 7 and the presence of carbonate radicals at pH 12 (Arslan *et al.*, 2002, in press).

An explanation for the relatively higher reaction rates of dye ozonation generally obtained at pH 12 with and without soda ash as compared to pH 7 (no carbonate) might be the reaction of dye molecules with $O^{\bullet -}$, that is the ionised (de-protonated) form of $OH^{\bullet}$, that are present at the concentration $[O^{\bullet -}] \approx 2[OH^{\bullet}]$ at pH 12 (Buxton *et al.*, 1989) according to

$$OH^{\bullet} \leftrightharpoons O^{\bullet -} + H^+ \quad pK_a = 11.9 \tag{12.19}$$

$O^{\bullet -}$ radicals are known to react with organic compounds in an order of magnitude slower than $OH^{\bullet}$, and almost immeasurably slowly with inorganic ions, such as CO_3^{2-} or HCO_3^-. Thus, it probably functions as the dominant and selective oxidising agent at pH 12 ($k_{D,O^{\bullet -}} = 10^8$–$10^9 M^{-1}s^{-1}$; Buxton *et al.*, 1988). It may be inferred that for the thorough and rapid oxidation of the dye constituents in the spent dye-bath liquor, molecular ozone or at least the above-mentioned radical species, which are more stable than $OH^{\bullet}$, are required.

To study the individual effects of the common dye assisting chemicals, exhausted reactive dye-bath effluents have been simulated by mixing synthetically hydrolysed reactive dyes with proper amounts of salts and alkali. Some of these studies are explained below.

Neppolian *et al.* (2002) studied the effect of NaCl and Na_2CO_3 on solar light induced TiO_2-assisted photobleaching of Reactive Blue 24. Increasing the initial dye concentration had a negative effect on the apparent first-order decolourisation rate constants. TiO_2/UVA oxidation was not affected by H_2O_2 and pH values ranging between 5–25 mM and 5–10, respectively, beyond which interval photocatalytic dye degradation was seriously inhibited. Chloride and sodium carbonate in the range of 0.25–2 g L^{-1} exhibited a negative impact on colour and COD degradation.

12.5.3 Other common anions

The effects of common ions (acetate, sulphate, oxalate, nitrate, biphosphate and also carbonate) on the photocatalytic oxidation of the textile dye Astrazone Orange G was studied by Sökmen and Özkan (2002). They observed that decolourisation rates were significantly retarded in the presence of anions due to competitive adsorption on the catalyst surface and their valence band hole (h^+) and $OH^\bullet$ scavenging effect.

Xu and Li (2000) investigated the UV/Fe^{3+} oxidation of the reactive azo dye Brilliant Red X-3b. The inhibition in the presence of SO_4^{2-} and Cl^- was mainly attributed to dual $OH^\bullet$ scavenging and complex formation effects. In contradiction to literature data, the inhibition by SO_4^{2-} was much more pronounced than by Cl^- probably as an indirect result of the different rates $(OH)^{2+}$ UV photolysis rates in the presence of anion complexation variations.

12.6 MERITS OF AOPs

The photodegradation of aqueous organic pollutants is an energy-intensive process, thus an appropriate 'figure-of-merit' definition based on the energy input seems logical. In order to compare treatment efficiencies for the different oxidative processes, it is important to use a general expression and relate the treatment efficiencies directly to operational treatment (here: electrical energy) costs. For this purpose, the EE/O defined as kWh electrical energy in terms of kWh required to degrade a contaminant or environmental group parameter by one order of magnitude in m^3 of contaminated water or wastewater was calculated for the photochemical treatment process (Bolton and Cater, 1994). Since the EE/O concept assumes first-order degradation kinetics with respect to pollutant concentration for batch treatment the following formula can be applied:

$$\text{EE/O} = \frac{P \times t}{\log(D_0/D) \times V} = \frac{P \times 38.40}{V \times k_d} \tag{12.20}$$

where

EE/O = electrical energy per order ($kWh\,m^{-3} \times$ order),
P = power of the oxidation unit (kW),
38.4 = a conversion factor,
V = volume of treated water or wastewater (L),
D_0, D = initial and final concentration of the pollutant (dye) ($mol\,L^{-1}$),
k_d = pseudo-first order degradation (here: decolourisation) rate constant for the pollutant (dye) in question (min^{-1}).

The appropriate figure-of-merit for zero order TOC abatement is the electrical energy per mass (EE/M) defined as the kWh of electrical energy required to remove 1 kg of the contaminant or organic carbon (Bolton and Cater, 1994):

$$\text{EE/M} = \frac{P \times t}{V \times (C_i - C_t)} \tag{12.21}$$

where

EE/M = electrical energy per mass ($kWh\,kg^{-1}$ TOC),
P = power input of the UV lamp (kW),
t = treatment time (h),
V = volume of treated aqueous pollutant or wastewater (m^3)
C_i, C_t = initial and final pollutant concentrations or collective environmental parameter (i.e. COD, TOC, in $kg\,m^{-3}$), respectively.

Calculation of the electrical energy requirements has proven to be very useful for AOPs employing artificial light sources (UV lamps) because a major fraction of the operating costs of AOPs are the electrical energy costs. CCOT (1995) demonstrated that AOPs, such as their patented H_2O_2/UV and Fenton/UV, provide an effective, economically justifiable treatment alternative for poor strippers and poor adsorbers, in other words non-volatile compounds and very hydrophilic compounds, such as water soluble textile dyes. Generally, speaking an EE/O value $\leqslant 10\,kWh\,m^{-3}$ per order is regarded as an economically acceptable power requirement.

Practical considerations when performing a cost comparison analysis are given below:

- At concentrations exceeding $10\,mg\,L^{-1}$ adsorbate, activated may become more expensive than the UV/oxidant system. Operating costs for AOPs are much less dependent on pollutant concentration.
- Regardless of concentration, costs for UV/oxidant systems (2–6 US$/1000 US gal for $10\,mg\,L^{-1}$ contaminant) are much less than for poor carbon absorbers (>25 US$/1000 US gal).
- For average carbon absorbers (1–25 US$/1000 US gal for $10\,mg\,L^{-1}$ contaminant), the most economical technology must be chosen on a case by case basis.
- AOP capital costs are typically 2–3 times higher than for GAC columns. Therefore, longer-term projects favour UV/oxidant systems whenever cumulative savings in operating costs offset the higher capital expense.

Table 12.3 Treatment of textile dyes with photochemical AOPs; comparison of process performances, EE/O and EE/M values.

Reference	Textile dye	AOP type	k_d (min^{-1}) or colour removal rate and time	COD removal incl. time (%)	TOC removal incl. time (%)	EE/O ($kWh\,m^{-3}\,order^{-1}$)	EE/M ($kWh\,kg^{-1}$ TOC)
Hustert and Zepp (1992)	Acid Orange 7	TiO_2/UVA	0.017		80 in 3 h	849	1609.5
Shu *et al.* (1994)	Acid Red 1	H_2O_2/UVC	0.717			19	
Shu *et al.* (1994)	Acid Yellow 23	H_2O_2/UVC	0.183			11	
Marechal *et al.* (1997)	Reactive Red 120	H_2O_2/UVC	70% in 2 min		94 in 6 min	17	22
Ince and Gönenç (1997)	Reactive Black 5	H_2O_2/UVC	0.88, 100% in 4 min	74 in 20 min	90 in 20 min	1.5	11
Ince and Tezcanli, (1999)	Reactive Red 141	Fe(II)/H_2O_2/UVC	0.62		79 in 20 min	1.0	7.4
Colonna *et al.* (1999)	Acid Red 1	H_2O_2/UVC	0.063			204	
Arslan and Balcioglu (1999)	Reactive Black 5	Fe(III)/ H_2O_2/UVA	0.54	81 in1 h		1.4	
Arslan and Balcioglu (1999)	Reactive Blue 21	TiO_2/UVA	81% in 2 h		68 in 2 h	56	81
Poulios and Aesopoulou (1999)	Reactive Orange 16	TiO_2/UVA	80% in 15 min			107	
Poulios and Tsachpinis (1999)	Reactive Black 5	TiO_2/UVA				104	
Tanaka *et al.* (2000)	Acid Orange 7	TiO_2/UVA	0.024		78 in 90 min	32,000	$>2 \times 10^6$
Xu and Lu (2000)	Reactive Brilliant Red X3b	TiO_2/UVA	75% in 50 min			1038	
Arslan-Alaton and Balcioglu (2001)	Reactive Black 5	H_2O_2/UVC	0.15	72 in 2 h	60 in 2 h	2.6	36
Arslan-Alaton and Balcioglu (2001)	Reactive Black 5	TiO_2/UVA	0.085	80 in 2 h	75 in 2 h	9.1	66
Chen *et al.* (2001)	Reactive Red 15	TiO_2/UVA/H_2O_2	0.024			3.0	
Galindo *et al.* (2001)	Acid Blue 74	TiO_2/UVA	90% in 5 min			2.5	
Galindo *et al.* (2001)	Acid Blue 74	TiO_2/UVA	80% in 12 min			10.3	
Diaper *et al.* (2002)	Reactive Black 5	TiO_2/UVC	0.0238		100% in 24 h	1807.05	4480
Diaper *et al.* (2002)	Acid Orange 7	TiO_2/UVC	0.0114		100% in 24 h	3772.63	4480
Sökmen and Özkan (2002)	Astrozone Orange G	TiO_2/UVA	60% in 4 min			9.72	

Herein, these empirical formulae have been applied to estimate electrical energy requirements on the basis of kg of TOC, COD or orders of colour removed during the treatment with UVC and UVA-driven photochemical AOPs. Table 12.3 summarises EE/O and EE/M values based on literature data available for dye degradation via photochemical (UV-light driven) AOPs, wherever these calculations have been possible. From Table 12.3 it is evident that although approximately the same initial dye concentration ($<100\,\mathrm{mg\,L^{-1}}$) has been used in all case studies, there are immense gaps between calculated EE/O and EE/M values as a consequence of the dye-type/structure and $OH^{\bullet}$ production method employed. Particularly the merits of TiO_2/UV-type AOPs (heterogeneous photocatalytic systems) appear to be to energy intensive for practical applications. Finally speaking, the high EE/O and EE/M values calculated for the TiO_2/UVA process show support that the fact TiO_2/UVA has only the potential for commercialisation if an alternative treatment system using solar (natural) light can be realised.

12.7 CONCLUSIONS

Treatment of textile dyes in an aqueous dye stream has attracted the attention of environmentalists, technologists and entrepreneurs because of its socio-economic and political dimensions. Generally speaking, there is no simple, single stage method available for treating recalcitrant textile dyes. The inherent advantages of AOPs over more conventional advanced treatment techniques, such as their simple operation, relatively fast oxidation rates and destructive nature have to be carefully outweighed against their high operation costs as well as the difficulties encountered in the control of accumulating oxidation products. An important disadvantage of AOPs is the virtually impossible control of the formation of reaction products being in some cases more toxic than the parent dye molecules as a result of the extremely rapid free radical reactions.

For instance in the case of supercritical water oxidation processes, qualified, experienced staff (i.e. post-graduate researchers) are required that would be always confronted with explosion hazards as a consequence of the extremely severe reaction conditions. However, high temperature, pressurised oxidation systems have been installed in the Far East and seem to be an alternative solution for highly concentrated dye wastes. Alternatively, the use of heterogeneous catalysts may reduce treatment times and the harshness of reaction conditions immensely.

In terms of COD and TOC removal efficiencies, the treatment performance of ambient temperature AOPs, such as Fenton's reaction, ozonation combined with H_2O_2 and/or UVC does not appear to be sufficient to make the processes practicable.

Particularly for photochemical AOPs, where usually UVC light irradiation is employed, the energy efficiency of the lamps as well as operating costs associated with artificial light sources is of prime importance. Currently, the best free radical production per unit of energy input does not exceed 25%, and the energy efficiency of solar light systems (TiO_2/UVA or TiO_2/near-UV–Vis) is typically less than 1%. Supported TiO_2 systems (fixed film beds) seriously suffer from reduced photocatalytic activity, although a costly separation procedure by means of microfiltration could be avoided.

The use of aqueous Fe^{2+}, Fe^{3+} and other transition metal salt solutions applied as homogenous catalysts will lead to a secondary pollution problem (sludge production after pH re-adjustment, residual ferrous ion in the treated effluent).

Common for all AOPs including pressurised oxidation systems is that a certain inert TOC fraction in the form of low molecular weight carboxylic acids (acetic, formic, maleic, etc.) remains in the treated effluent and does further degrade with extended treatment time. Mineralisation also increases the salinity (dissolved solids) content of the final effluent.

Since one of the major concerns in AOPs is the formation of undesirable reaction intermediates, a quick mineralisation of an organic compound suppressing the accumulation of dye intermediates (i.e. benzene, nitrobenzene, aminophenols) should be one of the major goals aside a biotreatability improvement. A combinative AOP treatment system could serve to tackle such an issue and hence deserves careful consideration in future efforts of environmental remediation.

Conclusively, although an immense number of AOPs have been proposed for textile dye treatment recently, a somewhat deeper elucidation of these processes shows clearly that total oxidation is technically and economically not yet feasible.

REFERENCES

Acero, J.L. and Van Gunten, U. (2000) Influence of carbonate on the ozone/hydrogen peroxide based advanced oxidation processes for drinking water treatment. *Ozone Sci. Eng.* **22**, 305–328.

Alvares, A.B.C., Diaper, C. and Parsons, S.A. (2001) Partial oxidation by ozone to remove recalcitrance from wastewaters – a review. *Environ. Technol.* **22**, 409–427.

Anbar, M. and Neta, P. (1967) A compilation of specific bimolecular rate constants for the reactions of hydrogen atoms and hydroxyl radicals with inorganic and organic compounds in aqueous solution. *Int. J. Appl. Radiat. Isot.* **16**, 493–523.

Aplin, R. and Waite, T.D. (2000) Comparison of three advanced oxidation processes for the degradation of textile dyes. *Water Sci. Technol.* **42**(5–6), 345–354.

Arslan, I. (2000) *Treatment of Reactive Dye-bath Effluents by Heterogeneous and Homogenous Advanced Oxidation Processes*. PhD Thesis, Bogaziçi University, Institute of Environmental Sciences, Istanbul, Turkey.

Arslan, I. (2001) Treatability of a simulated disperse dye-bath by ferrous iron coagulation, ozonation and ferrous iron-catalyzed ozonation. *J. Haz. Mat.* **85**(3), 229–241.

Arslan, I. and Akmehmet Balcioglu, I.A. (1999) Degradation of commercial reactive dyestuffs by heterogeneous and homogenous advanced oxidation processes: a comparative study. *Dyes Pigments* **43**, 189–195.

Arslan, I. and Akmehmet Balcioglu, I. (2000) Effect of common reactive dye auxiliaries on the ozonation of vinylsulphone and aminochlorotriazine containing dyehouse effluents. *Desalination* **130**, 61–71.

Arslan, I. and Akmehmet Balcioglu, I. (2001) Degradation of Remazol Black B dye and its simulated dyebath wastewater by advanced oxidation processes in heterogeneous and homogenous media. *Col. Tech.* **117**, 19–23.

Arslan-Alaton, I. and Akmehmet Balcioglu, I. (2001) Photochemical and heterogeneous photocatalytic degradation of waste vinylsulphone dyes: a case study with hydrolyzed reactive black 5. *J. Photochem. Photobiol. A* **141**, 247–254.

Arslan-Alaton, I. and Ferry, J.L. (2002) Application of polyoxotungstates as environmental catalysts: wet air oxidation of acid dye Orange II. *Dyes Pigments* (in press).

Arslan-Alaton, I., Kornmueller, A. and Jekel, M.R. (2002) Ozonation of spent reactive dye-baths: effect of HCO_3^-/CO_3^{2-} alkalinity. *ASCE J. Environ. Eng.* **128** (in press).

Bauer, C., Jacques, P. and Kalt, A. (2001) Photooxidation of an azo dye induced by visible light incident on the surface TiO_2. *J. Photochem. Photobiol. A* **140**, 87–92.

Beszedits, S. (1980) Ozonation to decolour textile effluents. *Am. Dyest. Rep.* **69**, 37–55.

Bolton, J.R. and Cater, S.R. (1994) Homogenous photodegradation of pollutants in contaminated water: an introduction. In: *Aquatic and Surface Photochemistry*, Heltz, G.R., Zepp, R.G. and Crosby, D.G. (eds). Lewis Publishers, Florida, pp. 467–490.

Buckley, C.A. (1992) Membrane technology for the treatment of dyehouse effluents. *Water Sci. Technol.* **25**(10), 203–209.

Buxton, G.V., Greenstock, C.L., Helman, W.P. and Ross, A.B. (1988) Critical review of rate constants for reaction of hydrated electrons, hydrogen atoms and hydroxyl radicals ($^{\bullet}OH/^{\bullet}O^-$) in aqueous solution. *J. Phys. Chem. Ref. Dat.* **17**, 513–886.

CCOT (1995). *The AOT Handbook*. CCOT, Markham, Ontario.

Chen, G., Lei, L. and Yue, P.L. (1999) Wet oxidation of high-concentration reactive dyes. *Ind. Eng. Chem. Res.* **38**, 1837–1843.

Chu, W. and Ma, C.W. (2000) Quantitative prediction of direct and indirect ozonation kinetics. *Water Res.* **34**(12), 3153–3160.

Chun, H., Yizhong, W. and Hongxiao, T. (2001) Preparation and characterization of surface bond-conjugated TiO_2/SiO_2 and photocatalysis for azo dyes. *Appl. Cat. B* **30**, 277–285.

Chuchley, J.H. (1995) The water company's view. In: *Colour in Dyehouse Effluent*, Cooper, P. (ed.). The Society of Dyers and Colourists, Alden Press, Oxford, pp. 31–43.

Colonna, G.M., Caronna, T. and Marcandalli, B. (1999) Oxidative degradation of dyes by ultraviolet radiation in the presence of hydrogen peroxide. *Dyes Pigments* **41**, 211–220.

Correia, V.M., Stephenson, T. and Judd, S.J. (1994) Characterization of textile wastewaters – a review. *Environ. Technol.* **15**, 917–929.

Crigeé, R. (1975) Mechanismus der Ozonolyse. *Angewandte Chemie* **87**, 765.

Diaper, C.D., Jarvis, P. and Parsons, S.A. (2002) Combined membrane photocatalysis process for dye waste treatment: adsorption and degradation studies. *Advanced Oxidation Processes for Water and Wastewater Treatment*, Cranfield, UK, April 3.

Donlagic, J. and Levec, J. (1997) Oxidation of an azo dye in subcritical aqueous solutions. *Ind. Eng. Chem. Res.* **36**, 3480–3486.

Donlagic, J. and Levec, J. (1998) Comparison of catalyzed and non-catalyzed oxidation of azo dye and effect on biodegradability. *Environ. Sci. Technol.* **32**(9), 1294–1302.

Easton, J.R. (1995) The dye maker's view. In: *Colour in Dyehouse Effluent*, Cooper, P. (ed.). The Society of Dyers and Colourists, Alden Press, Oxford, pp. 6–21.

Eilbeck, W.J. and Mattock, G. (1985) *Chemical Processes in Wastewater Treatment*. Ellis Horwood Ltd. Publishers, John Wiley and Sons, Inc., Chichester, pp. 117–123.

Galindo, C. and Kalt, A. (1999) UV/H_2O_2 oxidation of azo dyes in aqueous media: evidence of a structure – degradability relationship. *Dyes Pigments* **42**, 199–207.

Gaehr, F., Hermanutz, F. and Oppermann, W. (1994) Ozonation – an important technique to comply with new german laws for textile wastewater treatment. *Water Sci. Technol.* **30**(3), 255–263.

Georgiou, D., Melidis, P., Aivasidis, A. and Gimouhopoulus, K. (2002) Degradation of azo reactive dyes by ultraviolet radiation in the presence of hydrogen peroxide. *Dyes Pigments* **52**, 69–78.

Glaze, W.H. and Kang, J.W. (1989) Advanced oxidation processes: test of a kinetic model for the oxidation of organic compounds with ozone and hydrogen peroxide in a semibatch reactor. *Ind. Eng. Chem. Res.* **28**(11), 1580–1587.

Glaze, W.H., Kang, J.W. and Chapin, D.H. (1987) The chemistry of water treatment processes involving ozone, hydrogen peroxide and ultraviolet radiation. *Oz. Sci. Eng.* **9**, 335–352.

Gonzalves, M.S.T., Oliveira-Campos, A.M.F., Pinto, E.M.M., Plasencia, P.M.S. and Queiroz, M.J.R.P. (1999) Photochemical treatment of solutions of azo dyes containing TiO_2. *Chemosphere* **39**(5), 781–786.

Gould, J.P. and Groff, K.A. (1987) The Kinetics of ozonolysis of synthetic dyes. *Ozone Sci. Eng.* **9**, 153–164.

Herrera, F., Kiwi, J., Lopez, A. and Nadtochenko, V. (1999) Photochemical decolouration of Remazol Brilliant Blue and Uniblue A in the presence of Fe^{3+} and H_2O_2. *Environ. Sci. Technol.* **33**, 3145–3151.

Imamura, S. (1999) Catalytic and noncatalytic wet oxidation. *Ind. Eng. Chem. Res.* **38**, 1743–1753.

Imamura, K., Hiramatsu, A., Imada, M., Sakiyama, T., Tanaka, A., Yamada, Y. and Nakahashi, K. (2000) Development of a novel advanced oxidation system using combined UV/H_2O_2 technique and kinetic analysis for decomposition of dye solutions. *J. Chem. Eng. Jpn.* **33**(2), 253–261.

Ince, N.H. and Gönenç, D.T. (1997) Treatability of a textile azo dye by UV/H_2O_2. *Environ. Technol.* **18**, 179–185.

Ince, N.H. and Tezcanlý, G. (1999) Treatability of textile dye-bath effluents by advanced oxidation: preparation for reuse. *Water Sci. Technol.* **40**(1), 183–190.

Ince, N.H. and Tezcanli, G. (2001) Reactive dyestuff degradation by combined sonolysis and ozonation. *Dyes Pigments* **49**, 145–153.

Kiriakidou, F., Kondarides, D.I. and Verykios, X.E. (1999) The effect of operational parameters and TiO_2 doping on the photocatalytic degradation of azo-dyes. *Catal. Today* **54**, 119–130.

Kiwi, J., Lopez, A. and Nadtochenko, V. (2000) Mechanism and kinetics of the OH-radical intervention during Fenton Oxidation in the presence of a significant amount of radical scavenger (Cl^-). *Environ. Sci. Technol.* **34**(11), 2162–2168.

Kuo, W.G. (1992) Decolourizing dye wastewater with Fenton's Reagent. *Water Res.* **26**, 881–886.

Law, S.E., Wu, J., Hitchcock, D.R. and Eitemann, M.A. (1996) Ozone decolourization of cotton dyehouse wastewater. In: *International ASAE Meeting*, Phoenix, Arizona.

Legrini, O., Oliveros, E. and Braun, A.M. (1993) Photochemical processes. *Chem. Rev.* **93**, 671–698.

Lei, L., Chen, G., Hu, X. and Yue, P.L. (2000) Homogenous catalytic wet-air oxidation for the treatment of textile wastewater. *Water Environ. Res.* **72**, 147–151.

Marechal, M.L., Slokar, Y.M. and Taufer, T. (1997) Decolouration of chlorotriazine reactive azo dyes with H_2O_2/UV. *Dyes Pigments*, **33**, 181–298.

Masten, S.J. and Davies, H.R. (1993) The use of ozone and other strong oxidants for hazardous waste management. In: *Environmental Oxidants*, Nriagen, J.O. and Simmons, M.S. (eds). John Wiley and Sons, New York.

Morrison, C., Bandara, J. and Kiwi, J. (1996) Sunlight induced decolouration/degradation of non-biodegradable Orange II dye by advanced oxidation technologies in homogenous and heterogeneous media. *J. Adv. Oxid. Technol.* **1**(2), 160–169.

Nam, S., Renganathan, V. and Tratnyek, P.G. (2001) Substituent effects on azo dye oxidation by the Fe(III)-EDTA-H_2O_2 system. *Chemosphere* **45**, 59–65.

Namboodry, C.G. and Walsh, W.K. (1995) Decolourizing spent dyebath with hot peroxide. *Am. Dyest. Rep.* **9**, 86–95.

Namboodry, C.G. and Walsh, W.K. (1996) Ultraviolet light/hydrogen peroxide system for decolourizing spent reactive dyebath waste water. *Am. Dyest. Rep.* **85**(3), 15–25.

Neppolian, B., Choi, H.C., Sakthivel, S., Arabindoo, B. and Murugesan, V. (2002) Solar light induced and TiO_2-assisted degradation of textile dye reactive blue 4. *Chemosphere* **46**, 1173–1181.

Ollis, D.F., Pelizzetti, E. and Serpone, N. (1989) Heterogeneous photocatalysis in the environment: application to water purification. In: *Photocatalysis: Fundamentals and Applications*, Serpone, N. and Pelizetti, E. (eds). John Wiley and Sons, Inc., New York, pp. 603–637.

O'Neill, C., Hawkes, F.R., Hawkes, D.L., Lourencio, N.D., Pinheiro, M. and Deleé, W. (1999) Colour in textile effluents – sources, measurement, discharge consents and simulation: a review. *J. Chem. Techol. Biotechnol.* **74**, 1009–1018.

Pagga, U. and Brown, D. (1986) The degradation of dyestuffs: part ii; behavior of dyestuffs in aerobic biodegradation tests. *Chemosphere* **15**, 479–491.

Peralta-Zamora, P., Kunz, A., Gomes de Moraes, S., Pelegrini, R., Moleiro, P.C., Reyes, J. and Duran, N. (1988) Degradation of reactive dyes 1. a comparative study of ozonation, enzymatic and photochemical processes. *Chemosphere* **38**(4), 835–852.

Poon, C.S., Huang, Q. and Fung, P.C. (1998) Degradation kinetics of Cuprophenyl Yellow RL by UV/H_2O_2/ultrasonication process in aqueous solution. *Water Sci. Technol.* **38**(5), 1005–1014.

Poulios, I. and Aetopoulou, I. (1999) Photocatalytic degradation of the textile dye Reactive Orange 16 in the presence of TiO_2 suspensions. *Environ. Technol.* **20**, 479–487.

Poulios, I. and Tsachpinis, I. (1999) Photodegradation of the textile dye Reactive Black 5 in the presence of semiconducting oxides. *J. Chem. Technol. Biotechnol.* **74**, 349–37.

Prevot, A.B., Baiocchi, C., Brussino, M.C., Pramauro, E., Augugliaro, V., Marci, G. and Palmisano, L. (2001) Photocatalytic degradation of Acid Blue 80 in aqueous solution containing TiO_2 dispersions. *Environ. Sci. Technol.* **35**(5), 971–976.

Raffaner, I.I. and von Rohr, P.R. (2001) Promoted wet oxidation of the azo dye Orange II under mild conditions. *Ind. Eng. Chem. Res.* **40**(4), 1083–1089.

Reife, A. (1993) Dyes, environmental chemistry. In: *Kirk – Othmer Encyclopedia of Chemical Technology.* John Wiley and Sons, Inc., New York, pp. 753–784.

Reife, A. and Freeman, H.S. (1996) Ozonation. In: *Environmental Chemistry of Dyes and Pigments.* John Wiley and Sons, Inc., New York, pp. 43–60.

Reife, A., Betowski, D. and Freeman, H.S. (1998) Dyes and Pigments, Environmental Chemistry. In: *Encyclopedia of Environmental Analysis and Remediation*, Meyers, R.A. (ed.). John Wiley & Sons, Inc., New York, pp. 1442–1464.

Reife, A. and Freeman, H.S. (2000) Pollution prevention in the production of dyes and pigments. *AATCC J.* **32**(1), 56–60.

Rice, R.G. (1996) Application of ozone for industrial wastewater treatment – a review. *Ozone Sci. Eng.* **18**, 477–515.

Robinson, T., McMullan, G., Marchant, R. and Nigam, P. (2001) Remediation of dyes in textile effluent: a critical review on current treatment technologies with a proposed alternative. *Biores. Technol.* **77**, 247–255.

Rys, P. and Zollinger, H. (1972) *Fundamentals of Chemistry and Application of Dyes.* John Wiley & Sons, Inc., USA.

Shen, Y.S. and Wang, D.K. (2002) Development of photoreactor design equation for the treatment of dye wastewater by UV/H_2O_2 process. *J. Haz. Mat.* **B89**, 267–277.

Shende, R.V. and Mahajani, V.V. (1994) Wet air oxidation of anthraquinone and phthalocyanine class of reactive dyes. *Am. Dyest. Rep.* **83**(6), 40–46.

Shore, J. (1995) *Cellulosics Dyeing.* Society of Dyers and Colourists, Alden Press, Oxford.

Shu, H.Y. and Huang, C.R. (1995) Degradation of commercial azo dyes in water using ozonation and UV enhanced ozonation process. *Chemosphere* **39**, 3813–3825.

Shu, H.Y., Huang, C.R. and Chang, M.C. (1994) Decolourization of mono-azo dyes in wastewater by advanced oxidation process: a case study of Acid Red 1 and Acid Yellow 23. *Chemosphere* **29**, 2597–2607.

Slokar, Y.M. and Marechal, A.M. (1998) Methods of decolouration of textile wastewaters. *Dyes Pigments* **33**(4), 335–356.

Snowdon-Swan, L.J. (1995) Pollution prevention in the textile industries. In: *Industrial Pollution Prevention Handbook*, Freeman, H.M. (ed.). McGraw-Hill, Inc., New York, pp. 829–845.

So, C.M., Cheng, M.Y., Yu, J.C. and Wong, P.K. (2002) Degradation of the azo dye Procion Red MX-5B by photocatalytic oxidation. *Chemosphere* **46**, 905–912.

Solozhenko, E.G., Soboleva, N.M. and Goncharuk, V.V. (1995) Decolourization of azo dye solutions by Fenton's oxidation. *Water Res.* **29**, 2206–2210.

Sökmen, M. and Özkan, A. (2002) Decolourizing textile wastewater with modified titania: the effect of inorganic ions on the photocatalysis. *J. Photochem. Photobiol. A* **147**, 77–81.

Spadaro, J.T., Isabelle, L. and Renganathan, V. (1994) Hydroxyl radical mediated degradation of azo dyes: evidence for benzene generation. *Environ. Sci. Technol.* **28**(7), 1389–1393.

Tang, W.Z., Zhang, Z., An, H., Quintana, M.O. and Torres, D.F. (1997) TiO_2/UV photodegradation of azo dyes in aqueous solutions. *Environ. Technol.* **18**, 1–12.

Tanaka, K., Kanjana, P. and Teruaki, H. (2000) Photocatalytic degradation of commercial azo dyes. *Water Res.* **34**(1), 327–333.

Uygur, A. and Kök, E. (1999) Decolourization treatments of azo dye wastewaters including dichlorotriazinyl reactive groups by using advanced oxidation method. *JSDC* (now *Colouration Technology*) **115**, 350–354.

Vandevivere, P.C., Bianchi, R. and Verstraete, W. (1998) Treatment and reuse of wastewater from the textile wet-processing industry: review of emerging technologies. *J. Chem. Technol. Biotechnol.* **72**, 289–302.

Vinodgopal, K., Peller, J., Makogon, O. and Kamat, P.V. (2000) Ultrasonic mineralization of a reactive textile azo dye, Remazol Black B. *Water Res.* **32**(12), 3646–3650.

Wang, Y. (2000) Solar photocatalytic degradation of eight commercial dyes in TiO_2 suspension. *Water Res.* **34**(3), 990–994.

Westerhoff, P., Song, R., Amy, G. and Minear, R. (1997) Applications of ozone decomposition models. *Ozone Sci. Eng.* **19**, 55–73.

Xu, Y.M. and Lu, H.Q. (2000) Degradation of the dye X-3B by UV/Fe(III)-generated hydroxyl radicals in aqueous solution. *J. Photochem. Photobiol. A* **136**, 73–77.

Zepp, R.G. and Hoigné, J. (1992) Hydroxyl radical formation in aqueous reactions (pH 3–8) of Iron(II) with hydrogen peroxide. *Environ. Sci. Technol.* **26**, 313–319.

Zhan, H. and Tian, H. (1998a) Photocatalytic degradation of acid azo dyes in aqueous TiO_2 suspension I. The effect of substituents. *Dyes Pigments* **37**(3), 231–239.

Zhan, H. and Tian, H. (1998b) Photocatalytic degradation of acid azo dyes in aqueous TiO_2 suspension III. Spectral characterization of the excited state. *Dyes Pigments* **37**(3), 249–254.

13

Water treatment applications

Simon A. Parsons and Anthony Byrne

13.1 INTRODUCTION

The availability of a reliable supply of water is one of the most important determinants of our health. Historically, improvements in our health have been related to improvements in our water supply system from source to tap. The quality of water we receive in the UK today (99.82% of 2.8 million samples met the standards) is achieved through improvements in source protection, water treatment, operation and maintenance, quality monitoring, and training and education. To achieve these standards, a wide range of treatment processes are used. The treatment processes operated at each of the UK's 2200 water treatment works are driven by the quality of the incoming raw waters and the requirement of the 400,000 km of water mains supplying the water to our taps.

An example of a typical treatment scheme is shown below (Figure 13.1) and consists of two stages of solids separation, clarification and filtration, and the combination of ozone and activated carbon adsorbtion process to remove dissolved organics, such as pesticides. Before the water is put into supply, chlorine is added

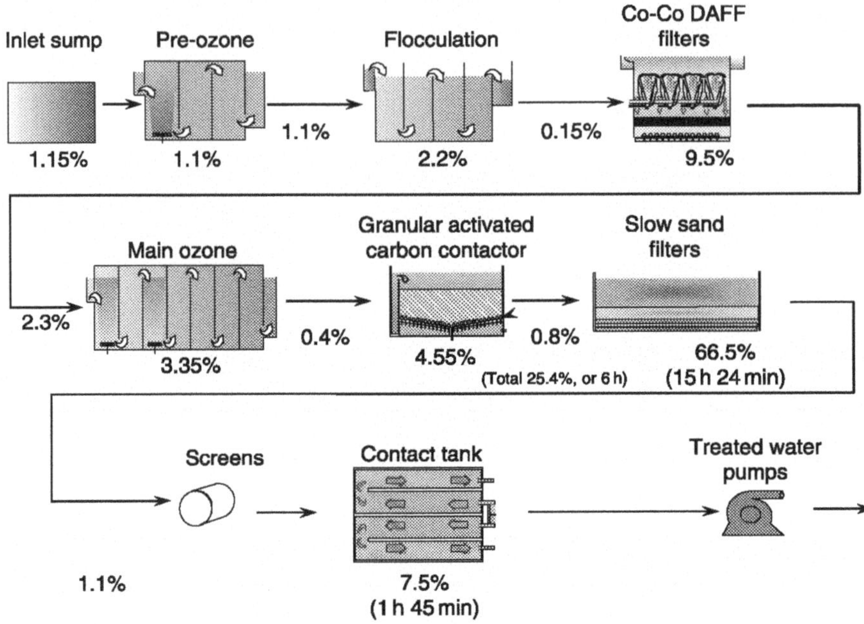

Figure 13.1 Example flowsheet of a modern surface water treatment works showing treatment processes and retention times (courtesy of Thames Water).

to disinfect the water and maintain a residual level through the distribution system and into supply.

13.2 OPPORTUNITIES FOR AOPs IN WATER TREATMENT

The number of publications dealing with advanced oxidation processes (AOPs) in the treatment of water is overwhelming, so while a number of the previous chapters have identified possible applications of individual AOPs in water treatment, this chapter aims to give the reader an insight into the use of AOPs in dealing with some current problems in drinking water treatment. Three main areas of opportunity are

1. removal of disinfection by-products (DBPs) precursors;
2. removal of micropollutants including pesticides, endocrine disrupting chemicals (and mimics), pharmaceuticals and biotoxins;
3. disinfection resistant pathogenic micro-organisms (e.g. *Cryptosporidium*).

13.2.1 DBPs

Water throughout the world contains natural organic matter (NOM) as a result of the interactions between the hydrological cycle and the biosphere and geosphere.

These interactions are responsible for the diverse nature of NOM, as the organic content of a particular water body is dependent on the surrounding environments of the biogeochemical cycles. NOM is a complex mixture of organic material and has shown to consist of organics as diverse as humic acids, hydrophilic acids, proteins, lipids, hydrocarbons and amino acids. The range of organic components in NOM varies from water to water and seasonally; this consequently leads to variations in the reactivity of NOM with chemical disinfectants, such as chlorine (Goslan *et al.*, 2002). NOM affects potable water quality as it

- enhances transportation and distribution of organic micropollutants;
- promotes DBP formation;
- provides an undesirable biological growth substrate;
- lowers the efficiency of treatment processes.

The conventional treatment process for waters containing NOM is coagulation/flocculation using inorganic coagulants, such as ferric chloride or aluminium sulphate. These processes remove NOM by adsorption onto flocs and can typically achieve 50–80% removal of DOC (Parsons *et al.*, 2002). Increasing coagulant dose will increase removal, but the process is limited when trying to reach TOC levels less than 1 mg L^{-1} (Croué *et al.*, 1993; Nowack *et al.*, 1999). The NOM problem is an intractable one, as even at these residual levels of TOC, the proposed DBP standards are difficult to achieve. The DBPs can be in the form of trihalomethanes (THMs), haloacetic acids (HAAs) and many other halogenated compounds, some of which are of major concern to the water treatment industry, as tests have shown links between cancer in laboratory animals and THMs. To control the risk, US and UK legislation has tightened in recent years to dictate the amount of THMs allowed in drinking water, the current consent in the UK is 100 $\mu g\,L^{-1}$ for total THMs (TTHMs) and 80 $\mu g\,L^{-1}$ in the US (Drinking Water Inspectorate UK, 1998; US EPA, 1998). An example of the impact increasing NOM levels has on colour and THM levels is shown in Figure 13.2.

Increasing DOC removal by increasing coagulant dose will also lead to an accompanying increase in the amount of plant residuals (sludge) generated and associated increases in costs and operators' time. In the US, enhanced coagulation is required by the Stage 1 D/DBPR and where it has been implemented sludge production increases when the coagulant dose is increased (Carlson *et al.*, 2000). The sludge produced during the coagulation of NOM is also difficult to dewater because of metal ion and organic content (Harbour *et al.*, 1999).

A range of AOPs have been tested for treating water containing commercial humic acid or NOM. The reported rate constant for the reaction between hydroxyl radicals and humic acid is 3×10^4 mg L^{-1} DOC s^{-1} and is significantly faster than ozone (Fukushima *et al.*, 2001). A recent study on UV-photo-oxidation of a reservoir water rich in NOM showed how as UV dose increased from 1000 to 26,000 mJ cm^{-2}, there was little direct oxidation of the NOM (Thompson *et al.*, 2002). There was though a substantial increase in biodegradability of the residual organic carbon showing that UV could be used as a pre-treatment to biological oxidation. Kleiser and Frimmel (2000) showed that water containing 2 mg L^{-1} of NOM was resistant

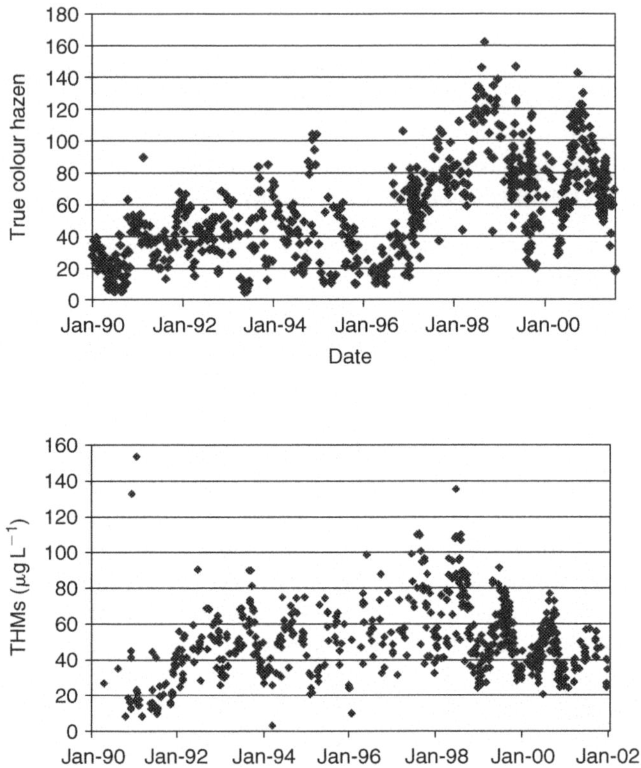

Figure 13.2 Example of annual changes in colour and THM levels as a result of elevated NOM levels.

to oxidation by ozone, but treatment with hydroxyl radicals, generated by the combination of H_2O_2 and UV ($8\,mg\,L^{-1}$), reduced the same DOC by 25% after 1050 min of treatment. In comparison, Wang *et al.* (2000) reported an 80% DOC removal from a $5\,mg\,L^{-1}$ solution of a soil humic acid in 120 min, although this was at significantly higher concentration (0.5%) of H_2O_2.

A comparison of UV/H_2O_2, Fenton's and photo-Fenton's for treating an upland catchments reservoir water (DOC $7.5\,mg\,L^{-1}$) showed how both Fenton's processes could remove excess of 90% of DOC and UV adsorbing species. This was compared to the ~70% achieved by conventional coagulation. Both Fenton's processes remove excess of 70% of raw water UV_{254} after 1 min with both achieving a maximum removal of 95%, PFR after 5 min compared to 30 min for FR, significantly quicker than UV/H_2O_2 (Figure 13.3).

Photocatalysis with UV/TiO_2 has shown significant potential for treating NOM and offers an alternative degradation treatment method, where the oxidation to CO_2 occurs via a complicated series of intermediates (Eggins *et al.*, 1997). Bekbolet (2002) also has been actively researching this area in recent years.

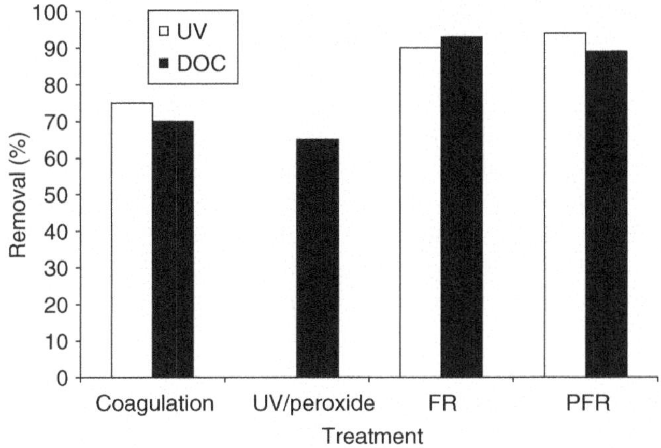

Figure 13.3 Maximum % UV_{254} and DOC removal from raw water with Fenton's, photo-Fenton's, H_2O_2/UV oxidation and conventional coagulation.

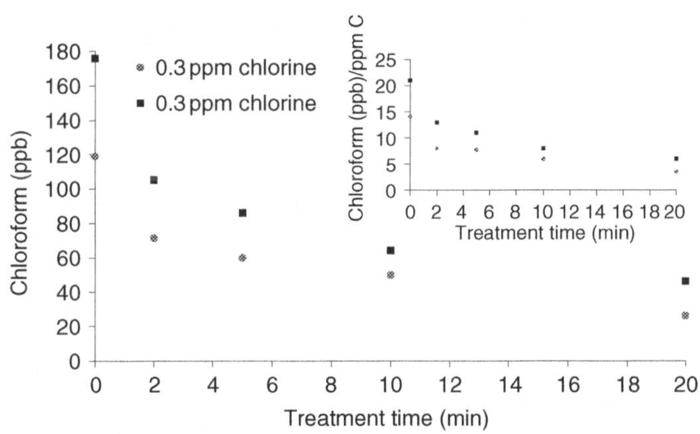

Figure 13.4 The reduction in chloroform formation after the chlorination (two doses) of humic waters treated by semiconductor photocatalysis (for different times). Insert shows the chloroform to dissolved organic carbon ratio over time.

One important aspect of the work carried out by Palmer (1998) involved the determination of the chlorinated by-products following semiconductor photocatalysis. The photocatalytic degradation of humic acids, was in the main, investigated using *Degussa P25* TiO_2 and humic acid (*Aldrich*) in a slurry type reactor. Figure 13.4 shows the reduction in chloroform formation after the chlorination (two doses) of humic waters treated by semiconductor photocatalysis (for different times).

Operational parameters, that is concentration, temperature, oxygen, light intensity and pH, have also been investigated with respect to their effect on the adsorption, rate of initial degradation and mineralisation of humic acid (Palmer *et al.*, 2002). Obviously, it is important to determine the extent of mineralisation which would be satisfactory for drinking water standards. However, another possibility would be to use photocatalysis as a pre-treatment in order to increase the bio-assimilable component, thus allowing the more effective use of biological methods. This approach has been shown to give bulk DOC removals as high as 87% (Fahmi *et al.*, 2003).

High-energy ultrasound has been investigated for reducing the THMFP of commercial fulvic and humic acids (Naffrechoux *et al.*, 2003). The combination of ultrasound and UV showed great potential in treating humic rich waters, in this case reducing THM levels from 1000 to 26 $\mu g\,L^{-1}$.

13.2.2 Micropollutants

While there are obvious issues regarding the reaction of organics with chlorine, there are a number of other questions relating to what organic are present in the water passing through conventional water treatment processes. The material is often polar in nature and is a heterogeneous mixture of organic molecules characterised by high solubility in water, and examples of chemical groups identified in this fraction include alcohols, ketones, esters, alkylhalides and aromatics. Few individual compounds have been identified, but any organic compound with a high water solubility (often characterised by a low K_{ow} values) has the potential to be in this fraction. A review of the literature has shown that a number of current and emerging pollutants fit into the same group: (i) a third of all pharmaceutical products previously identified in surface waters; (ii) four of the most popular 12 pesticides products used on arable crops in the UK; (iii) a number emerging pollutants, such as MTBE, vinyl chloride and HAAs. Recently, both US and German researchers have shown how these hydrophilic organics are poorly removed during wastewater treatment and can enter surface and groundwaters from wastewater discharges and then on into potable supply (Drewes *et al.*, 2003; Heberer, 2002).

13.2.2.1 Pharmaceuticals and endocrine disruptors compounds

There is increasing concern about the environmental effects of oestrogens and xenoestrogens in water (Wiese and Kelce, 1997). Oestrogenic substances act as endocrine disruptors (Kavlock *et al.*, 1996) and reported effects include the perturbation of sexual differentiation in embryos (Fry and Toone, 1981), cancers in reproductive organs (Hileman, 1994) and altered glucose and fat metabolism (Mathiessen, 1997). The three main human oestrogen compounds are 17-β-oestradiol (Figure 13.5), oestriol and oestrone, of which the former is the most potent (Montagnani *et al.*, 1996). These compounds can exhibit biological effects at very low levels of concentration (*ca.* 0.1 $ng\,dm^{-3}$) and detection at these concentrations is not possible by normal analytical methods (Routledge and Sumpter, 1996). Oestrogens may enter the aquatic environment from contraceptive pill residues, hormone replacement therapy residues and diethylstillbestrol residues (used to

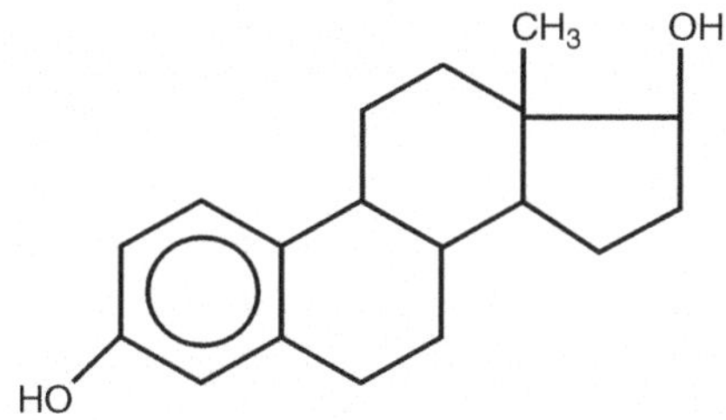

Figure 13.5 Structure of 17-β-oestradiol.

Table 13.1 Unit processes and operations used for micropollutant removal [adapted from Yoon *et al.* (2002)].

Classification	AC	AOP	UV	Cl_2/ClO_2	Coag./Floc.
Pesticides	>90	20–90	>90	70–90	<20
Steroids	>90	>90	>90	<20	<20
Industrial chemicals	>90	>90	>90	>90	<20–40
Antibiotics	40–90	20–90	40–90	20–90	<20–40
Antidepressants	70–90	20–90	40–90	20–70	<20–40
Anti-inflamants	>90	20–90	70–90	20–70	<20
Lipid regulators	>90		>90	20–70	<20
X-ray contrast media	70–90	70–90	20–90	20–70	<20–40
Synth. musks	70–90	70–90	>90	20–70	<20–40
Antimicrobials	70–90	70–90	40–90	20–70	<20–40
Surfactants	>90	>90	40–90	<20	<20–40

promote livestock growth) (Bitman and Cecil, 1970). Also giving cause for concern are xenoestrogen substances, a wide range of diverse compounds which mimic the biological effects of oestrogens. Substances classed as xenoestrogens include several pesticides, for example, DDT analogues and polychlorinated biphenyls (PCBs) (Ahel *et al.*, 1994), and also alkylphenol polyethoxylates which are surfactants used in detergents (Giger *et al.*, 1984). Although not as potent as the oestrogens, xenoestrogens are more abundant in the aquatic environment. Normal methods of water and sewage treatment are not completely effective in removing oestrogenic substances (Ahel and Giger, 1985). Although there is debate about the prevalence of EDCs in drinking water supplies, these substances are effective at such low levels that we must be pro-active in developing effective treatment technologies.

Table 13.1 compares the performance in removing micropollutants including EDCs with conventional treatment process, such as coagulation/flocculation and activated carbon with AOPs and UV processes (Yoon *et al.*, 2002). It is clear that both UV and AOPs have great potential in treating the whole range of chemicals listed, typically better than coagulation with metal salts and comparable with activated carbon.

There has been considerable interest in the fate and removal of EDCs in both water and wastewater treatment works and more recently in the use of AOPs.

Nagalwa *et al.* (2002) compared ozone and AOPs and found both to be efficient treatment. Lopez *et al.* (2003) showed how both UV and UV/H_2O_2 were very effective (>99% removal in 2.7 min) in removing pharmaceutical intermediates. Common drugs, such as paracetamol and aspirin, have been found in the effluent from sewage treatment works and, therefore, have the potential to travel into downstream water treatment works. Andreozzi *et al.* (2003) showed that UV/H_2O_2 was an effective treatment for paracetamol, removing the parent compound and reducing overall TOC levels by 40%. Electrochemical oxidation has been investigated for treatment of aspirin, where after 2 h 100% removal was possible with a platinum electrode and 90% for a carbon electrode (both at 50 mA cm^{-2}). Electrochemical oxidation has also shown potential for nonylphenol, where Kuramitz *et al.* (2002) showed 100% removal in 10 min.

Initial studies have reported the effective photocatalytic destruction of the parent compounds, that is 17-β-oestradiol, oestriol and oesterone (Coleman *et al.* 2000). However, perhaps more important is the removal of the oestrogenic activity. To this end experiments have been undertaken to determine the oestrogenic activity, following photocatalytic treatment, using a yeast screen bioassay (Routledge and Sumpter, 1996). In this bioassay, the DNA sequence of the human oestrogen receptor (hER) is integrated into the main chromosome of the yeast genome. The yeast cell also contains expression plasmids carrying oestrogen-responsive elements (EREs) controlling the expression of the reporter gene *lac-Z* (encoding for the enzyme β-galactosidase), which is used to measure the receptors' activity. Thus, in the presence of oestrogens (or chemicals with oestrogen activity), β-galactosidase is synthesised and secreted into the medium, where it breaks down the yellow coloured chromogenic substrate, chlorophenyl red-β-galactopyranoside (CPRG) to a red product, which can then be measured by absorbance at 540 nm.

In this work, the TiO_2 (*Degussa P25*) was immobilised onto Ti-alloy using an electrophoretic coating method (Byrne *et al.*, 1998). In the example shown in Figure 13.6, the reactor contained 8 cm^3 of 10 μg dm^{-3} 17-β-oestradiol. The graph clearly shows the efficient removal of the oestrogenic acitivity of the 17-β-oestradiol within the first 30 min of photocatalysis. Further work also demonstrated the effective destruction of the oestrogenic activity of oestrone and 17-α-ethinyl oestradiol (Coleman, 2000). These experiments clearly demonstrate that TiO_2 photocatalysis is a promising technology for the treatment of water contaminated with endocrine disrupting chemicals.

Pesticides Pesticides can be classified by their use either as an insecticide, herbicide or fungicide. The use of pesticides has developed enormously since the 1950s. They are particularly hazardous as they are chemically developed to be toxic and to some extent persistent in the environment. In permeable soils, they are able to infiltrate to groundwater and in more impermeable soils they find their way into surface watercourses via drainage networks. Pesticide levels in drinking waters taken from rivers tend to be seasonal phenomena, although they can be found at relatively high concentrations throughout the year. In groundwaters, they acccumulate over the years, so that the overall concentration is slowly but constantly increasing, without any discernable large seasonal variation. In Europe, the Drinking

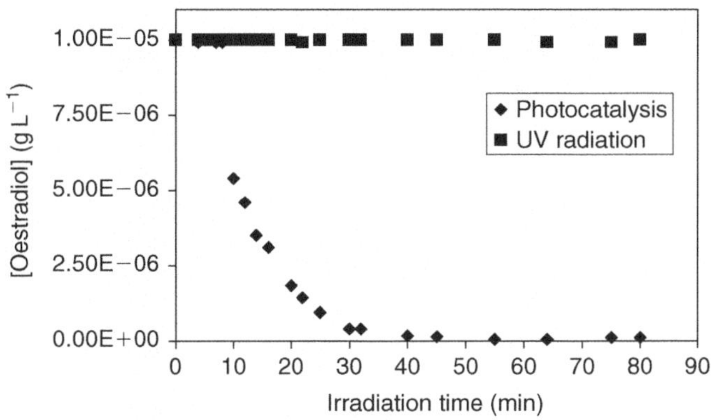

Figure 13.6 The removal of the oestrogenic activity of 17-β-oestradiol by photocatalysis (as determined by the yeast screen bioassay).

Water Directive sets a Maximum Allowable Concentration (MAC) value for total pesticides and any individual pesticide compound of 0.5 and 0.1 ppb, respectively. Total pesticides is expressed as 'pesticides and pesticide related products', which is defined as insecticides (persistent organochlorine compounds and carbamates), herbicides, fungicides, PCBs and polychlorinated terphenyls (PCTs). The current practise is often to determine the individual concentration of just three principal substances, parathion, BHC (lindane or HCH) and dieldrin, and to report the additive total as 'total pesticides'. This term is obviously misleading as only a small fraction of the real total is measured.

Significant work has been carried out on the oxidation of pesticides in water. The main problem with the destructive treatment of these substances using AOPs is that several intermediate species are produced which may in some cases be more stable and/or exhibit greater toxicity than the parent compound. Advanced analytical techniques, such as GCMS or LCMS, are required for the identification of such intermediates and the levels of pesticides in real waters is very low (ppb) adding to the analytical problems. Photocatalysis has been reported to be effective in the degradation of amide herbicides, bypyridinium herbicides, carbamate insecticides, chloronicotinoid insecticides, chlorophenol pesticides, dinitroaniline pesticides, halobenzonitrile pesticides, organochlorine insecticides, organophosphorus pesticides, phenol-based pesticides, pyrimidine pesticides, thiocarbamate herbicides and triazine herbicides (Burrow *et al.*, 2002). Of all the pesticides studied, the triazine herbicides are the only group which are reported to be resistant to complete mineralisation. Atrazine and simazine are triazine herbicides and are commonly found in surface and groundwater.

To compare the performance of a range of AOPs, an organic contaminant (Carbofuran) was selected from the literature and the rate of reactions compared. While this assessment does not take into account reactor configuration or starting concentrations for the purpose of this chapter it is an adequate comparison of performance (Figure 13.7). Carbofuran ($C_{12}H_{15}NO_3$) is a broad spectrum insecticide,

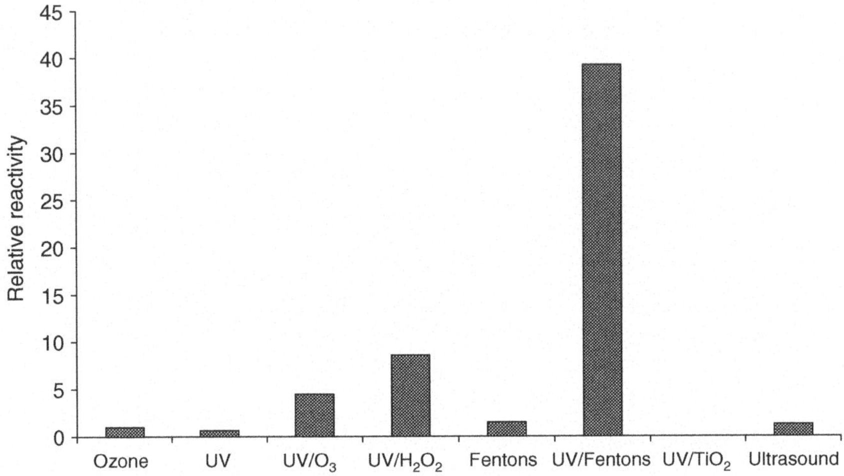

Figure 13.7 Comparison of AOPs performance in treating the pesticide Carbofuran [data taken from Hua and Pfalzer-Thompson (2001) and Benitez *et al.* (2002)].

which has been widely used and has been identified as a problem pollutant in many groundwaters. The comparison shows that all the AOPs perform better than both ozone and direct UV photolysis. The combination of UV with ozone improves the performance over ozone alone by 4.5 times, while the combination of UV and hydrogen peroxide increase the rate of oxidation by 8.5 times. Fentons reagent slightly improves the rate of oxidation; but when combined with UV light, the rate of oxidation increases by over 26 times. No comparative data was available for TiO_2/UV, but Tennakone *et al.* (1997) did report complete mineralisation of a $300\,mg\,L^{-1}$ solution of Carbofuran on TiO_2 coated glass slides in 15 h.

13.2.3 Disinfection of water

The disinfection of drinking water, using chemicals such as chlorine, has successfully protected public health against waterborne disease since the early 1900s (Gray, 1994). However, the reaction between chlorine and organic compounds, contained in the source water, can result in the production of DBPs, some of which have been shown to have mutagenic properties (Jolley *et al.*, 1990). The use of chemical disinfectants can also cause problems with taste and odour, leading to customer complaints (Gray, 1994). The emergence of waterborne pathogens, which are resistant to traditional chemical disinfection, such as *Cryptosporidium* and *Giardia*, has led to a reappraisal of traditional disinfection practise (Bouchier, 1998). Alternative treatment technologies are required to meet the demands of the consumers, the water companies and the regulators, without compromising public health and safety, but which maintain scientific credibility.

The use of UV in disinfection is well known and widely practised (see Chapters 2 and 3). For the other AOPs, the inactivation of micro-organisms is a relatively new area, with the majority of the research being carried out within the last decade and much of the work has been undertaken using photocatalysts.

Matsunaga *et al.* (1985) first demonstrated photocatalytic disinfection of *Lactobacillus acidophilus*, *Saccharomyces cerevisiae* and *Escherichia coli*. Subsequent research has primarily been directed towards the inactivation of bacterial micro-organisms, with few studies towards the inactivation of viral or protozoan organisms. The current state of photocatalytic disinfection has been recently comprehensively reviewed by Blake *et al.* (1999). A wide range of Gram positive and Gram negative bacteria have been subjected to photocatalytic treatment. *E. coli* has been extensively studied as it is used as the major faecal indicator organism in the water industry and due to its routine culture in microbiological laboratories. The photocatalytic inactivation of bacterial spores has also been reported (Pham *et al.*, 1995; Pham *et al.*, 1997; Tone *et al.*, 1993). Only a small number of viral organisms have been examined in photocatalytic disinfection studies. Poliovirus 1 (Watts *et al.*, 1995), phage MS2 (Sjogren and Sierka, 1994), *Lactobacillus* phage PL-1 (Kashige, 2001) and phage Qβ (Lee *et al.*, 1997, 1998; Otaki *et al.*, 2000), have all been inactivated in both slurry and immobilised reactors using UVA or UVC illumination sources. *Cryptosporidium parvum* is the only protozoan micro-organism to have currently undergone disinfection using photocatalysis. Otaki *et al.* (2000) reported 4 log inactivation using a UVC illuminated slurry TiO_2 reactor as determined by an *in vitro* excystation viability analysis. Curtis *et al.* (2002) reported 27% inactivation using photoelectrochemical disinfection, viability determined using vital dye exclusion assays. The application of a small electrical bias to an immobilised TiO_2 film has been shown to increase the rate of degradation of *E. coli* and *Clostridium perfringens* cells (Butterfield *et al.*, 1997; Dunlop *et al.*, 2002; Harper *et al.*, 2001).

A number of studies have, however, used bacterial samples from environmental origins, for example, secondary wastewater from municipal sewage treatment plants (Dillert *et al.*, 1998; Li *et al.*, 1996; Melian *et al.*, 2000; Watts *et al.*, 1995), water sourced from a pond (Ireland *et al.*, 1993) and river water (Wist *et al.*, 2002).

Mechanisms for the bactericidal action of TiO_2 photocatalysis involving reactive oxygen species (ROS) have been proposed by a number of authors (Kikuchi *et al.*, 1997; Maness *et al.*, 1999; Matsunaga *et al.*, 1985; Saito *et al.*, 1992; Sunada *et al.*, 1998; Zheng *et al.*, 2000) and reviewed by Blake *et al.* (1999). The production of hydrogen peroxide and the superoxide ion, in the presence of catalytic amounts of ferric ions, via the Haber–Weiss reaction and the generation of hydroxyl radicals, via Fenton reactions, at the surface or within the bacterial cell structure could also play a significant role in the inactivation process.

Regardless of the oxidative species responsible, there is substantial evidence that the cell membrane is the crucial site of attack. Zheng *et al.* (2000) designed studies to examine the order of cellular inactivation using *E. coli* as a model test organism. Damage to the cell membrane showed a good correlation with a decrease in cell viability. The mechanism of photocatalytic inactivation was proposed as the attack of the cell wall, followed by the cell membrane and finally oxidative attack of the cytoplasmic contents. The oxidation of bacterial cell membranes was studied by Maness *et al.* (1999). They concluded that alterations in the membrane structure, caused by lipid peroxidation, ultimately lead to conformational changes in membrane-bound proteins and electron mediators with the loss of cell membrane integrity. Once the cell membrane is damaged, cell respiration ceases. The loss of cell energy production mechanisms and subsequent oxidative attack of internal cellular components, could ultimately result in cell death.

Ultrasound has been used in the control of two problem organism in water treatment and supply, molluscs and algae. Workers at the University of Bucharest have investigated the removal of the water molluscs *Driessena polymorpha* and showed that a 3 min exposure to 2 W cm^{-2} lead to a 90% decrease in concentration. The control of *Microcystis* was repressed by ultrasonic radiation and has the potential for treating floating cells in surface waters.

13.3 PROSPECTS

It is clear from this chapter and many of the preceeding ones that AOPs have proven performance in removing many organic, inorganic and microbiological pollutants a modern water treatment works is expected to face. So the question is why are AOPs not widely used in existing works or often considered in new works designs? Many of the processes discussed in this book are available at full scale (Table 13.2), and some such as UV and UV/H_2O_2/O_3 are used at full scale but typically only when there is a specific micropollutant, such as NDMA or MTBE that requires removal (see Chapters 3 and 4). Suty *et al.* (2003) listed some of the disadvantages of AOPs as being:

- unspecific reactivity of hydroxyl radicals;
- production of more hydrophilic (and hence more difficult to treat) by-products than the original product;
- little knowledge of by-product toxicity;
- production of biodegradable dissolved organic carbon (BDOC) and the impact of treatment on biofilm re-growth in distribution systems.

Table 13.2 Comparison of AOP options for potable water treatment.

Process	Scale	Process	Barriers
UV	Full	Reactor	Performance/by-products
UV/H_2O_2/O_3	Full	Reactor, chemical dosing and destruction	Costs/by-products
Fenton's reagent	Full	Chemical dosing	Residuals
Photo-Fenton's	Laboratory/full	Reactor, chemical dosing	Residuals/costs/contact time
TiO_2/UV	Laboratory/pilot	Delivery method, reactor	Costs/recovery
Ultrasound	Laboratory/pilot	Reactor	Costs/equipment
Radiolysis	Laboratory/full	Reactor	Costs/equipment

On the plus side though Suty *et al.* (2003) and many others have noted the potential AOPs have over processes such as

- powerful oxidant leading to treatment of organics not attacked by chlorine or ozone;
- faster rate of reaction leading to smaller more compact processes;
- low chemical dose leading to lower residuals (and where applicable no residual processes).

Many of the issues regarding wider take up of technologies, AOPs are due to the process being higher in capital investment and operating costs and should, therefore, not be employed where cheaper technologies are effective (Esplugas *et al.*, 2002). One option for identifying where AOPs should be considered is to define the efficiency of commonly used processes, such as activated carbon adsorption. Activated carbon treatment is a well understood process whose operation is usually very simple and requires little manned attention. The capital cost of a carbon system will almost certainly be lower than that of an AOP, although if DOC levels are above 10 ppm, carbon treatment tends to become uncompetitive against AOP, because of the cost associated with change out and regeneration or disposal of the carbon. In addition, opportunities for AOPs exist for micropollutants that carbon will not adsorb pollutants, such as NDMA, chloroform and 1,4-dioxane. Table 13.1 above showed how AOPs and activated carbon were comparable processes for many micropollutants. Carbon adsorption like many conventional water treatment processes is not destructive and is merely a phase transfer, where the contaminant is rendered immobile on the carbon surface (Ince and Apikyan, 2000).

The ability to completely mineralise the pollutant must, therefore, offer significant benefits in water treatment or the ability of processes, such as TiO_2/UV, to act as an adsorbent like activated carbon, but with the potential for regeneration not off-site using heat but on-site using UV light. This can be seen in Figure 13.8 below, where a TiO_2 coated media is introduced to a $10\,mg\,L^{-1}$ solution of commercial humic acid, the adsorption of humic acid on to the media is clear and as reported above is an effective treatment method for waters containing NOM. The right hand image is the same media after irradiation with UV light and it can be seen that the adsorbed humic material has been oxidised and the adsorbent capacity of the media restored. Once a suitable, stable delivery method for TiO_2 has been commercially developed, the applications of this specific process in municipal water treatment are sure to follow. Photocatalysis may be a process for the future, but other process offer readily available solutions to treatment problems now. If, for example, a high coagulant dose is required to meet THM targets and sludge disposal costs is an issue, then the reduced metal ion dose possible with Fenton's reagent would be a switch on switch of solution for periods of elevated organic levels (Murray and Parsons, 2004).

The future of AOPs is predicted to have an annual growth rate of almost 15% in the UK by the year 2005 (Kaiser, 1994). The processes discussed in this chapter and through this book, while not the universal panacea for water treatment, either alone, or as part of a hybrid system are capable of producing cost effective treatment

Figure 13.8 Visual example of the adsorption and oxidation of humic acid on TiO_2.

in some applications with little or no competition. As with any treatment process, though there are certain areas that require further development include:

- the understanding of the degradation mechanisms, reaction and intermediates involved;
- the use of solar-based techniques for UV supply, since this energy, although not necessarily reliable in the UK, is free and plentiful in other areas;
- the use of catalysts to reduce the amount of reagent chemical required, and hence the associated cost;
- detailed evaluation and development of electrochemical and sonochemical techniques eliminating the need for on-site chemicals.

REFERENCES

Ahel, M. and Giger, W. (1985) Determination of alkylphenols and alkylphenol monoethoxylate and diethoxylate in environmental-samples by high-performance liquid-chromatography. *Anal. Chem.* **57**(8), 1577–1583.

Ahel, M., Giger, W. and Koch, M. (1994) Behavior of alkylphenol polyethoxylate surfactants in the aquatic environment. 1. Occurrence and transformation in sewage-treatment. *Water Res.* **28**(5), 1131–1142.

Andreozzi, R., Caprio, V., Marotta, R. and Vogna, D. (2003) Paracetamol oxidation from aqueous solutions by means of ozonation and H_2O_2/UV system. *Water Res.* **37**, 993–1004.

Back, A.P. and Christman, R.F. (1963) Characteristics of coloured water surface waters. *J. Am. Wat. Works Assoc.* **55**, 753–770.

Benitez, J.F., Acero, J.L. and Real, F.J. (2002) Degradation of carbofuran by using ozone, UV radiation and advanced oxidation processes. *J. Hazard. Mater.* **89**, 51–65.

Blake, D.M., Maness, P.C., Huang, Z., Wolfrum, E.J., Huang, J. and Jacoby, W.A. (1999) Application of the photocatalytic chemistry of titanium dioxide to disinfection and the killing of cancer cells. *Separ. Purif. Method.* **28**(1), 1–50.

Bouchier, I. (1998) *Cryptosporidium in Water Supplies*. Third Report of the Group of Experts, Department of Health/Department of the Environment, Published by HMSO.

Burrow, H.D., Canle, M., Santaballa, J.A. and Steenken, S. (2002) Reaction pathways and mechanisms of photodegradation of pesticides. *J. Photoch. Photobio. B-Biol.* **67**, 71–108.

Butterfield, I.M., Christensen, P.A., Curtis, T.P. and Gunlazuardi, J. (1997) Water disinfection using an immobilised titanium dioxide film in a photochemical reactor with electric field enhancement. *Water Res.* **31**(3), 675–677.

Byrne, J.A., Eggins, B.R., Brown, N.M.D., McKinney, B. and Rouse, M. (1998) Immobilisation of titanium dioxide for the treatment of polluted water. *Appl. Catal. B-Environ.* **17**, 25–36.

Coleman, H.M., Eggins, B.R., Byrne, J.A., Palmer, F.L. and King, E. (2000) Photocatalytic degradation of 17-β-oestradiol on immobilised TiO_2. *Appl. Catal. B-Environ.* **24**(1), L1–L5.

Croué, J.-P., Lefebvre, E., Martin, B. and Legube, B. (1993) Removal of dissolved hydrophobic and hydrophilic organic substances during coagulation/flocculation of surface waters. *Water Sci. Technol.* **27**, 143–152.

Curtis, T.P., Walker, G., Dowling, G.M. and Christensen, P.A. (2002) Fate of cryptosporidium oocysts in an immobilised titanium dioxide reactor with electric field enhancement. *Water Res.* **36**(9), 2410–2413.

Dillert, R., Siemon, U. and Bahnemann, D. (1998) Photocatalytic disinfection of municipal wastewater. *Chem. Eng. Technol.* **21**(4), 356.

Drewes, J.E., Reinhard, M. and Fox, P. (2003) Comparing microfiltration-reverse osmosis and soil-aquifer treatment for indirect potable reuse of water. *Water Res.* **37**, 3612–3621.

Dunlop, P.S.M., Byrne, J.A., Manga, N. and Eggins, B.R. (2002) The photocatalytic removal of bacterial pollutants from drinking water. *J. Photoch. Photobio. A-Chem.* **148**(1–3), 355–363.

Eggins, B.R., Palmer, F.L. and Byrne, J.A. (1997) Photocatalytic treatment of humic substances in drinking water. *Water Res.* **31**(5), 1223–1226.

Esplugas, S., Giménez, J., Contreras, S., Pascual, E. and Rodríguez, M. (2002) Comparison of different advanced oxidation processes for phenol degradation. *Water Res.* **36**, 1034–1042.

Fahmi, W.N., Mukaidani, T. and Okada, M. (2003) OC removal by multi-stage ozonation-biological treatment. *Water Res.* **37**, 150–154.

Fox, M.A. and Dulay, M.T. (1993) Heterogeneous photocatalysis. *Chem. Rev.* **93**(1), 341–357.

Fry, D.M. and Toone, C.K. (1981) Ddt-induced feminisation of gull embryos. *Science* **213**(4510), 922–924.

Fukushima, M., Tatsumi, K. and Nagao, S. (2001) Degradation characteristics of humic acid during photo-Fenton processes. *Environ. Sci. Technol.* **35**, 3683–3690.

Gianturco, F., Chiodaroli, C.M., Bellobono, I.R., Raimondi, M.L., Moroni, A. and Gawlik, B. (1997) Pilot-plant photomineralisation of atrazine in aqueous solution, by photocatalytic membranes immobilising titanium dioxide and promoting photocatalysts. *Fresen. Environ. Bull.* **6**, 461–468.

Giger, W., Brunner, P.H. and Schaffner, C. (1984) 4-Nonylphenol in sewage-sludge – accumulation of toxic metabolites from nonionic surfactants. *Science* **225**(4662), 623–625.

Goslan, E.H., Fearing, D.A., Banks, J., Wilson, D., Hillis, P., Campbell, A.T. and Parsons, S.A. (2002) Seasonal variations in the disinfection by-product precursor profile of a reservoir water. *J. Wat. Serv. Res. Tec.-AQUA* **51**, 475–482.

Gray, N.F. (1994) *Drinking Water Quality, Problems and Solutions.* John Wiley & Sons, Chichester, UK.

Harper, J.C., Christensen, P.A., Egerton, T.A., Curtis, T.P. and Gunlazuardi, J. (2001) Effect of catalyst type on the kinetics of the photoelectrochemical disinfection of water inoculated with *E-coli*. *J. Appl. Electrochem.* **31**(6), 623–628.

Heberer, T. (2002) Tracking persistent pharmaceutical residues from municipal sewage to drinking water. *J. Hydrol.* **266**, 175–189.

Hequet, V., Gonzalez, C. and Cloirec, P. (2001) Photochemical processes for atrazine degradation: methodological approach. *Water Res.* **35**, 4253–4260.

Hileman, B. (1994) Environmental estrogens linked to reproductive abnormalities, cancer. *Chem. Eng. News* **72**(5), 19–23.

Hoffmann, M.R., Martin, S.T., Choi, W.Y. and Bahnemann, D.W. (1995) Environmental applications of semiconductor photocatalysis. *Chem. Rev.* **95**(1), 69–96.

Hua, I. and Pfalzer-Thompson, U. (2001) Ultrasonic irradiation of Carbofuran: decomposition kinetics and reactor characterization. *Water Res.* **35**, 1445–1452.

Hustert, K. and Moza, P.N. (1991) Photocatalytic degradation of s-triazine herbicides. *Toxicol. Environ. Chem.* **31–32**, 97–102.

Ireland, J.C., Klostermann, P., Rice, E.W. and Clark, R.M. (1993) Inactivation of *Escherichia coli* by titanium-dioxide photocatalytic oxidation. *Appl. Environ. Microbiol.* **59**(5), 1668–1670.

Jolley, Johnson, Minear, Jacobs, Condie, Katz and Mattice (1990) Water chlorination. chemistry, environmental impact and health effects. In: *Water Chlorination: The Challenge*, C., J.J. and L., J.R. (eds). Lewis Publishers, Inc., Chicago, Vol. 6, pp. 21–27.

Kashige, N., Kakita, Y., Nakashima, Y., Miake, F. and Watanabe, K. (2001) Mechanism of the photocatalytic inactivation of *Lactobacillus casei* phage PL-1 by titania thin film. *Curr. Microbiol.* **42**(3), 184–189.

Kavlock, R.J., Daston, G.P., DeRosa, C., FennerCrisp, P., Gray, L.E., Kaattari, S., Lucier, G., Luster, M., Mac, M.J., Maczka, C., Miller, R., Moore, J., Rolland, R., Scott, G., Sheehan, D.M., Sinks, T. and Tilson, H.A. (1996) Research needs for the risk assessment of health and environmental effects of endocrine disruptors: a report of the US EPA-sponsored workshop. *Environ. Health Persp.* **104**, 715–740.

Kikuchi, Y., Sunada, K., Iyoda, T., Hashimoto, K. and Fujishima, A. (1997) Photocatalytic bactericidal effect of TiO_2 thin films: dynamic view of the active oxygen species responsible for the effect. *J. Photoch. Photobiol. A-Chem.* **106**(1–3), 51–56.

Kleiser, G. and Frimmel, F.H. (2000) Removal of precursors for disinfection by-products (DBPs) – differences between ozone- and OH-radical induced oxidation. *Sci. Total Environ.* **256**, 1–9.

Konstantinou, L.K., Sakellarides, T.M., Sakkas, V.A. and Albanis, T.A. (2001) Photocatalytic degradation of selected s-triazine herbicides and organophosphorus insecticides over aqueous TiO_2 suspensions. *Environ. Sci. Technol.* **35**, 398–405.

Kuramitz, H., Saitoh, J., Hattori, T. and Tanaka, S. (2002) Electrochemical removal of p-nonylphenol from dilute solutions using a carbon fiber anode. *Water Res.* **36**, 3323–3329.

Lee, S., Nishida, K., Otaki, M. and Ohgaki, S. (1997) Photocatalytic inactivation of phage Q beta by immobilized titanium dioxide mediated photocatalyst. *Water Sci. Technol.* **35**(11–12), 101–106.

Lee, S., Nakamura, M. and Ohgaki, S. (1998) Inactivation of phage Q beta by 254 nm UV light and titanium dioxide photocatalyst. *J. Environ. Sci. Health Part A-Toxic/Hazard. Subst. Environ. Eng.* **33**(8), 1643–1655.

Li, X.Z., Zhang, M. and Chua, H. (1996) Disinfection of municipal wastewater by sensitized photo-oxidation. *Water Sci. Technol.* **33**(3), 111–118.

Macarthy, J.F. (1989) Aquatic humic substances: influence on fate and treatment of pollutants. In: *American Chemical Society*, Suffet, I.H. and MaCarthy, P. (eds), Vol. pp. 263–269.

Maness, P.C., Smolinski, S., Blake, D.M., Huang, Z., Wolfrum, E.J. and Jacoby, W.A. (1999) Bactericidal activity of photocatalytic TiO_2 reaction: toward an understanding of its killing mechanism. *Appl. Environ. Microbiol.* **65**(9), 4094–4098.

Mathiessen, P. (1997) *Oestrogenic Substances in the Environment.* Royal Society of Chemistry (Analytical Division) Meeting, London.

Matsunaga, T., Tomoda, R., Nakajima, T. and Wake, H. (1985) Photoelectrochemical sterilization of microbial-cells by semiconductor powders. *FEMS Microbiol. Lett.* **29**(1–2), 211–214.

Melian, J.A.H., Rodriguez, J.M.D., Suarez, A.V., Rendon, E.T., do Campo, C.V., Arana, J. and Pena, J.P. (2000) The photocatalytic disinfection of urban waste waters. *Chemosphere* **41**(3), 323–327.

Mills, A. and Le-Hunte, S. (1997) An overview of semiconductor photocatalysis. *J. Photoch. Photobio. A-Chem.* **108**(1), 1–35.

Mills, A., Davies, R.H. and Worsley, D. (1993) Water-purification by semiconductor photocatalysis. *Chem. Soc. Rev.* **22**(6), 417–425.

Montagnani, D.B., J., P., Davie, T.J.A. and Vinson, G.P. (1996) Environmental persitant oestrogen – like substances in UK river systems. **10**, 399–406.

Murray, C.M. and Parsons, S.A. (2004) Removal of NOM from drinking water: Fenton's and photo Fenton's processes. *Chemosphere* **54**, 1017–1023.

Naffrechoux, E., Combet, E., Fanget, B. and Petrier, C. (2003) Reduction of chloroform formation potential of humic acid by sonolysis and ultraviolet irradiation. *Water Res.* **37**, 1948–1952.

Nowack, K.O., Cannon, F.S. and Arora, H. (1999) Ferric chloride plus GAC for removing TOC. *J. Am. Wat. Works Assoc.* **91**, 65–78.

Otaki, M., Hirata, T. and Ohgaki, S. (2000) Aqueous microorganisms inactivation by photocatalytic reaction. *Water Sci. Technol.* **42**(3–4), 103–108.

Palmer, F.L. (1998) *Semiconductor Photocatalytic Degradation of Humic Waters.* DPhil Thesis, Faculty of Science, University of Ulster.

Palmer, F.L., Eggins, B.R. and Coleman, H.M. (2002) The effect of operational parameters on the photocatalytic degradation of humic acid. *J. Photochem. Photobiol. A-Chem.* **148**(1–3), 137–143.

Parsons, S.A., Goslan, E.H., Jarvis, P.R., Hurst, A.M. and Fearing, D.A. (2002) NOM – the relationship between character and treatability. *Natural Organic Mater Characterisation and Treatment*, Cranfield, UK, May 15.

Penuela, G.A. and Barcelo, D. (2000) Comparative photodegradation study of atrazine and desethlyatrazine in water samples containing titanium dioxide/hydrogen peroxide and ferric chloride/hydrogen peroxide. *J. AOAC Int.* **83**, 53–60.

Pham, H.N., McDowell, T. and Wilkins, E. (1995) Photocatalytically-mediated disinfection of water using TiO_2 as a catalyst and spore-forming bacillus-pumilus as a model. *J. Environ. Sci. Health Part A-Environ. Sci. Eng. Toxic Hazard. Subst. Contr.* **30**(3), 627–636.

Pham, H.N., Wilkins, E., Heger, K.S. and Kauffman, D. (1997) Quantitative analysis of variations in initial Bacillus pumilus spore densities in aqueous TiO_2 suspension and design of a photocatalytic reactor. *J. Environ. Sci. Health Part A-Environ. Sci. Eng. Toxic Hazard. Subst. Contr.* **32**(1), 153–163.

Reuter, J.H. and Perdue, E.M. (1977) Importance of heavy metal–organic matter interactions in natural waters. *Geochim. Cosmochim. Acta* **41**, 325–334.

Rook, J.J. (1977) Chlorination reactions of fulvic acids in natural water. *J. Environ. Sci. Technol.* **11**, 478–482.

Routledge, E.J. and Sumpter, J.P. (1996) Estrogenic activity of surfactants and some of their degradation products assessed using a recombinant yeast screen. *Environ. Toxicol. Chem.* **15**(3), 241–248.

Saito, T., Iwase, T., Horie, J. and Morioka, T. (1992) Mode of photocatalytic bactericidal action of powdered semiconductor TiO_2 on mutans streptococci. *J. Photochem. Photobiol. B-Biol.* **14**(4), 369–379.

Schulten, H.R., Plage, B. and Schnitzer, M. (1991) A chemical-structure for humic substances. *Naturwissenschaften* **78**(7), 311–312.

Sjogren, J.C. and Sierka, R.A. (1994) Inactivation of phage Ms2 by iron-aided titanium-dioxide photocatalysis. *Appl. Environ. Microbiol.* **60**(1), 344–347.

Sunada, K., Kikuchi, Y., Hashimoto, K. and Fujishima, A. (1998) Bactericidal and detoxification effects of TiO_2 thin film photocatalysts. *Environ. Sci. Technol.* **32**(5), 726–728.

Suty, H., De Traversay, C. and Coste, M. (2003) Applications of advanced oxidation processes: present & future. In: *Proceedings of the 3rd Conference on Oxidation Technologies for Water and Wastewater Treatment*, Goslar, Germany, May 18–22, p. 8.

Tennakone, K., Tilakaratne, C.T.K. and Kottegoda, I.R.M. (1997) Photomineralization of carbofuran by TiO_2-supported catalyst. *Water Res.* **31**, 1909–1912.

Thompson, J., Roddick, F.A. and Drikas, M. (2002) UV photooxidation facilitating biological treatment for the removal of NOM from drinking water. *J. Wat. Suppl.: Res. Technol.- AQUA* **51**(6), 297–306.

Tone, S., Taya, M., Kato, S., Horie, Y., Ashikaga, Y. and Joo, H.K. (1993) Kinetic-analysis of photochemical sterilization of thermoduric bacterial-spores in slurry of semiconductor catalyst particles with aeration. *Kag. Kog. Ronbunshu* **19**(6), 1149–1156.

Wang, G.-S., Hsieh, S.-T. and Hong, C.-S. (2000) Destruction of humic acid in water by UV light – catalysed oxidation with hydrogen peroxide. *Water Res.* **34**(15), 3882–3887.

Watts, R.J., Kong, S.H., Orr, M.P., Miller, G.C. and Henry, B.E. (1995) Photocatalytic inactivation of coliform bacteria and viruses in secondary waste-water effluent. *Water Res.* **29**(1), 95–100.

Wiese, T.E. and Kelce, W.R. (1997) An introduction to environmental oestrogens. *Chem. Ind.* **16**, 648–653.

Wist, J., Sanabria, J., Dierolf, C., Torres, W. and Pulgarin, C. (2002) Evaluation of photocatalytic disinfection of crude water for drinking-water production. *J. Photochem. Photobiol. A-Chem.* **147**(3), 241–246.

Zheng, H., Maness, P.C., Blake, D.M., Wolfrum, E.J., Smolinski, S.L. and Jacoby, W.A. (2000) Bactericidal mode of titanium dioxide photocatalysis. *J. Photochem. Photobiol. A-Chem.* **130**(2–3), 163–170.

Index

Page numbers in bold type refer to figures and tables.

IWA Publishing's authorised EU representative for General Product Safety Regulations is Diane D'Arras, 15 rue Duret, 75116 Paris, France, e-mail: safety@iwap.co.uk.

Printed and bound by CPI Group (UK) Ltd, Croydon, CR0 4YY
12/05/2026
02108864-0007